Post-Fermentation and -Distillation Technology

Stabilization, Aging, and Spoilage

Post-Fermentation and -Distillation Technology

Stabilization, Aging, and Spoilage

Edited by
Matteo Bordiga

CRC Press
Taylor & Francis Group
Boca Raton London New York

CRC Press is an imprint of the
Taylor & Francis Group, an **informa** business

CRC Press
Taylor & Francis Group
6000 Broken Sound Parkway NW, Suite 300
Boca Raton, FL 33487-2742

Printed on acid-free paper

International Standard Book Number-13: 978-1-4987-7869-5 (Hardback)

Visit the Taylor & Francis Web site at
http://www.taylorandfrancis.com

and the CRC Press Web site at
http://www.crcpress.com

FSC www.fsc.org MIX Paper from responsible sources FSC® C013056

Printed and bound in Great Britain by
TJ International Ltd, Padstow, Cornwall

To My Son

Contents

Preface

This book focuses on the post-fermentation and -distillation technology applied to wine, beer, vinegar, and distillates in a broad spectrum. Stabilization, aging, and spoilage represent the three major sections of the text. The book provides a comprehensive overview of all of the post-fermentation operations related to these products, focusing on the complex issue of their stability. Wines, for example, must be stable against microbial activity as well as undesirable chemical and physical–chemical reactions occurring in the bottle. There are five goals of "finishing" a wine: clarity, stability, compositional adjustment, style development, and packaging. It is important, especially in white wines, that the wine at the point of consumption not be cloudy or contain any haze or precipitate. Currently, haze represents a visual defect associated with spoilage in the eyes of the consumer.

It is also important to prevent unwanted microbial growth from occurring in the wine after the primary fermentation is complete, as this affects the flavor and aroma profile in unpredictable ways. *Saccharomyces* autolysis will replenish nutrients in the wine, making them available for other organisms. *Saccharomyces* do not consume all possible bacterial energy sources. Many spoilage organisms are obligate aerobes, so the wine must be protected against exposure to air once the carbon dioxide blanket generated during fermentation has dissipated.

Similarly, the discussion also involves beer, vinegar, and distillates. This represents a novel approach, not limiting the book to the issues of aging, stabilization, and spoilage of wine. Covering three other main fermentation categories certainly makes this project interesting. New technologies as well as new materials are included in the discussion (e.g., square barrels, synthetic closures, tetra pak®). The book is a good mix of referenced research with practical applications, also reporting case studies of these various applications of novel technologies. Consumer packaged goods companies are forced to re-evaluate packaging formats and materials, as products and their packaging are inextricably linked to the trends taking place within the consumption habits. At the same time, as consumers grow more concerned about energy consumption and the environment as a whole, they are also increasingly demanding more eco-friendly and socially conscious products and packaging.

Matteo Bordiga

Editor

Dr. Matteo Bordiga received his PhD in food science from the Università del Piemonte Orientale, Novara, Italy in 2010. He received his MS in chemistry and pharmaceutical technologies from the same university. Dr. Bordiga's main research activity concerns food chemistry, investigating the different classes of polyphenols from analytical, technological, and nutritional points of view. More recently, he moved his research interests towards wine chemistry, focusing his attention on the entire production process—from vine to glass. He has published more than 30 research papers in peer-reviewed international and national journals. Since 2013, he has been an editorial board member of the *International Journal of Food Science & Technology*. In 2014, he was appointed editor-in-chief for *Wine Studies*, a peer-reviewed international open-access journal focused on wine-related topics. He served as editor of the book, *Valorization of Wine Making By-Products* (CRC Press/Taylor & Francis Group, 2015). All his research activities have been developed through important collaborations with foreign institutions (Foods Science and Technology Department and Foods for Health Institute, University of California, Davis; Fundación Parque Científico y Tecnológico de Albacete, and Instituto Regional de Investigación Científica Aplicada, Universidad de Castilla-La Mancha, Ciudad Real, Spain).

Contributors

Roger Barth
West Chester University
West Chester, Pennsylvania

Tommaso Bonciani
University of Modena and Reggio Emilia
Reggio Emilia, Italy

Matteo Bordiga
Università del Piemonte Orientale
Novara, Italy

Federico Casassa
California Polytechnic State University
San Luis Obispo, California

Thomas Collins
Washington State University
Richland, Washington

Giuseppe Corradini
University of Modena and Reggio Emilia
Reggio Emilia, Italy

Teresa Garde-Cerdán
Institute of Vine and Wine Sciences (CSIC-CAR-UR)
Logroño, Spain

Paolo Giudici
University of Modena and Reggio Emilia
Reggio Emilia, Italy

Ana Gonzalo-Diago
Institute of Vine and Wine Sciences (CSIC-CAR-UR)
Logroño, Spain

Paul S. Hughes
Oregon State University
Corvallis, Oregon

Julie Kellershohn
Ryerson University
Toronto, Ontario, Canada

Federico Lemmetti
University of Modena and Reggio Emilia
Reggio Emilia, Italy

Juana Martínez
Institute of Vine and Wine Sciences (CSIC-CAR-UR)
Logroño, Spain

Tiziana Nardi
Council for Agricultural Research and Economics
Research Center for Viticulture and Enology (CREA-VE)
Conegliano, Italy

Dawid Ostrowski
Ostra Distillers
Robinvale, Victoria, Australia

Eva P. Pérez-Álvarez
Institute of Vine and Wine Sciences (CSIC-CAR-UR)
Logroño, Spain

Pilar Rubio-Bretón
Institute of Vine and Wine Sciences (CSIC-CAR-UR)
Logroño, Spain

Inge Russell
Heriot-Watt University (ICBD)
Edinburgh, Scotland

Frank Vriesekoop
Harper Adams University
Newport, Shropshire, United Kingdom

1 Fermentation Process

Tiziana Nardi and Matteo Bordiga

CONTENTS

1.1 WINE

1.1.1 Alcoholic Fermentation

Winemaking represents the process of wine production, from the selection of grapes to the bottling of finished wine. Grapes must be healthy, without defects due to bacterial and/or fungal infections. Winemaking must be performed as soon as possible to prevent wine defects. If grapes are damaged during harvest and/or transport to the winery, issues or defects may arise due to the activity of acetic and/or lactic bacteria. It is possible to obtain about 70 L of must from 100 kg of grapes. The stem, generally separated from the grapes and discarded, represents a lignified vegetable structure containing a high concentration of polyphenols. Polyphenols and aroma compounds are present in the skin of grapes. Acids (tartaric, malic, and citric—about 0.5%–1.5% w/v), sugars (fructose and glucose—about 15%–30% w/v), and small quantities of aromatic compounds and polyphenols are present in the pulp.

Grape crushing represents the first activity of winemaking. Usually, grapes are treated with a mechanical crusher/destemmer. Once crushed, berries are then transferred to the fermentation tank. For white wine production, grape must be treated very carefully because the compounds present in the skin and stem must not pass into the must. Usually, for this typology, crushing is generally obtained with a simple grape pressing. When the must has been prepared, selected yeasts, sulfur dioxide, and nutrient substances are added. Yeasts are generally already present on the grapes and in the winery environment but they can give detrimental results. For this reason, selected commercial strains of yeast are added to the must. Sulfur dioxide is used in winemaking (generally at 50–100 mg/L) due to its inhibitory activity against the natural microflora (bacteria and yeasts) in the juice, thus facilitating the action of the selected yeasts added to the must. Another activity related to this compound acts to inhibit oxidative enzymes responsible for wine browning. Ammonium salts and vitamins (biotin and thiamin) are often added as nutrients for the yeast. If the sugar concentration is low, sugar may be added. However, this addition is strictly subject to local regulations. For example, in Italy it is only possible to add concentrated and rectified must, not sucrose. After about 10 hours from inoculation with yeast, the primary, alcoholic fermentation (AF) starts. This fermentation generally lasts for about 8 days and, during this period, the yeast cells metabolize the sugars in the must, producing carbon dioxide gas and alcohol. The temperature during the fermentation affects the taste of the product. The proper temperature for red wines is typically from 25°C to 28°C, while for white wines it is from 20°C to 25°C. Other substances (minor products) are produced during alcoholic fermentation (e.g., glycerol, acetic acid, higher alcohols, and acetaldehyde). Wine quality is also defined by low concentrations of acetic acid, higher alcohols, and acetaldehyde.

1.1.2 Malolactic Fermentation

Malolactic fermentation (MLF) represents the "second fermentation" of wine that in traditional winemaking used to take place during storage. In modern winemaking the timing of occurrence of MLF is advancing from springtime to autumn (in the Northern Hemisphere), taking place after alcoholic fermentation or during alcoholic fermentation, due both to climate change and to an increasingly careful management of this step. Nevertheless, MLF is the microbial transformation that, more than others, affects postfermentation stages such as aging, stabilization, and possible spoilage. In this context, MLF management has been recently recommended by the Organisation Internationale de la Vigne et du Vin (OIV), together with other biological methods, as a good winemaking practice to avoid wine spoilage that causes major economic losses (OIV, 2014).

1.1.2.1 Lactic Acid Bacteria in Grapes and Wine

Lactic acid bacteria (LAB) perform malolactic fermentation and constitute a ubiquitous group of bacteria that occur in a range of environments, including many foods and beverages. In the oenological environment, they can be found throughout all stages of winemaking. LAB can be isolated on many surfaces and environments including grapevine leaves, grapes, various winery equipment, and barrels. At harvest, low numbers (fewer than 100 cells/g) of LAB can be found on grapes, although acetic bacteria and yeast are found in much higher numbers (Fugelsang, 1997). Lactic acid bacteria display high morphological and physiological diversity; in fact, the term lactic acid bacteria emerged at the beginning of the twentieth century to describe a heterogeneous group of bacteria that are currently defined as spherical (cocci) or rod-shaped (bacilli), Gram-positive, catalase-negative, immobile, nonsporulating, anaerobic, aerotolerant producers of lactic acid as a primary metabolite of sugar (Fugelsang, 1997). Concerning the specific environment of winemaking, the most studied and well-known species of LAB is *Oenococcus oeni*s because it is the predominant species conducting the biological transformation called malolactic fermentation. Nevertheless, the LAB composition of grape must at the beginning of the alcoholic fermentation is much more variable (the most common isolates are part of the genera *Lactobacillus*, *Pediococcus*, *Leuconostoc*, and *Oenococcus*) and generally dominated by *Lactobacillus* species (Betteridge et al., 2015). *Pediococcus* can be found mostly after the MLF as well as in wines of higher pH. Indeed, wine does not differ from other foodstuffs in which LAB are responsible for a critical fermentation step (here, MLF), but they can also cause changes that adversely affect the organoleptic properties of the final product. Therefore, a relevant part of studies about lactic acid bacteria in foods is the study of the negative effects they have on the quality and composition of the final product. A number of recent molecular identification studies have detected new species of lactic acid bacteria in both musts (*Lactobacillus bobalius*, Mañes-Lázaro et al., 2008a; and *Lactobacillus uvarum*, Mañes-Lázaro et al., 2008b) and wines (*Lactobacillus nagelii*, Edwards et al., 2000; *Lactobacillus vini*, Rodas et al., 2006; and *Lactobacillus oeni*, Mañes-Lázaro et al., 2009). Moreover, in recent studies, genera and species not frequently detected in wines such as *Weissella*, *Fructobacillus*, and species such as *Oenococcus kitaharae* and *Lactobacillus fabifermentans* were identified during AF (González-Arenzana et al., 2016) or in grape marc after fermentation (Maragkoudakis et al., 2013). The density of LAB in the initial phases of winemaking (the must phase and first stages of AF) ranges from approximately 103 to 104 colony forming units/mL. Afterwards, only a few species are capable of surviving in grape must and fermenting wine because of the hostile conditions that they encounter: mainly low pH, a lack of nutrients, and the presence of ethanol.

In an interesting recent work, molecular advanced techniques—next-generation sequencing (NGS)—have been used to obtain a more complete picture of how microbial communities change during grape fermentation and how different fermentation techniques might affect the microbial community composition, including bacteria with undesirable effects on wine quality (Piao et al., 2015). A better understanding of the microbial dynamics and their effect on the final product will likely be of great importance to help winemakers produce wine styles of consistent and high quality in the future. Both implantation effectiveness and bacterial metabolism can differ widely depending on species and strains involved in MLF. In general, lacking heme-linked cytochromes and catalase, all LAB obtain energy from carbohydrates by fermentative metabolism (Kandler, 1983). Overall, the LAB group can utilize a wide range of carbohydrates, including the hexoses (glucose, fructose, mannose, and galactose), as well as other pentoses, polyols, and oligosaccharides. This capability is dependent on the species and strains involved, as well as the pH of the medium. Moreover, since malic acid cannot be used by wine LAB as a sole carbohydrate source (see next paragraph), the availability and utilization of fermentable carbohydrates in wine by LAB is essential to enable the onset of bacterial growth and the occurrence of MLF. Furthermore, recent studies have clearly demonstrated that grape-derived phenolic glycosides also significantly stimulate the growth of *O. oeni* (Schopp et al., 2013).

1.1.2.2 The Chemistry of MLF

The term "malolactic fermentation" describes the enzymatic conversion of L-malic acid to L-lactic acid and CO_2 by LAB (Wibowo et al., 1985); thus it does not technically refer to a fermentation process, but rather to a decarboxylation reaction (Figure 1.1). Malic acid is a major acid in wines; the conversion of a dicarboxylic acid (L-malic acid) into a monocarboxylic acid (L-lactic acid) increases pH and modifies the sensory profile of wine. In 1901, Seifert could already describe for the first time that lactic acid bacteria from wine transformed L-malic acid into L-lactic acid and carbon dioxide via a direct reaction, meaning that the intermediate pyruvic acid was not formed during this conversion (Munoz et al., 2011). This reaction is catalyzed by the enzyme malate decarboxylase—often referred to as the malolactic enzyme—that requires cofactors NAD+ and Mn++ (Lonvaud-Funel and De Saad, 1982; Spettoli et al., 1984). The malolactic enzyme, which was purified for the first time in *Lactobacillus plantarum*, has been found in all species of lactic acid bacteria isolated in wine (Munoz et al., 2011).

Malic and citric acids do not serve as the sole energy sources for the growth of LAB (Liu et al., 1995). Consequently, malolactic bacteria require sugars as a carbon source. However, under conditions of limiting sugar availability or of low pH, which inhibit sugar metabolism, energy (ATP) generated from MLF is beneficial to cell growth (Henick-Kling, 1993). Another, minor (<1%) activity of the malolactic enzyme has also been suggested to stimulate the metabolic activity and initial growth rates of wine LAB (Morenzoni, 2006). All wine lactic acid bacteria are able to perform the malolactic reaction. However, *Oenococcus oeni* is the major actor that carries out this fermentation since it is the best, well-adapted wine–associated species (especially when pH is beyond 3.5) and, for a long time, it has been used almost exclusively for the induction of malolactic fermentation in wines (Bartowsky et al., 2015; Wibowo et al., 1985). Among LAB, *Oenococcus* was separated from the genus *Leuconostoc* by 16S rDNA sequence analysis in 1995 (Dicks et al., 1995). For a long time, this genus only included one species: *O. oeni*. In 2006, *Oenococcus kitahareae* (Endo and Okada, 2006) was described, isolated from a composting distilled shochu residue and, more lately, a third species was ascribed to the genus *Oenococcus alcoholitolerans*, isolated from cachaça and ethanol fermentation processes (Badotti et al., 2014). In recent years, a number of scientific studies have focused on *Lactobacillus plantarum* as a potential starter for MLF, also leading to industrial interest in this genus (Berbegal et al., 2016; Iorizzo et al., 2016; Spano and Capozzi, 2011; du Toit et al., 2011). These studies open a perspective on the potential of *Lactobacillus* as the next generation of MLF starter cultures, discussing their occurrence during winemaking,

$$\text{L-malic acid} \xrightarrow[\text{Malolactic enzyme}]{NAD^+,\ Mn^{2+}} \text{L-lactic acid} + CO_2$$

(*malate NAD⁺ decarboxylase*)

FIGURE 1.1 Reaction of decarboxylation, describing the enzymatic conversion of L-malic acid to L-lactic acid and CO_2 by lactic acid bacteria (LAB).

major metabolic activities, factors influencing their growth, and their performance in conducting MLF that will probably be further developed in the near future.

1.1.3 Sensory Impacts of Malolactic Fermentation

1.1.3.1 Acidity Reduction

The decarboxylation of malic acid to lactic acid is a biological deacidification reaction, well recognized as one of the main metabolic capabilities of LAB in wine, and its conduct is of major commercial importance to the winemaking process. Depending on the style of a wine, this enzymatic conversion can be beneficial or detrimental. It is a fact that the sensory impact of deacidification is particularly detectable and appreciated in wines with high acidity; thus the development of strategies to favor a microbial deacidification of low pH wines is particularly critical. Wines produced from grapes cultivated in cool climate areas contain a naturally high level of malic acid (up to 8 g/L) and are considered to benefit from an acidity reduction (Lasik, 2013). Moreover, in such wines the induction of MLF is often difficult to achieve due to harsh conditions (mainly, low pH), so it is important to prevent sluggish or stuck fermentation. Thus, many recent investigations focused on wines with high acidity (cool-climate whites, most of the sparkling base wines, and other varieties), improving knowledge in this field and proposing and optimizing techniques as the use of bacterial starters with early inoculation as efficient methods for achieving MLF (Guzzon et al., 2016; Knoll et al., 2012; Ruiz et al., 2012). On the other hand, wines produced from grapes grown in warm to hot regions have lower total acidity, and a further reduction in acidity from MLF can have a negative impact on their quality, causing a flat taste. Nevertheless, in wines characterized by high pH, biological deacidification is often desired because it results in improved microbial stability. In these cases, the impact of MLF is rather indirect, although the lack of acidity caused by this transformation is not strongly perceivable or even detrimental to freshness. It is usually accepted in order to remove nutrients (malic acid as a possible carbon substrate, and other compounds consumed by LAB) that would leave place for late, uncontrolled, and possibly incomplete malolactic fermentations (eventually ascribed to *Pediococcus* spp.) or for other microbial transformations including development of *Brettanomyces* in wines during aging (Morenzoni, 2006). Furthermore, a very recent frontier research approach introduced the perspective of using *Lactobacillus* strains with ability to induce biological acidification in low acidity grape musts to obtain more acidic wines, together with the achievement of malolactic fermentation. This recent study shows a selection of *Lactobacillus* strains that can grow in must, carry out MLF, and at the same time acidify grape must by synthesizing lactic acid from sugars. Indeed, homofermentative or facultative heterofermentative bacteria such as *Lactobacilli* are good candidates to be used as acidifying starters since they can synthesize only lactic acid from grape must sugars by lactic fermentation and have no danger of acetic acid synthesis (Lucio et al., 2016).

1.1.3.2 Production of Flavor-Active Compounds

In addition to the deacidification reaction that characterizes MLF, it is becoming increasingly recognized that a diverse range of other metabolic activities are associated with the growth and development of LAB in wine, which can have a significant influence on wine quality. The complex aroma and flavor compounds found in wine largely originate from the grape, from yeast metabolism during alcoholic fermentation, and from oak when used. Bacterial metabolism during malolactic fermentation might contribute to wine flavor by the formation of additional compounds and the modification of grape-, yeast-, and oak-derived compounds (Swiegers et al., 2005). Indeed, MLF affects the final aroma and taste balance by modifying fruit-derived aromas and producing aroma-active compounds. Research has shown that LAB have the potential to impact the aroma profile of wine by

i. The production of volatile secondary metabolites
ii. The modification of grape- and yeast-derived metabolites including ethyl esters, acetate esters, primary terpene alcohols, glycoside-related aroma compounds, acids and alcohols
iii. The removal of existing flavor compounds by metabolism and adsorption to the cell wall (Bartowsky and Henschke, 1995; Bartowsky and Pretorius, 2009; Bartowsky et al., 2015; Lonvaud-Funel, 1999; Nielsen and Richelieu, 1999; de Revel et al., 1999; Ugliano et al., 2003)

In the same studies, flavor attributes imparted by MLF are usually described as buttery, lactic, nutty, yeasty, oaky, sweaty, and earthy; MLF may also impact fruity and vegetative aromas, as well as the mouthfeel of wine. Many of these alterations are strain dependent; however, the vinification technique can also affect the final wine aroma profile and these flavor impacts of individual bacterial strains are of great interest for winemakers (Lerm et al., 2010).

From a molecular point of view, over the last 15 years, research studies have deepened the characterization of *O. oeni* diverse array of secondary metabolic activities during MLF, which can modify the sensory properties of wine (Cozzolino, 2016). These secondary activities include the metabolism of organic acids, carbohydrates, polysaccharides and amino acids, and numerous enzymes such as glycosidases, esterases, and proteases, which generate volatile compounds well above their odor detection threshold. Recent studies using array-based comparative genome hybridization and genome sequencing of *O. oeni* strains have revealed the large genomic diversity within this species, confirming that phenotypic variation between *O. oeni* strains is central for producing different wine styles (Bartowsky and Borneman, 2011; Liu et al., 2016).

From an applicative point of view, results based on wide surveys usually show that lactic acid bacteria modify the fruity notes of wines but without a specific trend (Antalick et al., 2012), since the impact is strongly dependent on the strain carrying out the fermentation and on its inoculation timing. For example, in recent studies comparing aromatic compounds released from natural precursors by selected *Oenococcus oeni* strains during malolactic fermentation, some bacteria resulted in good candidates to be used when floral wines are desired, while others were the best preserving the fruity aroma of wines (Pérez-Martín et al., 2015; Sumby et al., 2013). Other works comparing the effects of inoculating grape must with malolactic bacteria at various stages of alcoholic fermentation (usually, the beginning of alcoholic fermentation [co-inoculation with yeast], midalcoholic fermentation, and/or postalcoholic fermentation) on the wine chemical composition showed important differences. This often suggested that co-inoculation is a worthwhile alternative for winemaking if compared with traditional postalcoholic fermentation LAB inoculation or with spontaneous MLF (Abrahamse and Bartowsky, 2012; Cañas et al., 2015; Guzzon et al., 2016; Knoll et al., 2012). In general, more and more findings illustrate that MLF is an effective and novel way of modulating the volatile and aroma compound profile of wine.

1.1.3.3 Risk of Organoleptic Defects

Many secondary metabolites produced by bacteria are volatile and potentially affect negatively wine sensory qualities. Lactic acid bacteria can spoil wine during winemaking if MLF is not controlled, takes place too early, or gets stuck or, afterwards, during maturation and aging (Lonvaud-Funel, 1999). In the first case, deviations of metabolism by the fermenting bacteria (usually *O. oeni* or *L. plantarum*) during malolactic fermentation can affect wine quality. In the second case, growth of *Lactobacillus*, *Pediococcus*, and even some *Oenococcus* species in wine, happening after malolactic fermentation, can lead to numerous spoilage scenarios, as they can form undesirable aromas and flavor compounds. Overall, the chemicals involved in LAB-caused wine depreciation include compounds ranging from acetaldehyde to acetic acid, diacetyl, tetrahydropyridines, acrolein, beta-glucans, and biogenic amines, as summarized by Bartowsky (2009). Concerning organic acids, not only malic acid but also citric acid in the wine is metabolized by *O. oeni* and by numerous genera of the lactic acid bacteria, resulting in the production of acetic acid and diacetyl. One of the intermediary compounds in the metabolism of citric acid, diacetyl, is considered one of the most important flavors produced during MLF. When present at a concentration above the sensory threshold, diacetyl gives the wine an aroma that can be characterized as buttery or nutty. It has been demonstrated that threshold values in different wines vary from 0.2 mg/L in Chardonnay wine to 2.8 mg/L in Cabernet Sauvignon. The compound can add pleasant aromas and complexity to wine at concentrations below 4 mg/L, but above this, it becomes unpleasant with overt buttery notes (Bartowsky, 2009; Nielsen and Richelieu, 1999; Swiegers et al., 2005). In general, wines that have undergone MLF have higher concentrations of diacetyl (Liu, 2002).

A number of factors, including some that the winemaker can control, particularly during malolactic fermentation, affect the final level of diacetyl in wine. The bacterial strain used, oxygen exposure, fermentation temperature and duration of malolactic fermentation, wine type, and sulfur dioxide impact diacetyl production (Bartowsky, 2009). It should be pointed out that diacetyl is formed chemically from the oxidative decarboxylation of α-acetolactate, an unstable intermediary compound produced during citrate metabolism. Its formation and degradation by bacteria are directly related to growth and to the metabolism of sugar and malic acid together with citric acid, since it is formed as an intermediate metabolite in the reductive decarboxylation of pyruvic acid (Swiegers et al., 2005). Moreover, previous studies observed the utilization of diacetyl by *Oenococci*, which is not surprising, since many LAB contain diacetyl reductase that converts the flavorful diacetyl to the much less flavorful acetoin and 2,3-butanediol; therefore some strains can decrease the amount of this molecule (Liu, 2002).

Certain strains of lactic acid bacteria (particularly *Lactobacillus* strains) are also capable of degrading tartaric acid, although this capacity is much less common than that of malic and citric acid metabolism. Tartaric acid is only degraded in certain conditions after the metabolism of other organic acids. The catabolism of this acid always alters wine by causing a slight reduction in fixed acidity and an increase in volatile acidity (Munoz et al., 2011). Acetaldehyde, a highly volatile compound with an apple-like and nutty aroma, is one of the most important sensory carbonyl compounds formed during vinification, constituting more than 90% of the total aldehyde content in wine, and originates mainly from yeast metabolism (Swiegers et al., 2005). At low levels, acetaldehyde gives a pleasant fruity aroma, but results in an undesirable aroma described as green, grassy, or apple-like when present in excess. The aroma can be masked by the addition of SO_2, but binding of SO_2 to acetaldehyde reduces its effectiveness as an antimicrobial compound and its antioxidant effect. The interaction of acetaldehyde with phenolics improves red wine color by forming stable polymeric pigments resistant to SO_2 bleaching, but it may also induce phenolic haze and eventual deposition of condensed pigments (Bauer and Dicks, 2004; Swiegers et al., 2005). Some strains of *Oenococcus* and *Lactobacillus* spp. can metabolize free and SO_2-bound acetaldehyde to acetic acid and ethanol (Osborne et al., 2000). Free SO_2 released from the degradation of SO_2-bound acetaldehyde by SO_2-sensitive strains of *O. oeni* may cause inhibition, resulting in stuck or sluggish MLF. The chemical and sensory impact of the ethanol and acetic acid formed by the metabolism of acetaldehyde by lactic acid bacteria is believed to be limited, but the reduction in the acetaldehyde pool in wine is believed to influence final wine (Osborne et al., 2006). By using efficient acetaldehyde-degrading strains to conduct MLF, the addition of SO_2 to reduce acetaldehyde aroma can be minimized (Bauer and Dicks, 2004).

Mousy wines result from the metabolism of ornithine and lysine, leading to the formation of extremely potent and unpleasant nitrogen-heterocylic compounds: tetrahydropyridines. These molecules are perceived on the back palate as a persistent aftertaste reminiscent of caged mice because of interactions with the mouth environment, since an increase in pH renders the compounds volatile (mousy taint is indeed described as a taste rather than a smell). Production of these compounds seems to be limited to the heterofermentative LAB (*O. oeni* and some species of *Lactobacillus*) (Costello et al., 2001). A wine with a viscous and thick texture is referred to as "ropy" because of the presence of excess exopolysaccharides such as B-D-glucan. In this context, the production of exopolysaccharides is almost exclusively because of *Pediococcus* growth in wine (Bartowsky, 2009).

Biogenic amine (BA) formation results from the decarboxylation of the corresponding amino acids by the action of microorganisms. Many bacterial genera are able to decarboxylate amino acids. This reaction is thought to favor growth and survival in acidic media, since it induces an increase in pH. In wine, several amino acids can be decarboxylated; as a result, histamine, tyramine, putrescine, cadaverin, and phenylethylamine are usually found, the first three being the most frequent (Lonvaud-Funel, 2001). The presence of these compounds is considered by some authors a fundamental parameter to the detriment of alcoholic beverages, for both sensory problems and consumer health issues (Marques et al., 2008). Since winemaking involves the growth of lactic acid bacteria for malolactic fermentation, biogenic amines may occur. However, not all bacterial strains are able to produce them. Since the early 2000s, it has been possible to detect the presence of undesirable histamine-producing strains by PCR test or DNA probe based on the presence of the gene encoding histidine decarboxylase, ornithine, and/or tyrosine decarboxylase. This problem usually arises in spontaneous MLF, where an unknown LAB population carries out the fermentation, whereas the absence of these genes is increasingly required as a selection criterion for selected starters (Lonvaud-Funel, 2001).

1.1.4 Current Knowledge on Malolactic Fermentation: Managing the Process

1.1.4.1 Physicochemical Parameters Affecting MLF Development

Various studies have reported many factors that influence the occurrence of LAB and MLF in wines. In addition to oxygen and CO_2, Henick-Kling (1993) listed carbohydrates, amino acids, vitamins and minerals, and organic acid content, as well as the alcohol level, pH, SO_2, the method of vinification, and interrelationships between LAB and other wine microorganisms to be the most influential factors to affect LAB growth. Much other research work over the last 20 years has deepened the impact of these factors, providing to winemakers an interpretation key to understanding MLF problems and some tools to manage the MLF process (Alexandre et al., 2004; Bauer and Dicks, 2004; Lerm et al., 2010; Liu et al., 2016; Terrade and de Orduña, 2009). Indeed, if some of the parameters affecting MLF feasibility are not easy to change (grape variety, alcohol or potential alcohol, pH, malic acid content), many others can be managed by the winemaker in order to minimize risks of stuck or sluggish fermentations (SO_2, yeast strain chosen for AF, temperature, nutrients to be added). Table 1.1 summarizes the main physicochemical factors affecting MLF and their potential impact, according to recent studies.

1.1.4.2 Use of Starter Cultures

Winemakers started to recognize the benefits of inoculating grape must or wine with commercial starter cultures of LAB to ensure the successful completion of MLF and to reduce the risks associated with spontaneous MLF in the 1980s (Davis et al., 1985). Potential risks include the presence of unidentified/spoilage bacteria that can produce undesirable or off-flavors, a delay in the onset or completion of MLF, the production of biogenic amines and the development of bacteriophages—all of which contribute to a decrease in the quality of the wine (Bauer and Dicks, 2004; Jaomanjaka et al., 2016; Lerm et al., 2010). Thus, while MLF can occur spontaneously, more and more winemakers, particularly in New World winemaking regions, prefer to minimize the danger of a failed or sluggish MLF by inoculating with a reliable, commercially available starter culture. By inoculating a commercial starter culture, most of which contain *O. oeni* as the single LAB culture, the winemaker can promote the rapid start and completion of MLF and also encourage a positive flavor contribution to the wine (Bartowsky et al., 2015; Lerm et al., 2010).

O. oeni is still the main species used in the many commercial starter cultures available today. Most of the cultures are prepared with single strains or a mixture of

TABLE 1.1
Main Factors Affecting MLF Effectiveness

Factor	Effect on MLF	Mainly Cross Linked with	Ref.
Temperature[a]	Temperature affects growth rate, length of the lag phase, and maximum population of malolactic bacteria. Optimal growth rate of *O. oeni*, in absence of ethanol, is around 25°C. Optimum growth at 10%–14% alcohol is between 18°C and 20°C.	Ethanol	Bauer and Dicks, 2004
Ethanol	Ethanol decreases optimal growth temperature of LAB; ethanol tolerance is decreased at elevated temperatures due to membrane fluidity. Generally *O. oeni* strains are able to survive and proliferate in 10% ethanol, concentrations > 14% (v/v) inhibit *O. oeni* growth. Ethanol tolerance in LAB is strain dependent in *Oenococcus* and *Lactobacillus* species.	Temperature, SO_2	Bauer and Dicks, 2004; Henick-Kling, 1993; Lerm et al., 2010
pH	Wines of pH 3.3 and above generally exhibit few problems, whereas at lower pH, difficulties may arise. *O. oeni* usually represents the dominant species in wine below pH 3.5. At higher pH *Lactobacillus* and *Pediococcus* spp. may survive and grow. pH influences SO_2 impact on MLF.	SO_2	Bauer and Dicks, 2004; Henick-Kling, 1993
SO_2[a]	Antimicrobial activity of SO_2 strongly affects growth of LAB cells and influences malolactic activity. pH has crucial effect on the form of SO_2 present. The lethal level of molecular SO_2 for most wine LAB is <0.3 mg/L (corresponding, e.g., to 5 mg/L free SO_2 at pH 3.2 and 13% ethanol, 15 mg/L free SO_2 at pH 3.6 and 13% ethanol). To a lesser extent, also, bound inhibits MLF (bound SO_2 at 30 mg/L delays the growth of LAB, bound SO_2 > 50 mg/L reduces malolactic activity by 50%). Some yeast strains are capable of producing rather large amounts of SO_2 during AF.	pH, yeast	Bauer and Dicks, 2004; Henick-Kling, 1993; Krieger and Silvano, 2016; Lerm et al., 2010
Malic acid	L-malate stimulates growth and biomass production by *O. oeni* up to 3 g/L. Higher concentrations may show inhibitory effect. At low pH, L-malate is metabolized at a high rate, whereas carbohydrate metabolism proceeds very slowly.	pH	Bauer and Dicks, 2004; Krieger and Silvano, 2016; Lerm et al., 2010

(*Continued*)

TABLE 1.1 (CONTINUED)
Main Factors Affecting MLF Effectiveness

Factor	Effect on MLF	Mainly Cross Linked with	Ref.
Nutrients[a]	Depending on strains, some amino acids (4 to 9) are essential for growth, while others are required for optimum growth of *O. oeni*. Nicotinic acid, riboflavin, pantothenic acid, and either thiamine or pyridoxine are necessary vitamins, manganese an essential ion. Yeast can consume these nutrients; yeast autolysis releases nutrients that stimulate LAB growth and malolactic activity.	Yeast	Krieger and Silvano, 2016; Terrade and de Orduña, 2009
Yeast strain[a]	Inhibitory and stimulatory effects differ between strains: AF with SO_2 producing yeast strain results in wine inhibitory to MLF and medium chain fatty acid production by yeast negatively affects LAB growth and reduce ability to metabolize malic acid. Antagonism between yeast and LAB during alcoholic fermentation may be also explained by nutrient depletion and production of specific inhibitory peptides. New nontargeted studies recently provided insights into other molecules belonging to classes of oligopeptides, carbohydrates, amino acids.	Nutrients, SO_2	Alexandre et al., 2004; Bauer and Dicks, 2004; Lerm et al., 2010; Liu et al., 2016
Phenolic compounds	Polyphenolic compounds impact the growth of bacteria in wine, but research results are very confusing, with some studies describing the phenolic compounds as activators, and others as inhibitors. In a recent study, the addition of grape tannins and red wine extracts to a white wine showed a positive influence on the growth, survival, and malolactic activity of *O. oeni* strains; a 3-O-galloyl esterase and gallate decarboxylase have been found in *L. plantarum*. Results showed different malolactic behaviors in relation to wine phenolic compositions for *O. oeni* and *L. plantarum*. Diversity was found within each group.		Chasseriaud et al., 2015; du Toit et al., 2011

[a] Factor that the winemaker can manage during the vinification process.

two or three strains. Nowadays, many commercial starter cultures are available to induce malolactic fermentation; most consist of strains of lactic acid bacteria, which have a high malolactic activity and a high tolerance of low pH and high ethanol content. These starter cultures have been commercialized in various forms, including fresh, frozen, and lyophilized cultures (Gonzalez et al., 2011; Munoz et al., 2011). Since the introduction of malolactic starter cultures for improving the induction of MLF, there has been considerable research and development aimed at optimizing inoculation strategies and strain selection to further enhance MLF efficiency.

One of the major considerations has been to determine the optimal time point for inoculation. Starter cultures can be co-inoculated with yeast (at the beginning or towards the end of alcoholic fermentation) or inoculated sequentially (after alcoholic fermentation). Generally, relative to sequential inoculation, co-inoculation reduces overall vinification time. This has important consequences for the wine industry: Speeding up vinification rate leads to more rapid wine stabilization and reduces the risk of spoilage (Bartowsky et al., 2015). Indeed, grape must is an environment more suitable than wine for microbial growth because it does not contain some of the limiting factors; in these conditions, better adaptation and activity of malolactic bacteria are expected. The obstacles to microbial activity, in particular ethanol and eventually some yeast-produced molecules, accumulate gradually during alcoholic fermentation, allowing time for bacterial biomass adaptation and ensuring a greater chance of survival for lactic acid bacteria (Guzzon et al., 2016).

Nonetheless, there has been some reluctance by industry to adopt co-inoculation as a practice. One possible concern for using this tool is that *O. oeni* is heterofermentative. This means that under certain conditions, one of the products of its sugar metabolism is acetic acid. Thus it might be assumed that *O. oeni* has the potential to produce wines with elevated volatile acidity. Considering that the consumption of sugar and malic acid can occur simultaneously, the bacteria utilized in this kind of fermentation must be tailored specifically to avoid spoilage phenomena associated with the consumption of sugar by lactic acid bacteria via heterolactic fermentation. It has been demonstrated by several works that, at least under winemaking conditions and with careful management of fermentations, *O. oeni* survives better and does not produce acetic acid when grown in grape juice at low pH (Zapparoli et al., 2015a,b). Under these conditions, indeed, it preferentially utilizes organic acids (malic and citric acids) rather than sugars (Henick-Kling, 1993).

There has been growing interest internationally in studying and industrially using co-inoculation in the production of many red and some white wines. Co-inoculation strategies have been found to benefit production of a wide range of grape varieties providing reliable malolactic fermentation (as reviewed by Bartowsky et al., 2015), as well as in different contexts of harsh situations including high alcohol (Zapparoli et al., 2009) and low pH (Guzzon et al., 2013, 2016; Knoll et al., 2012; Pan et al., 2011), and to be appreciated for their aromatic contribution to the final product, as discussed in Section 1.1.3.2. As the pH increases, there is a shift to a preference for sugar utilization of *Oenococcus oeni*, thus increasing the risk of acetic acid accumulation in case of early inoculation (Bartowsky et al., 2015). However, in many spontaneous MLFs in wines exhibiting a pH over 3.5, *Pediococcus damnosus* and certain unknown *Lactobacillus* strains may dominate, creating the risk of organoleptic defects. The application of proper microbiological selection criteria to wines in this category has recently led to the isolation of bacterial strains capable of inducing a quality MLF in high-pH wines. In this context, one of the emerging trends in the application of malolactic fermentation is to use alternatives to *O. oeni* as starter cultures, with a particular interest in *Lactobacillus plantarum*. This bacterium is homofermentative for hexoses such as glucose; it will produce only lactic acid and not acetic acid when it metabolizes glucose, thus eliminating any potential risk of producing volatile acidity.

In addition, *L. plantarum* has a preference for malate as an energy source at low pH, making it suitable for MLF in a cofermentation or even prealcoholic fermentation, when it can begin decarboxylating malate before a yeast starter culture has been added (Bartowsky et al., 2015; du Toit et al., 2011). From a sensory point of view, *L. plantarum* produces a broader range of extracellular enzymes, including glycosidases and esterases, than *O. oeni*. These enzymes potentially play an important role in the development of wine sensory properties through the release of flavor molecules; thus *L. plantarum* may enhance wine sensory properties more largely than *O. oeni* (Bartowsky et al., 2015; du Toit et al., 2011). Last but not least, the wide genetic diversity of the *Lactobacillus* spp. and in particular of the *plantarum* species makes it interesting for recent projects aiming at enhancing regionality of wine through MLF (Iorizzo et al., 2016; Spano and Capozzi, 2011; Testa et al., 2014). Indeed, recent studies have shown how microbial activity is an integral part of regionally distinct wine characteristics (terroir), an important aspect of wine production and consumer appreciation. Wine production and grape and wine microbiota present regionally defined patterns associated with vineyard and climatic conditions and are objects of increasing interest (Bokulich et al., 2016) thanks to the high-throughput sequencing technologies that are providing the greatest advance to this sector of research. So, research has been conducted on regional microbial isolates and their potential application in winemaking, since grape resident microbial diversity forms an untapped reservoir of indigenous bacteria strains and may be primarily considered in an MLF starter selection scheme, and new findings are likely to be expected in this field.

1.1.4.3 Control of MLF to Avoid Further Microbial Spoilage

How best to avoid wine spoilage is not always easy to define. Even appropriate hygiene practices and the chemically harsh nature of wine cannot be relied on as a deterrent to unwanted bacteria. As an initial barrier, the high ethanol concentrations (up to 16% v/v) and high wine acidity (pH as low as 2.9) can inhibit development of bacterial populations; however, in wines with lower ethanol concentrations and low acidity (above pH 3.6), it can be challenging to arrest late bacterial growth, since, unlike the treatment of wort in beer brewing, grape must is not pasteurized prior to yeast inoculation (Bartowsky, 2009; Ribéreau-Gayon et al., 2006). Some kinds of risks are easier to avoid than others. For example, injudicious aeration during and/or after the winemaking process can result in the growth and activity of acetic acid bacteria, high volatile acidity, and a vinegary taint in wine. These bacteria are classified into the genera *Acetobacter*, *Acidomonas*, *Gluconobacter*, and *Gluconacetobacter*; of these, *Gluconobacter oxydans*, *Acetobacter aceti*, *Acetobacter pasteurianus*, *Gluconacetobacter liquefaciens*, and *Gluconacetobacter hansenii* are normally associated with grapes and wine, but can be easily avoided by limiting oxygen (Swiegers et al., 2005).

Sulfur dioxide (SO_2) is the key additive for the preservation of wines. Carbonyl and keto compounds in wine can bind to SO_2 and decrease its efficacy, resulting in higher total SO_2 requirements. Increased consumer demand for low sulfite and organic wines poses production challenges if SO_2 binders have not been properly managed during vinification. In this context, the correct management of fermentation stages (alcoholic and malolactic) can be critical to obtain the best ratio between free and bond SO_2. In particular, malolactic fermentation has been known to reduce bound SO_2 levels. Some studies suggest that microbiological wine stabilization 1 week after malic acid depletion is an effective strategy for maximum removal of SO_2 binders while reducing the risk of possible post-ML spoilage by *O. oeni* leading to the production acetic acid and biogenic amines (Jackowetz and Mira de Orduña, 2012; Osborne et al., 2006).

Another point on which MLF management can be important to prevent spoilage is biogenic amine formation. Previous works showed that most of the commercial malolactic bacteria did not produce BA, and that the application of commercial malolactic starters in wines is useful to reduce the BA amounts, since in the inoculated wines BA concentrations were significantly lower when compared with those not inoculated (Lonvaud-Funel, 2001). These results suggest that the use of selected malolactic starters can minimize BA production (Marques et al., 2008). When BA-producing strains are present in indigenous microflora, the winemaker is particularly encouraged to inoculate selected malolactic starters to replace the indigenous microflora. Nevertheless, when the dominance of starter cultures on the indigenous BA-producer microflora is not sufficient, this does not represent the definitive solution. Thus, a recent work reports the selection of autochthonous strains of *Lactobacillus plantarum* able to degrade BA and their suitability to be used as malolactic starter in wine production (Capozzi et al., 2015). This represents one more scenario in which BA could be controlled, in fermented foods, by modulating microbial resources as MLF.

Moreover, MLF management strongly affects the development of *Brettanomyces* during subsequent wine aging. Some studies showed that wines that underwent MLF inhibited the growth of *Brettanomyces*, resulting in a product containing little or no volatile phenols. Wines that did not undergo MLF or that underwent late spontaneous MLF that proceeded slowly allowed proliferation of *Brettanomyces*, resulting in a product containing more volatile phenols (Gerbaux et al., 2009; Nardi et al., 2014). Thus, early inoculation of wine with malolactic bacteria may be a tool for lowering the risk of volatile phenol production. In a recent resolution, the International Organisation of Vine and Wine drafted the "Code of good vitivinicultural practices in order to avoid or limit contamination by *Brettanomyces*" (OIV, 2014). The document attests that if MLF is delayed, the risk of production of volatile phenols increases because *Brettanomyces* can take advantage of the time between alcoholic and malolactic fermentation to multiply, benefiting from the absence of SO_2. Thus, the use of malolactic starters is proposed as a good way to limit *Brettanomyces* development. Moreover, co-inoculation or early sequential inoculation is presented as the best tool to prevent *Brettanomyces* contamination by reducing the lag phase in between AF and MLF, as shown in scientific studies (Coulon et al., 2010; Gerbaux et al., 2009). After malolactic fermentation, it is recommended to eliminate all microorganisms—particularly by adding SO_2.

In case of infection, the addition of chitosan is one alternative for controlling the growth of undesirable microorganisms, particularly *Brettanomyces* (as also reported by Nardi et al., 2014). In this context, some investigations may propose in the future new strategies to select *Oenococcus oeni* strains holding competitive advantages in surviving in wine, preventing microbial spoilage, and improving the wine organoleptic profile, thanks to their biofilm formation. Indeed, prior observations showed that *O. oeni* is able to resist several months in harsh wine conditions when adhered on oak barrels. Since biofilm is a prevailing microbial life style in natural environments, the capacity of *O. oeni* to form biofilms was recently investigated on winemaking material such as stainless steel and oak chips (Bastard et al., 2016). Promising results showed that biofilm could be considered as a novel approach for performing MLF and as an alternative way of adapting MLF starters to wine stress.

1.2 BEER

1.2.1 Microbial Aspect of Malting and Brewing

Starting from raw materials to stability, brewing beer involves microbial activity at every stage. Most of these activities are desirable, as beer is the result of a traditional food fermentation. Conversely, others represent threats to the quality of the final product and must be controlled actively through careful management. While beer fermentation itself represents a monocultural microbial phenomenon (with few exceptions), the complete process involves a succession of microbial constituents that dramatically influence the final product.

1.2.1.1 Barley

Brewing microbiology starts in the barley field, where plant–microbial interactions and the microbiological status of the grain may have serious implications for processing and beer quality. Even if these microbes do not survive the malting and brewing processes, secretory factors may persist, affecting quality. After harvest, barley may be stored for a time prior to malting to overcome dormancy. During this time, microbes continue to grow, interacting with the living grain. For this reason, conditions must be monitored carefully to ensure that the grain is stored in a low-moisture, low-temperature environment, minimizing microbial growth (Petters et al., 1988). Fungal pathogens (such as *Fusarium* spp.) of barley and other cereals are responsible for producing mycotoxins that survive the brewing process. Some mycotoxins have been detected in barley, including deoxynivalenol (DON), nivalenol, T-2 toxin, HT-2 toxin, and diacetoxyscirpenol. Fungal infection is also responsible for an issue with more immediate consequences to the consumer: gushing. Hydrophobins (hydrophobic fungal peptides) are the cause of this phenomenon. These compounds serve as nucleation sites for CO_2 bubbles in beer, resulting in the spontaneous release of gas and overfoaming once the container is opened (Deckers et al., 2011). The correlation between fungal growth and gushing is well established. A successful approach for suppression of *Fusarium* growth on barley (and consequently reduction of gushing) is represented by in-field application of LAB. Moreover, different plant pathogenesis factors may also precipitate premature yeast flocculation (PYF) during fermentation.

Yeast flocculation occurs when cell wall mannoproteins bind lectin-like glycoproteins on other cells (Stratford and Carter, 1993). In PYF, yeast aggregation and settling occur prior to full attenuation of sugar, resulting in significantly decreased beer quality (e.g., incomplete fermentation, off-flavors).

1.2.1.2 Malt

The process of malting consists of three primary steps: steeping, germination, and kilning. The successive steeping and aeration cycles promote more than plant growth, and although kilning diminishes viable counts of microbes, microbial activity during germination may affect beer quality downstream. After kilning, low-moisture conditions must be maintained with particular attention to avoiding microbial spoilage, especially as malt results are somewhat hygroscopic and rich in soluble nutrients at this stage. To change the steep liquor between air rests, reducing dissolved nutrients and reintroducing suspended biomass, and to control the steep temperature (microbial growth is limited at lower temperatures) represent a valuable approach to reducing the inhibitory effects of microbial growth on malt quality (Briggs and McGuinness, 1993). Previous studies have also recommended microbial inoculation of steep liquor to control the growth of detrimental microbiota during germination. *Wickerhamomyces anomalus* has been shown to inhibit *Fusarium* growth on malt when added during steeping, thus preventing hydrophobin production and beer gushing. Similarly, *Geotrichum candidum* and *Lactobacillus plantarum* have also been shown to diminish *Fusarium* growth on malt (Laitila et al., 2002).

1.2.1.3 Wort

During mashing, the microbial load diminishes, but thermotolerant microbes, especially homofermentative LAB, remain active in the nutrient-rich, high-moisture environment. However, bacterial growth may have beneficial consequences, and mash acidification by lactic acid bacteria can improve the extraction, fermentability, and nitrogen yield of wort. Similarly, foam stability, color, and flavor of beer may be improved (Lowe et al., 2005). Bacterial growth can also cause serious problems during extended mashing. For example, *Bacillus* spp. is responsible for excessive acidification and nitrosamine formation (by reduction of nitrate to nitrite). Growth of *Clostridium* can produce high levels of butyric acid, giving the beer a cheese-like aroma. Following the mash, wort is boiled for an extended period, thus effectively sterilizing it. However, wort is a nutrient-rich, high-pH (about 5.5) medium, so, if appropriate precautions are not taken to ensure rapid fermentation, once it leaves the kettle it is vulnerable to opportunistic spoilage agents. Gram-negative enterobacteria, especially species of *Klebsiella*, *Citrobacter*, *Enterobacter*, *Obesumbacterium*, and *Escherichia*, represent the most prevalent wort spoilers. In wort, these produce dimethyl sulfide, organic acids, and 2,3-butanediol, giving beer an unpleasant fruity or vegetal aroma.

1.2.1.4 Beer

After a cooling period, oxygenated wort is pumped to fermenters, where strains of *Saccharomyces* are added to rapidly convert it to beer through the fermentation of maltose and other sugars to ethanol and carbon dioxide. The resulting conditions are hostile to the growth of most microorganisms. The fermentation process of the young beer follows the process of the maturation in the lagering tanks, designed only for the second phase of the beer fermentation process. During this production phase, the beer maturates, or saturates with the carbon dioxide, which is formed from the remaining yeast activity. Maturation represents the longest stage of the beer production process. The chosen recipe, temperature, and set pressure in the maturation tank may determine the length of the beer maturation process. Maturation of bottom-fermented beer takes about 3 weeks and light lager at least 30 days; beer with special alcohol content (>12%) takes 60 days up to several months. Spatial orientation of the tanks (horizontal or vertical) represents a crucial issue in designing the breweries (Figures 1.2 and 1.3). Even if the use of the horizontal type shows an advantage in saving space, the vertical solution represents the standard type in modern breweries and this concept is preferred where small space is not a fundamental limitation. Together with the spatial orientation of the tanks, different cooling layouts appear fundamental in the development of a brewery. Different types of lagering tanks, according to the design of a cooling method, are reported next.

1.2.1.5 Beer Maturation Lagering Tanks with Single Jacket (Air Cooled)

Simple and cheap noninsulated tanks are cooled by cold air. This represents the traditional way of cooling in a lager cellar. The advantage is represented by low cost, but some disadvantages exist, such as high cooling energy demands (this may be affected by the quality of the walls' thermal insulation in the lager cellar). It must be highlighted that, for the brewery cellar with single-shell tanks to create a cooling box, it appears necessary to have a room with very good insulation of walls and ceiling and with a refrigerator door. The operating costs of cooling are particularly

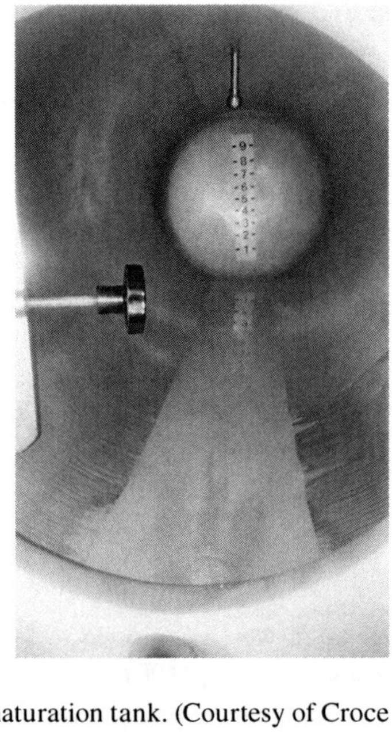

FIGURE 1.2 Horizontal maturation tank. (Courtesy of Croce di Malto S.r.l., Trecate, Italy.)

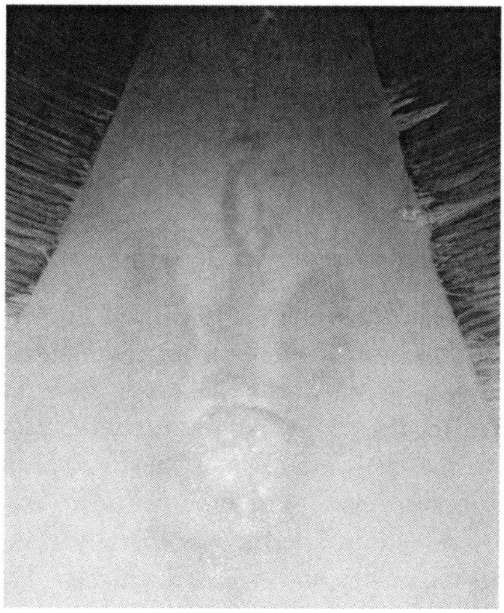

FIGURE 1.3 Yeast settled to the tank bottom. (Courtesy of Croce di Malto S.r.l., Trecate, Italy.)

representative during the summer months, significantly higher than the lager cellar with insulated tanks cooled by liquid. These increased costs related to dedicated building and operation usually outweigh the savings from buying a brewery that has cheaper technology.

1.2.1.6 Beer Maturation Lagering Tanks with Double Jacket (Liquid Cooled)

These tanks are more elaborate, cooled by glycol or ice water circulating in the double jacket duplicators—the modern way of cooling a lager cellar, representing the standard configuration for modern breweries. Besides the apparent disadvantage of the higher initial cost, these tanks have only advantages such as low cost of the cooling operation. Moreover, it is evident that this type of choice requires few constraints for installation (no need to insulate storage cellar, tanks may be placed in any room), helping the user with comfortable operations (in the room with tanks there is no need to regulate the temperature). There is also the possibility to control maturation in each tank separately.

1.2.1.7 Brewing Yeast

The strains of *Saccharomyces* employed in beer production, even though all produce ethanol as a fermentation end product, are classified into the categories of ale and lager yeasts. Ale yeasts (*Saccharomyces cerevisiae* strains) are the more diverse yeasts, basically isolated worldwide. These are often referred to as "top-fermenting" yeasts. This definition arises because in traditional open fermenters they rise to the surface of the vessel, facilitating their collection by skimming. However, hydrostatic pressure in modern fermenters tends to overcome this tendency of ale yeast, accordingly collecting in the cone of the tank. The "bottom-fermenting" nomenclature refers to lager yeast, because of its tendency not to rise to the surface under any set of fermentation conditions (*Saccharomyces pastorianus*). It must be highlighted that lager yeast is a more complex organism if compared to ale yeast, and it has been proposed that it arose in perhaps two separate steps involving the hybridization of *S. cerevisiae* with *S. bayanus* (Turakainen et al., 1994).

1.2.2 Reaction Mechanisms of Aging Processes in Beer

Beer represents a water–ethanol solution with a pH ~ 4.2 in which hundreds of different molecules are dissolved. These originate from raw materials such as water, malt, hops, and adjuncts followed by the wort production, fermentation, and maturation processes. Immediately after the fermentation and aging process is completed, the beer is filtered to remove the large quantity of yeast and hops (Figures 1.4 through 1.6). However, a freshly bottled beer, whose constituents are not in chemical equilibrium, represents a closed system. Thermodynamically, this system will thus strive to reach a status of minimal energy and maximal entropy. Therefore, molecules are subjected to many reactions mainly determined by the reaction rates (temperature dependent) under practical storage conditions.

1.2.2.1 Reactive Oxygen Species in Stored Beer

Oxygen, in particular, is responsible for starting some important aging reactions, thus causing a rapid deterioration of beer flavor. Oxygen in the ground state ($^3O_2$) is quite stable and not easily reactive with organic molecules. In the presence of ferrous

FIGURE 1.4 A 50 μm filter bag, typically used for removing yeasts/hop. The image highlights the inner surface (dirty, given the presence of residues of yeasts/hop) as the filter has been turned off. (Courtesy of Croce di Malto S.r.l., Trecate, Italy.)

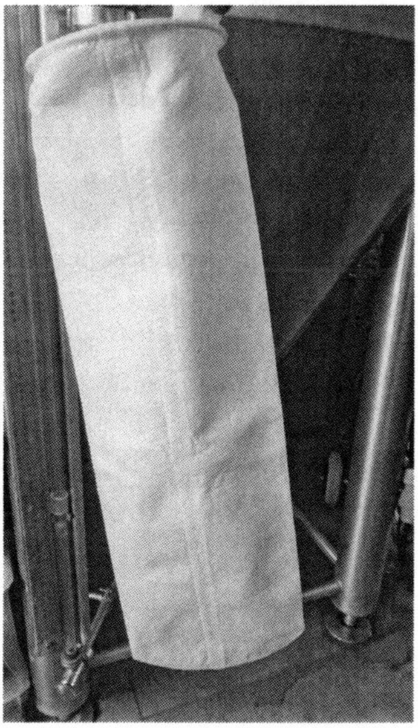

FIGURE 1.5 A new 10 μm filter bag. (Courtesy of Croce di Malto S.r.l., Trecate, Italy.)

FIGURE 1.6 Two different typologies of new 10 μm cartridge filters. (Courtesy of Croce di Malto S.r.l., Trecate, Italy.)

iron (Fe^{2+}) in beer, oxygen is able to capture an electron and form the superoxide anion $\left(O_2^-\right)$ and Fe^{3+}. Copper ions (Cu^+ is oxidized to Cu^{2+}) show the same behavior (Kaneda et al., 1999). Polyphenols, sugars, isohumulones, and alcohols might act as electron donors of a mixed function oxidation system, a part of which is represented by Cu^+/Cu^{2+} and Fe^{2+}/Fe^{3+} ions (Kaneda et al., 1992). The superoxide anion can be protonated to form the perhydroxyl radical ($OOH^•$), characterized by a much higher reactivity. Considering both the pH of beer and the pKa of this reaction (4.8), the majority of the superoxide results in the perhydroxyl form. The superoxide anion can also be reduced by iron/copper ions to the peroxide anion $\left(O_2^{2-}\right)$, quickly protonated to hydrogen peroxide (H_2O_2) in beer. Then hydroxyl radicals ($OH^•$) can be produced from H_2O_2 or, alternately, the superoxide anion (O^{2-}) due to a metal-induced reaction such as Fenton and the Haber-Weiss reaction. During beer aging, the concentration of free radicals increases with increasing iron/copper ion concentrations, with increasing oxygen concentrations or with higher storage temperatures (Kaneda et al., 1992). The reactive oxygen species (ROS) react with all kinds of organic molecules in beer, such as polyphenols and alcohols, resulting in several changes in the sensory profile. Among these, the hydroxyl radical, representing one of the most reactive species identified, is able to react with ethanol in beer. As previously reported by Andersen and Skibsted (1998), the 1-hydroxyethyl radical, as quantitatively the most important radical in beer, actually was derived from the reaction of ethanol with the hydroxyl radical.

1.2.3 Aging Reactions Producing Carbonyl Compounds

1.2.3.1 Oxidation of Higher Alcohols

Ethanol, 2-methyl-propanol, 2-methyl-butanol, 3-methyl-butanol, and 2-phenyl-ethanol represent the most important alcohols in beer. The corresponding aldehydes increase during beer aging due to exposure to higher oxygen levels. However, other parameters, such as high temperatures, low pH, and the supplementation of additional higher alcohols, may lead to higher concentrations of aldehydes.

1.2.3.2 Strecker Degradation of Amino Acids

Amino acids in stored beer may be a source of aldehydes. When either valine or leucine was added to beer (in the presence of oxygen), the formation of 2-methyl-propanal and 3-methyl-butanal increased significantly (Blockmans et al., 1975). This reaction (catalyzed by Fe and Cu ions) has been explained by a Strecker reaction between amino acids and α-dicarbonyl compounds. The reaction involves transamination, followed by decarboxylation of the subsequent α-ketoacid, resulting in an aldehyde. In beer, additional α-dicarbonyl compounds may be formed by the Maillard reaction. However, it must reported that Strecker degradation is only important at strongly increased amino acid contents, but not at the amino acid concentrations normally present in beer (±1 g/L) (Thum et al., 1995).

1.2.3.3 Aldol Condensation

Under the mild conditions existing in beer during storage, aldol condensation of carbonyl compounds may be possible. (E)-2-nonenal was formed by aldol condensation of acetaldehyde with heptanal following a model beer (containing 20 mmol/L of proline) stored for 20 days at 50°C (Hashimoto and Kuroiwa, 1975). In these reactions, the amino acids represents the basic catalysts through the formation of an imine intermediate. This pathway can produce carbonyl compounds, characterized by a lower flavor threshold, from carbonyls (less flavor active).

1.2.3.4 Degradation of Hop Bitter Acids

In beer, the degradation of hop bitter acids is able to decrease sensory bitterness. Moreover, there are indications that some products obtained from the degradation of hop bitter acids (iso-α-acids, α-acids, and β-acids) are involved in the appearance of aging flavors. During the fast degradation of iso-α-acids in the presence of ROS the *trans*-isomer appears much more sensitive than the *cis*-isomer (De Cooman et al., 2000). However, recent study indicates that, in the presence of suitable electron acceptors, electrons are released from iso-α-acids, not necessarily requiring the involvement of oxygen species (Huvaere et al., 2003). Degradation products of iso-α-acids represent carbonyl compounds with different chain lengths, such as 2-alkanones (C3 to C11), alkanals (C2 to C10), 2-alkenals (C4 to C7), and 2,4-alkedienals (C6 to C7) (Hashimoto and Eshima, 1979).

1.2.3.5 Oxidation of Unsaturated Fatty Acids

Linoleic acid (C18:2) and linolenic acid (C18:3) represent the only lipid substrates of significance in beer and wort arising from malted barley. These acids are mainly released from triacylglycerols due to the activity of barley and malt lipases. Hydrolysis of these compounds to fatty acids occurs mainly during mashing (Schwarz et al., 2002). During wort production, enzymatic and nonenzymatic oxidation generates (E)-2-nonenal potential initiating the (E)-2-nonenal formation during storage. This potential is defined as the potential of wort to release (E)-2-nonenal after its treatment for 2 h at 100°C, at pH 4.0 (argon atmosphere) (Drost et al. 1990). There are two possible oxidation routes: auto-oxidation or an enzymatic oxidation with lipoxygenases (LOX). Auto-oxidation of a fatty acid is initiated by the slower reacting peroxy radical ($ROO^{\cdot}$), abstracting the most weakly bound H-atom in the fatty acid. Germinated barley contains two lipoxygenase enzymes, namely LOX-1 and LOX-2.

The first is present in raw barley and increases during germination, whereas the second only develops during germination. These may oxidize fatty acids to their hydroperoxy acids. Both enzyme results are very heat sensitive; therefore, during kilning, most activity is destroyed and the remaining activity in malt is mainly due to LOX-1.

1.2.3.6 Acetalization of Aldehydes

2,3-Butanediol (up to 280 mg/L in beer) usually undergoes a condensation reaction with different aldehydes such as acetaldehyde, isobutanal, 3-methylbutanal, and 2-methyl-butanal (Peppard and Halsey, 1982). The result of this reaction leads to the formation of the respective cyclic acetals (2,4,5-trimethyl-1,3-dioxolane; 2-isopropyl-4,5-dimethyl-1,3-dioxolane; 2-isobutyl-4,5-dimethyl-1,3-dioxolane; and 2-sec butyl-4,5-dimethyl-1,3-dioxolane). It must be noted that an equilibrium between 2,4,5-trimethyl-1,3-dioxolane, acetaldehyde, and 2,3-butanediol is reached quite rapidly. Consequently, in beer aging, the increase in the acetaldehyde concentration causes the 2,4,5-trimethyl-1,3-dioxolane concentration to increase similarly (Vanderhaegen et al., 2003).

1.2.3.7 Maillard Reactions

The complex reactions between reducing sugars and proteins, peptides, amino acids, or amines are all classified as Maillard reactions. Several heterocyclic compounds found in aged beers belong to this type of reaction, such as furfural and 5-hydroxymethyl-furfural. The latter represents, quantitatively, one of the most important heterocyclic staling compounds. The reaction starts with the formation of a Schiffs base, between a carbonyl group of a hexose sugar and an amino group, leading to an imine. This undergoes an Amadori rearrangement to a more stable 1-amino-1-deoxyketose (also called an Amadori product). However, because of the pH of beer, the Amadori product is subject to enolization, releasing an amine, which leads to 3-deoxy-2-hexosulose (3-DH). This reactive α-dicarbonyl compound can (among various secondary products) give rise to 5-hydroxymethyl-furfural (HMF). Similarly, following this pathway, furfural is produced starting from a pentose. At this stage HMF, furfural, and other compounds are merely intermediates of the Maillard reaction. They are subject to further reactions, such as condensation, dehydration, cyclization, and isomerization, thus forming brown pigments of high molecular weight (melanoidins).

1.2.3.8 Synthesis and Hydrolysis of Volatile Esters

Chemical condensation between ethanol and organic acids occurs at significant rates during beer storage. 3-Methyl-butyric acid and 2-methyl-butyric commonly lead to ethyl 3-methyl-butyrate and ethyl 2-methyl-butyrate. Conversely, some esters, such as iso-amyl acetate, may be hydrolyzed, decreasing their contribution to beer flavor.

1.2.3.9 Degradation of Polyphenols

During beer storage, phenolic polymers interact with proteins, thus forming insoluble complexes and hazes. Polyphenols easily react with ROS and free radicals. Simple polyphenols polymerize to tannins (high molecular weight), either by acid catalysis or by oxidative mechanisms. Polyphenols are first oxidized to quinones or semiquinone radicals, which interact with other analogous compounds. Moreover, acetaldehyde (formed by yeast or by ethanol oxidation) is able to induce polymerization reactions through the formation of ethyl bridges between flavanols (Saucier et al., 1997). Conversely, in oxidized phenols, ring opening has been proposed as an alternative degradation mechanism (Cilliers and Singleton, 1990).

1.2.4 Oxidative versus Nonoxidative Beer Aging

Following the previous considerations, it is clear that oxygen triggers the release of free radicals. These may react easily with many beer constituents, leading to rapid modifications in the flavor profile. Since oxygen is very detrimental, brewers have tried to minimize the oxygen pickup in finished beer. Due to modern filling equipment, total oxygen levels in the bottle may be kept within 0.1 mg/L. At such low oxygen levels, it is debatable whether the formation of ROS is the determining factor in the aging of these beers. However, other molecules have enough reactivity to interact and form staling compounds. Beer staling is regarded as the result of oxidation only, but nonoxidative processes may be responsible, especially due to the low oxygen levels reached in modern breweries. Nonoxidative reactions (e.g., esterification) may affect flavor quality. Even (E)-2-nonenal paradoxically seems to arise due to nonoxidative mechanisms in beer.

1.2.4.1 Antioxidants and Reducing Agents versus Pro-Oxidants

Most of the presumed antioxidants may have pro-oxidant capabilities as well (e.g., ascorbic acid, sulfhydryl compounds, chelating agents, polyphenols, and Maillard-reaction products) (Bamforth, 2001). Melanoidins are known to scavenge free and active oxygen, act as reducing agents, and chelate metals. However, their pro-oxidant activity have been reported as well (Ames, 2001). An antioxidant is defined as a compound present at low concentrations compared with those of an oxidizable one that significantly delays or prevents oxidation of that substrate. Nevertheless, the way antioxidants can exert their antioxidant activity differs between compounds. At beer pH, the scavenging activity of the polyphenols and phenolic acids is very low if compared to sulfite (Nakamura et al., 2003). However, some antioxidants (e.g., (+)-catechin) may have pro-oxidant properties at low concentrations. Sulfite is a potent inhibitor of flavor deterioration. This compound is able to stabilize the flavor in two ways: as antioxidant and as carbonyl scavenger in aldehyde–bisulfite adducts. Sulfite is able to effectively retard the formation of radicals in beer during storage. Yeasts represent potent antioxidative creatures. *Saccharomyces cerevisiae* is able to reduce a variety of carbonyls and seems to play a very important role in the reduction of beer staling compounds, precursors included (Shimizu et al., 2002).

1.3 VINEGAR

1.3.1 Overview

Vinegar represents the world's oldest cooking ingredient and food preservation method. Its use can be traced back over 10,000 years. Throughout the centuries its use ranged from basic ingredient (e.g., vinegars flavored with fruit, honey, malt, etc.) even up to medicinal purposes. Vinegar appears as an acid liquid produced from the fermentation of ethanol in a process that yields acetic acid as key ingredient. Its concentration typically ranges from 4% to 8% by volume for table vinegar and up to 18% for pickling. Small amounts of tartaric acid, citric acid, and other acids are commonly present in natural vinegars. Its primary use is directed to flavor and preserve foods and as an ingredient in salad dressings and marinades. Secondly, vinegar is also used as a cleaning agent. There are many different types of vinegars, usually classified on the raw material used for its production. Malt, wine, apple cider, balsamic, fruit, and many other types of vinegar exist in the market.

1.3.1.1 Wine Vinegar

Wine vinegar is produced from red or white wine. This represents the most commonly used vinegar in Mediterranean countries and Central Europe. Similarly to wine, there is a considerable range in quality. High-quality vinegars are matured in wood for up to 2 years, thus exhibiting a complex, mellow flavor. Generally, its acidity results are lower compared to white or cider vinegars. There are more expensive wine vinegars produced from individual varieties of wine (e.g., Champagne or Pinot).

1.3.1.2 Beer Vinegar

Typically, the United Kingdom, Germany, Austria, and the Netherlands produce vinegar starting from beer. Although its flavor depends on the particular type of beer from which it is made, generally it presents a malty taste. Only one type of fermentation takes place in beer vinegar production, similarly to wine vinegar: the conversion of ethanol into acetic acid. Other types of vinegars are grouped under a production process characterized by two types of fermentations: one for ethanol formation and another conversion of the ethanol into acetic acid.

1.3.1.3 Fruit Vinegars

These are made from fruit wines, usually without any additional flavoring. Apple, blackcurrant, raspberry, quince, and tomato represent common flavors of this vinegar typology. Usually, the flavors of the original fruits remain in the final product. Most fruit vinegars are produced in Europe, where there is a growing demand for high-quality vinegars produced solely from specific fruits. However, several varieties are also produced in Asia. Persimmon vinegar is popular in South Korea and jujube vinegar and wolfberry vinegar in China. Jamun sirka is vinegar produced starting from rose apple in India, considered to be medicinally valuable for the stomach.

1.3.1.4 Balsamic Vinegar

Balsamic vinegar is an aromatic, aged type of vinegar traditionally crafted in the Modena and Reggio Emilia provinces of Italy (Figure 1.7). It is produced using concentrated juice, or must, of white grapes (typically Trebbiano variety). It appears dark brown and its flavor is rich, sweet, and complex. This is the result of a long aging period, usually from 12 to 25 years, thus receiving the *protected designation of origin*. The aging process is designed by using a successive number of casks made

FIGURE 1.7 Possible configurations of the barrel sets. (Courtesy of Acetaia dei Bago, Azienda Agricola La Vittoria di Mislej Patrizia, Vignola, Italy.)

FIGURE 1.8 A series of different sized barrels used to make balsamic vinegar. Commonly known as "battery." (Courtesy of Acetaia dei Bago, Azienda Agricola La Vittoria di Mislej Patrizia, Vignola, Italy.)

of various types of wood (including oak, mulberry, chestnut, cherry, juniper, ash, and acacia) (Figure 1.8). The final product is really expensive due to the long and complex aging process. However, a cheaper form of balsamic vinegar became widely known and available in recent decades. The commercial balsamic vinegar sold in supermarkets is typically made with concentrated grape juice mixed with strong vinegar, with the addition of caramel and sugar.

1.3.1.5 Malt Vinegar

Malting barley, causing the starch in the grain to turn to maltose, may be used to make malt vinegar. Successively, ale is brewed from the maltose and allowed to turn into vinegar, which is then aged. It appears typically light brown in color.

1.3.1.6 Rice Vinegar

Rice vinegar is most popular in East and Southeast Asia. It is available in "white" (light yellow), red, and black varieties. The Japanese prefer light typology for the preparation of sushi rice and salad dressings. Red rice vinegar traditionally is colored using red yeast rice. Black rice vinegar (produced with black glutinous rice) is really popular in China. White rice vinegar shows a mild acidity and a not particularly complex flavor.

1.3.1.7 Cane Vinegar Production

Cane vinegar, produced from sugar cane juice, is most popular in the Philippines, although it is also produced in France and the United States. It ranges from dark yellow to golden brown and has a mellow flavor, similar in some respects to rice vinegar. Generally, vinegar production allows utilization of over-ripe fruits and cane by-products such as molasses and bagasse. It is possible to produce naturally fermented vinegar from sugarcane juice (also young canes may be used). If the juice extracted is below 15°Brix, small amounts of sugar or molasses may be added.

1.3.1.8 Others

Worldwide, it is possible to find many other different types of vinegars, such as coconut vinegar, palm vinegar, raisin vinegar, honey vinegar, and sugar cane vinegar. Coconut and palm vinegars are often used in the Philippines, where they are produced. Vinegar made from raisins is typically used in cuisines of the Middle East. Though vinegar produced starting from honey is rare, products are commercially available in Italy, France, Romania, and Spain.

1.3.2 Vinegar Production

1.3.2.1 Bacterial Cultures

Acetobacter acetii cultures are used for vinegar production. These work perfectly at a temperature of about 28°C with full air injection. The lowest temperature tolerated is 20°C and the maximum is about 33°C. Outside these ranges of temperatures, there is no conversion from alcohol into acetic acid. Usually, the starting alcohol should be lower than 7.5% (v/v) and free sulfites should not be present. In the commercial production, usually mother of vinegar is added as a source of *Acetobacter*. Mother of vinegar represents the gooey film appearing on the surface of the alcohol product as it is converted to vinegar. Once skimmed off the top, mother of vinegar is added to following batches of alcohol to speed the formation of vinegar. It consists of natural carbohydrate (cellulose) holding the highest concentration of *Acetobacter*. Herbs and fruits are often used to flavor vinegar. Commonly used herbs are tarragon, garlic, and basil; among fruits, raspberries, cherries, and lemons are used.

1.3.2.2 Processing Methods

Vinegar contains, mainly, acetic acid by weight and small quantities of other molecules such as alcohol, glycerol, esters, sugars, and salts. The transformation of wine or fruit juice to vinegar is a chemical process in which a partial oxidation of ethyl alcohol allows the formation of acetaldehyde. Then, the acetaldehyde is converted into acetic acid. Thus, it is clear that the production of vinegar involves two types of biochemical reactions: alcoholic fermentation and oxidation of alcohol into acid. The first crucial step in the production of vinegar is represented by alcoholic fermentation of carbohydrates (under anaerobic conditions). In this process sugar is fermented to alcohol by the action of yeast species. Oxidation of alcohol to acid by *Acetobacter* species represents the second major step (aerobic process).

Commercial vinegar is produced by either a fast or a slow process. In the slow (natural) process, vats of cider are allowed to sit open at room temperature. After several days to even months, the fruit juices ferment into alcohol and then oxidize into acetic acid. Slow methods generally are used with traditional vinegars; fermentation proceeds slowly, requiring weeks or months. The longer fermentation period allows for the accumulation and subsequent recovery of mother of vinegar. Conversely, fast methods add mother of vinegar to the source liquid before adding air (pump system) to promote oxygenation in order to obtain the fastest fermentation. Thus vinegar may be produced in a period ranging from 1 to 3 days. In modern production, the generator method and the submerged fermentation method are employed. These methods are designed with the aim of infusing as much oxygen as possible into the alcohol product. The three common methods used are (1) the generator or trickling method, (2) the submerged fermentation or the acetator method, and (3) the Orleans traditional method.

1.3.2.3 The Orleans Method

This process represents one of the oldest known methods for the production of vinegar. Developed in France, it is characterized by a slow, continuous process. High-grade vinegar is utilized as a starter culture, to which wine is added (in weekly intervals). The vinegar is fermented in 200 L capacity barrels. About 70 L, high-grade vinegar is added to the barrel along with 15 L of wine. After 1 week, a further 15 L of wine is added (this step is repeated at weekly intervals). After about 1 month, vinegar may be withdrawn from the barrel (about 15 L per week) as more wine is added to replace the vinegar. One of the issues related to this method concerns adding new wine to the barrel without disturbing the floating bacterial mat. This may be overcome by using a glass tube reaching the bottom of the barrel. Additional liquid is poured in through the tube, thus not disturbing the bacteria.

Manufacturing steps: The alcohol is poured into the barrel via long-necked funnels. Mother of vinegar is added. The barrel is filled to a level just below the holes on the ends. Netting is placed over the holes to prevent insects from getting into the barrels. The filled barrels are allowed to sit for several months. The room temperature is kept at about 29°C. When the alcohol has converted to vinegar, it is drawn off. About 15% of the liquid is left in the barrel to blend with the next batch.

1.3.2.4 Generator Method

Because the Orleans process is slow, other methods have been developed to speed up the process. This method uses a generator (a cylindrical tank with a perforated false bottom supporting beech-wood shavings) that is fitted with devices allowing the alcoholic solution to trickle down through the shavings in which the acetic acid bacteria are inoculated. An adjusted solution of alcohol acidified with acetic acid and special nutrients for the growth of acetic acid bacteria represents the mix used. The mix is fluxed from the top of the chamber, thus allowing it to trickle down over the shavings, and then this is collected at the bottom of the generator. The process continues until the desired strength is obtained. Oxidation of alcohol may result in the development of temperatures detrimental for the bacteria. In order to keep the temperature down to about 25°C, a cooling system is required. This method is quicker compared to others and the vinegar may be produced within a period of 10 days.

Manufacturing steps: Vats are filled with vinegar-moistened beech wood shavings. The alcohol product is poured into the top of the vat and slowly drips down through the fillings. An air compressor blows air through the holes. Vinegar is poured off from the bottom of the vat into storage tanks. The vinegar produced using this method has very high acetic acid content (often as high as 14%). This must be diluted with water to bring its acetic acid content to a range of about 5%.

1.3.2.5 Submerged Fermentation Method

A tank filled with alcohol is pumped with oxygen and maintained at warm temperature. The wine is kept at a temperature between 26°C and 38°C while nutrients and air are pumped into the mixture. This method is commonly used in the production of wine vinegars. Submerged vinegar systems are most commonly used by industry to produce high quantities of vinegar with a high content of acetic acid (Figure 1.9). This system works with the continuous aeration of the liquid. The vinegar bacteria are floating in the liquid, not producing a vinegar mother. Slime does not form in the machine, thus allowing the vinegar to appear exceptionally clean. Air may be pumped into the liquid using automatically controlled turbine systems. The vinegar

FIGURE 1.9 Industrial vinegar fermenter. (Courtesy of Ponti S.P.A., Ghemme, Italy.)

is pumped out of the machine, and the machine is then refilled with wine. These systems are designed for producers who plan to make more than 50,000 L of vinegar per year. Another air system commonly used is the Venturi. This represents a smaller and cheaper system compared to a turbine one. The processing is realized with a pumping system in a closed stainless steel tank. A Venturi air nozzle brings the air into the liquid. The time required for this process is around 30% longer than in a turbine system, but this is better for quality because of the lower aeration avoiding flavor loss. Currently, the Venturi air system is the most common processing system for small and medium vinegar makers worldwide.

Manufacturing steps: The acetators (large stainless steel tanks) are fitted with centrifugal pumps in the bottom. As the pump stirs the alcohol, nutrients are piped into the tank. The temperature is kept between 26°C and 38°C. Within a matter of hours, vinegar is produced. The vinegar is piped from the acetators to a filtering machine (Figures 1.10 and 1.11). The stainless steel plates press the alcohol through paper filters to remove any sediment (about 3% of the total product).

1.3.3 Quality Control

Two factors require particular attention when making vinegar: oxygen supply and temperature. Oxygen should be spread throughout the mixture and temperature of fermentation should be kept at around 28°C. Sometimes undesirable microorganisms might spoil the product. For example, *Turbatrix aceti* (vinegar nematode) may cause deterioration of flavor and appearance. Therefore, care should be taken to check contamination by undesirable bacteria or molds. The generator must be periodically disinfected to avoid contamination. Moreover, when the vinegar is fully fermented, it should be stored in separate containers or wooden barrels. The recommended place should be cool (under 10°C) and dark. The final quality of vinegars depends on the selection of suitable starter cultures (generally mixed cultures). However, other factors affect the quality, such as the starting material, production method, and aging process.

Generally, between products made by traditional methods and those by industrial scale, some differences arise. Aromatic compounds significantly effect on the quality of vinegars. The aroma represents a complex fraction, containing several compounds with a wide range of volatilities, polarities, and concentrations ranging from several milligrams per liter to nanograms per liter. To date, more than a hundred different chemical compounds are in the aroma of wine vinegar, including carbonyl compounds, ethers, acetals, lactones, acids, alcohols, phenols, and volatile esters (Callejon et al., 2009). During aging, the contact with wood produces a substantial increase in the aromatic complexity (Callejon et al., 2010). Polyphenolic compounds, in addition to their antioxidant activity, are responsible for the color and astringency of vinegar. Acetification is an aerobic process where oxygen appears crucial to the growth of the bacteria. Oxygen affects the classes of polyphenolic compounds to different degrees. For example, the flavonol content of vinegars is influenced by oxygen availability during submerged fermentation. Conversely, surface acetification vinegars do not affect phenolic aldehydes, which are released from wooden barrels (Garcıa-Parrilla et al., 1998). The aging process involves the reaction of compounds.

FIGURE 1.10 Tangential flow filtration system. (Courtesy of Ponti S.P.A., Ghemme, Italy.)

These include both polymerization and the release of compounds from the wood. The substances given depend on the type of wood and related degree of toasting. In addition, the contact surface ratio and the aging time affect vinegar quality. The evolution of phenolic compounds showed significant differences in compounds such as vanillin, syringaldehyde, coniferyl aldehyde, and cinnamic acid in vinegar after 90 days of aging in oak barrels (Tesfaye et al., 2002).

1.3.4 Maturation and "Oaking" of Vinegar

Considering that both the maturation and aging process in wood barrels (such as cherry, chestnut, oak, mulberry, ash, or juniper barrels) result in very high cost,

FIGURE 1.11 Storage tanks. (Courtesy of Ponti S.P.A., Ghemme, Italy.)

plastic tanks are often used. To imitate the renowned wooden flavor typically associated with top-quality vinegar, vinegar producers add oak chips or wood flavor distillates. To achieve good results, normally oak chips have to remain for at least 1 month in the vinegar. The use of ultrasound can speed up this process significantly. To sonicate vinegar with additive oak powder/chips may assist the extraction of flavor. Due to its highly energetic cavitational forces, sonication is able to break cell walls and makes the intracellular material available. Ultrasound creates alternating high-pressure and low-pressure cycles, thus ensuring a higher mass transfer between the plant cells and the vinegar. Moreover, oak powder is really well dispersed in the liquid medium because ultrasound is a very effective means of dispersal. The ultrasound technology also results in an efficient method for emulsifying and dissolving.

This is useful in vinegar production (e.g., caramel is added to the vinegar). Caramel (food additive E150) represents one of the oldest and most widely used food additives, giving the vinegar the desired dark brown color.

REFERENCES

Abrahamse, C.E., and Bartowsky, E.J. (2012). Timing of malolactic fermentation inoculation in Shiraz grape must and wine: Influence on chemical composition. *World J. Microbiol. Biotechnol.* 28, 255–265.

Alexandre, H., Costello, P.J., Remize, F., Guzzo, J., and Guilloux-Benatier, M. (2004). *Saccharomyces cerevisiae—Oenococcus oeni* interactions in wine: current knowledge and perspectives. *Int. J. Food Microbiol.* 93, 141–154.

Ames, J.M. (2001). Melanoidins as pro- or antioxidants. *Cerevisia* 26, 210–216.

Andersen, M.L., and Skibsted, L.H. (1998). Electron spin resonance spin trapping identification of radicals formed during aerobic forced aging of beer. *J. Agric. Food Chem.* 46, 1272–1275.

Antalick, G., Perello, M.-C., and de Revel, G. (2012). Characterization of fruity aroma modifications in red wines during malolactic fermentation. *J. Agric. Food Chem.* 60, 12371–12383.

Badotti, F., Moreira, A.P.B., Tonon, L.A.C., de Lucena, B.T.L., Gomes, F. de C.O., Kruger, R., Thompson, C.C., de Morais, M.A., Rosa, C.A., and Thompson, F.L. (2014). *Oenococcus alcoholitolerans* sp. nov., a lactic acid bacteria isolated from cachaça and ethanol fermentation processes. *Antonie Van Leeuwenhoek* 106, 1259–1267.

Bamforth, C.W. (2001). Oxido-reduction processes and active forms of oxygen in aqueous systems. *Cerevisia* 26, 149–154.

Bartowsky, E.J. (2009). Bacterial spoilage of wine and approaches to minimize it. *Lett. Appl. Microbiol.* 48, 149–156.

Bartowsky, E.J., and Borneman, A.R. (2011). Genomic variations of *Oenococcus oeni* strains and the potential to impact on malolactic fermentation and aroma compounds in wine. *Appl. Microbiol. Biotechnol.* 92, 441–447.

Bartowsky, E.J., and Henschke, P.A. (1995). Malolactic fermentation and wine flavour. *Aust. Grapegrow. Winemak. Aust.* 378, 83–94.

Bartowsky, E.J., and Pretorius, I.S. (2009). Microbial formation and modification of flavor and off-flavor compounds in wine. In *Biology of Microorganisms on Grapes in Must and in Wine*, Springer, pp. 209–231.

Bartowsky, E.J., Costello, P.J., and Chambers, P.J. (2015). Emerging trends in the application of malolactic fermentation. *Aust. J. Grape Wine Res.* 21, 663–669.

Bastard, A., Coelho, C., Briandet, R., Canette, A., Gougeon, R.D., Alexandre, H., Guzzo, J., and Weidmann, S. (2016). Effect of biofilm formation by *Oenococcus oeni* on malolactic fermentation and the release of aromatic compounds in wine. *Front. Microbiol.* 7, 613.

Bauer, R., and Dicks, L.M.T. (2004). Control of malolactic fermentation in wine. A review. *Afr. J. Enol. Vitic.* 25, 74–88.

Berbegal, C., Peña, N., Russo, P., Grieco, F., Pardo, I., Ferrer, S., Spano, G., and Capozzi, V. (2016). Technological properties of *Lactobacillus plantarum* strains isolated from grape must fermentation. *Food Microbiol.* 57, 187–194.

Betteridge, A., Grbin, P., and Jiranek, V. (2015). Improving *Oenococcus oeni* to overcome challenges of wine malolactic fermentation. *Trends Biotechnol.* 33, 547–553.

Blockmans, C., Devreux, A., and Masschelein, C.A. (1975). Formation de composés carbonyls et alteration du goût de la bière. *Proceedings of the European Brewery Convention Congress*, 699–713.

Bokulich, N.A., Collins, T.S., Masarweh, C., Allen, G., Heymann, H., Ebeler, S.E., and Mills, D.A. (2016). Associations among wine grape microbiome, metabolome, and fermentation behavior suggest microbial contribution to regional wine characteristics. *mBio* 7, e00631–16.

Briggs, D.E., and McGuinness, G. 1993. Microbes on barley grains. *J. Inst. Brew.* 99, 249–255.

Callejon, R.M., Tesfaye, W., Torija, M.J., Mas, A., Troncoso, A.M., and Morales, M.L. (2009). Volatile compounds in red wine vinegars obtained by submerged and surface acetification in different woods. *Food Chem.* 113(4), 1252–1259.

Callejon, R.M., Torija, M.J., Mas, A., Morales, M.L., and Troncoso, A.M. (2010). Changes of volatile compounds in wine vinegars during their elaboration in barrels made from different woods. *Food Chem.* 120(2), 561–571.

Cañas, P.M.I., Romero, E.G., Pérez-Martín, F., Seseña, S., and Palop, M.L. (2015). Sequential inoculation versus co-inoculation in Cabernet Franc wine fermentation. *Food Sci. Technol. Int.* 21, 203–212.

Capozzi, V., Russo, P., Ladero, V., Fernández, M., Fiocco, D., Alvarez, M.A., Grieco, F., and Spano, G. (2015). Biogenic amines degradation by *Lactobacillus plantarum*: Toward a potential application in wine. *Biog. Amines Fermented Foods* 3, 122.

Chasseriaud, L., Krieger-Weber, S., Déléris-Bou, M., Sieczkowski, N., Jourdes, M., Teissedre, P.L., Claisse, O., and Lonvaud-Funel, A. (2015). Hypotheses on the effects of enological tannins and total red wine phenolic compounds on *Oenococcus oeni*. *Food Microbiol.* 52, 131–137.

Cilliers, J.J.L., and Singleton, V.L. (1990). Autoxidative phenolic ringopening under alkaline conditions as a model for natural polyphenols in food. *J. Agric. Food Chem.* 38, 1797–1798.

Costello, P.J., Lee, T.H., and Henschke, P. (2001). Ability of lactic acid bacteria to produce N-heterocycles causing mousy off-flavour in wine. *Aust. J. Grape Wine Res.* 7, 160–167.

Coulon, J., Perello, M.C., Lonvaud-Funel, A., De Revel, G., and Renouf, V. (2010). *Brettanomyces bruxellensis* evolution and volatile phenols production in red wines during storage in bottles. *J. Appl. Microbiol.* 108, 1450–1458.

Cozzolino, D. (2016). Metabolomics in grape and wine: Definition, current status and future prospects. *Food Anal. Methods* 9, 2986–2997.

Davis, C.R., Wibowo, D., Eschenbruch, R., Lee, T.H., and Fleet, G.H. (1985). Practical implications of malolactic fermentation: A review. *Am. J. Enol. Vitic.* 36, 290–301.

Deckers, S.M., Lorgouilloux, Y., Gebruers, K., Baggerman, G., Verachtert, H., Neven, H., Michiels, C., Derdelinckx, G., Delcour, J.A., and Martens, J. 2011. Dynamic light scattering (DLS) as a tool to detect CO2-hydrophobin structures and study the primary gushing potential of beer. *J. Am. Soc. Brew. Chem.* 69, 144–149.

De Cooman, L., Aerts, G., Overmeire, H., and De Keukeleire, D. (2000). Alterations of the profiles of iso-alpha-acids during beer ageing, marked instability of trans-iso-alpha-acids and implications for beer bitterness consistency in relation to tetrahydroiso-alpha-acids. *J. Inst. Brew.* 106, 169–178.

de Revel, G., Martin, N., Pripis-Nicolau, L., Lonvaud-Funel, A., and Bertrand, A. (1999). Contribution to the knowledge of malolactic fermentation influence on wine aroma. *J. Agric. Food Chem.* 47, 4003–4008.

Dicks, L.M.T., Dellaglio, F., and Collins, M.D. (1995). Proposal to reclassify *Leuconostoc oenos* as *Oenococcus oeni* [corrig.] gen. nov., comb. nov. *Int. J. Syst. Evol. Microbiol.* 45, 395–397.

Drost, B.W., van den Berg, R., Freijee, F.J.M., van der Velde, E.G., and Hollemans, M. (1990). Flavor stability. *J. Am. Soc. Brew. Chem.* 48, 124–131.

du Toit, M., Engelbrecht, L., Lerm, E., and Krieger-Weber, S. (2011). *Lactobacillus*: The next generation of malolactic fermentation starter cultures—An overview. *Food Bioprocess Technol.* 4, 876–906.

Edwards, C.G., Collins, M.D., Lawson, P.A., and Rodriguez, A.V. (2000). *Lactobacillus nagelii* sp. nov., an organism isolated from a partially fermented wine. *Int. J. Syst. Evol. Microbiol.* 50, 699–702.

Endo, A., and Okada, S. (2006). *Oenococcus kitaharae* sp. nov., a non-acidophilic and non-malolactic-fermenting oenococcus isolated from a composting distilled shochu residue. *Int. J. Syst. Evol. Microbiol.* 56, 2345–2348.

Fugelsang, K.C. (1997). The Lactic Acid Bacteria. In *Wine Microbiology* (Springer US), pp. 3–47.

Garcıa-Parrilla, M.C., Heredia, F.J., and Troncoso, A.M. (1998). The influence of the acetification process on the phenolic composition of wine vinegars. *Sci. des Aliments* 18 (2), 211–221.

Gerbaux, V., Briffox, C., Dumont, A., and Krieger, S. (2009). Influence of inoculation with malolactic bacteria on volatile phenols in wines. *Am. J. Enol. Vitic.* 60, 233–235.

Gonzalez, R., Munoz, R., and Carrascosa, A.V. (2011). Production of wine starter cultures. *Mol. Wine Microbiol.* (V. Carrascosa, R. Munoz, and R. González, eds.) Elsevier Amst. Boston, pp. 279–302.

González-Arenzana, L., Santamaría, P., Gutiérrez, A.R., López, R., and López-Alfaro, I. (2016). Lactic acid bacteria communities in must, alcoholic and malolactic Tempranillo wine fermentations, by culture-dependent and culture-independent methods. *Eur. Food Res. Technol.* 243, 41–48.

Guzzon, R., Davide, S., Villegas, T.R., Malacarne, M., Larcher, R., Nardi, T., Vagnoli, P., and Krieger-Weber, S. (2016). Exploitation of simultaneous alcoholic and malolactic fermentation of Incrocio Manzoni, a traditional italian white wine. *South Afr. J. Enol. Vitic.* 37(2), 124–131.

Guzzon, R., Villega, T.R., Pedron, M., Malacarne, M., Nicolini, G., and Larcher, R. (2013). Simultaneous yeast–bacteria inoculum. A feasible solution for the management of oenological fermentation in red must with low nitrogen content. *Ann. Microbiol.* 63, 805–808.

Hashimoto, N., and Eshima, T. (1979). Oxidative degradation of isohumulones in relation to flavour stability of beer. *J. Inst. Brew.* 85, 136–140.

Hashimoto, N., and Kuroiwa, Y. (1975). Proposed pathways for the formation of volatile aldehydes during storage of bottled beer. *Proc. Am. Soc. Brew. Chem.* 33, 104–111.

Henick-Kling, T. (1993). Malolactic Fermentation. In *Wine Microbiology and Biotechnology* (G.H. Fleet, ed.) Harwood Academic Publishers, Switzerland.

Huvaere, K., Andersen, M.L., Olsen, K., Skibsted, L.H., Heyerick, A., and De Keukeleire, D. (2003). Radicaloid-type oxidative decomposition of beer bittering agents revealed. *Chem. Eur. J.* 9, 4693–4699.

Iorizzo, M., Testa, B., Lombardi, S.J., García-Ruiz, A., Muñoz-González, C., Bartolomé, B., and Moreno-Arribas, M.V. (2016). Selection and technological potential of *Lactobacillus plantarum* bacteria suitable for wine malolactic fermentation and grape aroma release. *LWT-Food Sci. Technol.* 73, 557–566.

Jackowetz, J.N., and Mira de Orduña, R. (2012). Metabolism of SO2 binding compounds by *Oenococcus oeni* during and after malolactic fermentation in white wine. *Int. J. Food Microbiol.* 155, 153–157.

Jaomanjaka, F., Claisse, O., Blanche-Barbat, M., Petrel, M., Ballestra, P., and Marrec, C.L. (2016). Characterization of a new virulent phage infecting the lactic acid bacterium *Oenococcus oeni*. *Food Microbiol.* 54, 167–177.

Kandler, O. (1983). Carbohydrate metabolism in lactic acid bacteria. *Antonie Van Leeuwenhoek* 49, 209–224.

Kaneda, H., Kano, Y., Koshino, S., and Ohyanishiguchi, H. (1992). Behavior and role of iron ions in beer deterioration. *J. Agric. Food Chem.* 40, 2102–2107.

Kaneda, H., Kobayashi, M., Takashio, M., Tamaki, T., and Shinotsuka, K. (1999). Beer staling mechanism. *Tech. Q. Master Brew. Assoc. Am.* 36, 41–47.

Knoll, C., Fritsch, S., Schnell, S., Grossmann, M., Krieger-Weber, S., Du Toit, M., and Rauhut, D. (2012). Impact of different malolactic fermentation inoculation scenarios on Riesling wine aroma. *World J. Microbiol. Biotechnol.* 28, 1143–1153.

Krieger, S., and Silvano, A. (2016). Environmental factors affecting malolactic fermentation, in: *Malolactic Fermentation—Importance of Wine Lactic Acid Bacteria in Winemaking*, Lallemand Montreal.

Laitila, A., Alakomi, H.L., Raaska, L., Mattila-Sandholm, T., and Haikara, A. 2002. Antifungal activities of two *Lactobacillus plantarum* strains against *Fusarium* moulds *in vitro* and in malting of barley. *J. Appl. Microbiol.* 93, 566–576.

Lasik, M. (2013). The application of malolactic fermentation process to create good-quality grape wine produced in cool-climate countries: A review. *Eur. Food Res. Technol.* 237, 843–850.

Lerm, E., Engelbrecht, L., and Du Toit, M. (2010). Malolactic fermentation: The ABC's of MLF. *South Afr. J. Enol. Vitic.* 31(2), 186–212

Liu, S.-Q. (2002). Malolactic fermentation in wine–Beyond deacidification. *J. Appl. Microbiol.* 92, 589–601.

Liu, S.Q., Davis, C.R., and Brooks, J.D. (1995). Growth and metabolism of selected lactic acid bacteria in synthetic wine. *Am. J. Enol. Vitic.* 46, 166–174.

Liu, Y., Forcisi, S., Harir, M., Deleris-Bou, M., Krieger-Weber, S., Lucio, M., Longin, C., Degueurce, C., Gougeon, R.D., Schmitt-Kopplin, P. et al. (2016). New molecular evidence of wine yeast-bacteria interaction unraveled by non-targeted exometabolomic profiling. *Metabolomics* 12, 1–16.

Lonvaud-Funel, A. (1999). Lactic acid bacteria in the quality improvement and depreciation of wine. In *Lactic Acid Bacteria: Genetics, Metabolism and Applications* (Springer), pp. 317–331.

Lonvaud-Funel, A. (2001). Biogenic amines in wines: Role of lactic acid bacteria. *FEMS Microbiol. Lett.* 199, 9–13.

Lonvaud-Funel, A., and De Saad, A.S. (1982). Purification and properties of a malolactic enzyme from a strain of *Leuconostoc mesenteroides* isolated from grapes. *Appl. Environ. Microbiol.* 43, 357–361.

Lowe, D.P., Ulmer, H.M., Barta, R.C., Goode, D.L., and Arendt, E.K. 2005. Biological acidification of a mash containing 20% barley using *Lactobacillus amylovorus* FST 1.1: Its effects on wort and beer quality. *J. Am. Soc. Brew. Chem.* 63, 96–106.

Lucio, O., Pardo, I., Krieger-Weber, S., Heras, J.M., and Ferrer, S. (2016). Selection of *Lactobacillus* strains to induce biological acidification in low acidity wines. *LWT-Food Sci. Technol.* 73, 334–341.

Mañes-Lázaro, R., Ferrer, S., Rodas, A.M., Urdiain, M., and Pardo, I. (2008a). *Lactobacillus bobalius* sp. nov., a lactic acid bacterium isolated from Spanish Bobal grape must. *Int. J. Syst. Evol. Microbiol.* 58, 2699–2703.

Mañes-Lázaro, R., Ferrer, S., Rosselló-Mora, R., and Pardo, I. (2008b). *Lactobacillus uvarum* sp. nov., a new lactic acid bacterium isolated from Spanish Bobal grape must. *Syst. Appl. Microbiol.* 31, 425–433.

Mañes-Lázaro, R., Ferrer, S., Rosselló-Mora, R., and Pardo, I. (2009). *Lactobacillus oeni* sp. nov., from wine. *Int. J. Syst. Evol. Microbiol.* 59, 2010–2014.

Maragkoudakis, P.A., Nardi, T., Bovo, B., D'Andrea, M., Howell, K.S., Giacomini, A., and Corich, V. (2013). Biodiversity, dynamics and ecology of bacterial community during grape marc storage for the production of grappa. *Int. J. Food Microbiol.* 162, 143–151.

Marques, A.P., Leitão, M.C., and San Romão, M.V. (2008). Biogenic amines in wines: Influence of oenological factors. *Food Chem.* 107, 853–860.

Morenzoni, R. (2006). *Malolactic Fermentation in Wine: Understanding the Science and the Practice.* Lallemand, Montreal.

Munoz, R., Moreno-Arribas, M., and de las Rivas, B. (2011). Lactic acid bacteria. *Mol. Wine Microbiol.* 1st ed., Elsevier Acad. Press Lond. UK, pp. 191–226.

Nakamura, Y., Franz, O., and Back, W. (2003). *Proceedings of the European Brewing Congress*, Vol. 65, Dublin, pp. 612–620.

Nardi, T., Vagnoli, P., Minacci, A., Gautier, S., and Sieczkowski, N. (2014). Evaluating the impact of a fungal-origin chitosan preparation on *Brettanomyces bruxellensis* in the context of wine aging. *Wine Stud.* 3(1), 13–15.

Nielsen, J.C., and Richelieu, M. (1999). Control of flavor development in wine during and after malolactic fermentation by *Oenococcus oeni*. *Appl. Environ. Microbiol.* 65, 740–745.

OIV (2014). Oenology resolutions. OIV/OENO 462/2014.

Osborne, J.P., de Orduña, R.M., Pilone, G.J., and Liu, S.-Q. (2000). Acetaldehyde metabolism by wine lactic acid bacteria. *FEMS Microbiol. Lett.* 191, 51–55.

Osborne, J.P., Dubé Morneau, A., and Mira de Orduna, R. (2006). Degradation of free and sulfur-dioxide-bound acetaldehyde by malolactic lactic acid bacteria in white wine. *J. Appl. Microbiol.* 101, 474–479.

Pan, W., Jussier, D., Terrade, N., Yada, R.Y., and Mira de Orduña, R. (2011). Kinetics of sugars, organic acids and acetaldehyde during simultaneous yeast-bacterial fermentations of white wine at different pH values. *Food Res. Int.* 44, 660–666.

Peppard, T.L., and Halsey, S.A. (1982). The occurrence of 2 geometrical-isomers of 2,4,5-trimethyl-1,3-dioxolane in beer. *J. Inst. Brew.* 88, 309–312.

Pérez-Martín, F., Izquierdo-Cañas, P.M., Seseña, S., García-Romero, E., and Palop, M.L. (2015). Aromatic compounds released from natural precursors by selected *Oenococcus oeni* strains during malolactic fermentation. *Eur. Food Res. Technol.* 240, 609–618.

Petters, H.I., Flannigan, B., and Austin B. (1988). Quantitative and qualitative studies of the microflora of barley malt production. *J. Appl. Bacteriol.* 65, 279–297.

Piao, H., Hawley, E., Kopf, S., DeScenzo, R., Sealock, S., Henick-Kling, T., and Hess, M. (2015). Insights into the bacterial community and its temporal succession during the fermentation of wine grapes. *Food Microbiol.* 6, 809.

Ribéreau-Gayon, P., Dubourdieu, D., Donèche, B., and Lonvaud, A. (2006). *Handbook of Enology, the Microbiology of Wine and Vinifications.* New York, John Wiley & Sons.

Rodas, A.M., Chenoll, E., Macian, M.C., Ferrer, S., Pardo, I., and Aznar, R. (2006). *Lactobacillus vini* sp. nov., a wine lactic acid bacterium homofermentative for pentoses. *Int. J. Syst. Evol. Microbiol.* 56, 513–517.

Ruiz, P., Izquierdo, P.M., Seseña, S., García, E., and Palop, M.L. (2012). Malolactic fermentation and secondary metabolite production by *Oenoccocus oeni* strains in low pH wines. *J. Food Sci.* 77, M579–M585.

Saucier, C., Bourgeois, G., Vitry, C., Roux, D., and Glories, Y. (1997). Characterization of (+)-catechin-acetaldehyde polymers: A model for colloidal state of wine polyphenols. *J. Agric. Food Chem.* 45, 1045–1049.

Schopp, L.M., Lee, J., Osborne, J.P., Chescheir, S.C., and Edwards, C.G. (2013). Metabolism of nonesterified and esterified hydroxycinnamic acids in red wines by *Brettanomyces bruxellensis*. *J. Agric. Food Chem.* 61, 11610–11617.

Schwarz, P., Stanley, P., and Solberg, S. (2002). Activity of lipase during mashing. *J. Am. Soc. Brew. Chem.* 60, 107–109.

Shimizu, C., Ohno, M., Araki, S., Furusho, S., Watari, J. and Takashio, M. (2002). Effect of reduction of carbonyl compounds by yeast on flavor stability of Happoshu. *J. Am. Soc. Brew. Chem.* 60, 122–129.

Spano, G., and Capozzi, V. (2011). Food microbial biodiversity and "microbes of protected origin. *Front. Microbiol.* 2, 237.

Spettoli, P., Nuti, M.P., and Zamorani, A. (1984). Properties of malolactic activity purified from *Leuconostoc oenos* ML34 by affinity chromatography. *Appl. Environ. Microbiol.* 48, 900–901.

Stratford, M., and Carter, A.T. 1993. Yeast flocculation—Lectin synthesis and activation. *Yeast* 9, 371–378.

Sumby, K.M., Jiranek, V., and Grbin, P.R. (2013). Ester synthesis and hydrolysis in an aqueous environment, and strain specific changes during malolactic fermentation in wine with *Oenococcus oeni. Food Chem.* 141, 1673–1680.

Swiegers, J.H., Bartowsky, E.J., Henschke, P.A., and Pretorius, I.S. (2005). Yeast and bacterial modulation of wine aroma and flavour. *Aust. J. Grape Wine Res.* 11, 139–173.

Terrade, N., and de Orduña, R.M. (2009). Determination of the essential nutrient requirements of wine-related bacteria from the genera *Oenococcus* and *Lactobacillus. Int. J. Food Microbiol.* 133, 8–13.

Tesfaye, W., Morales, M.L., Garcìa-Parrilla, M.C., and Troncoso, A.M. (2002). Evolution of phenolic compounds during an experimental aging in wood of sherry vinegar. *J. Agric. Food Chem.* 50(24), 7053–7061.

Testa, B., Lombardi, S.J., Tremonte, P., Succi, M., Tipaldi, L., Pannella, G., Sorrentino, E., Iorizzo, M., and Coppola, R. (2014). Biodiversity of *Lactobacillus plantarum* from traditional Italian wines. *World J. Microbiol. Biotechnol.* 30, 2299–2305.

Thum, B., Miedaner, H., Narziss, L., and Black, W. (1995). Bildung von Alterundscarbonylen—Mogeliche Mechanismen und Bedeutung bei der Bierlagerung. *Proceedings of the European Brewery Convention Congress*, pp. 491–498.

Turakainen, H., Kristo, P., and Korhola, M. (1994). Consideration of the evolution of the *Saccharomyces-Cerevisiae* Mel gene family on the basis of the nucleotide-sequences of the genes and their flanking regions. *Yeast* 10, 1559–1568.

Ugliano, M., Genovese, A., and Moio, L. (2003). Hydrolysis of wine aroma precursors during malolactic fermentation with four commercial starter cultures of *Oenococcus oeni. J. Agric. Food Chem.* 51, 5073–5078.

Vanderhaegen, B., Neven, H., Coghe, S., Verstrepen, K.J., Derdelinckx, G., and Verachtert, H. (2003). Bioflavoring and beer refermentation. *Appl. Microbiol. Biotechnol.* 62, 140–150.

Wibowo, D., Eschenbruch, R., Davis, C.R., Fleet, G.H., and Lee, T.H. (1985). Occurrence and growth of lactic acid bacteria in wine: A review. *Am. J. Enol. Vitic.* 36, 302–313.

Zapparoli, G., Torriani, S., Malacrino, P., Suzzi, G., and Dellaglio, F. (2015a). Interactions between *Saccharomyces* and *Oenococcus oeni* strains from Amarone wine affect malolactic fermentation and wine composition. *VITIS-J. Grapevine Res.* 42, 107.

Zapparoli, G., Tosi, E., Azzolini, M., Vagnoli, P., and Krieger, S. (2009). Bacterial inoculation strategies for the achievement of malolactic fermentation in high-alcohol wines. *Afr. J. Enol. Vitic.* 30, 49–55.

Zapparoli, G., Tosi, E., and Krieger, S. (2015b). Influence of the pH of Chardonnay must on malolactic fermentation induced by bacteria co-inoculated with yeasts. *VITIS-J. Grapevine Res.* 45, 197.

2 Distillation Processes and Distillates

Frank Vriesekoop and Dawid Ostrowski

CONTENTS

2.1 DISTILLING PROCESSES

2.1.1 Distilling: Art versus Science

Distillation is the art and science of separating alcohol and its accompanying volatiles from its fermented base, purifying, and, ultimately, refining the spirit fraction (predominantly alcohol) for its intended use. The art and science of distillation is based on a combination of skills and experience, and an understanding of the processes that occur within a distillation vessel. The science of distillation is the process of separating a mixture of liquids with distinctively different boiling points through the process of controlled evaporation and the collection of various fractions through the subsequent condensation of vapors. The science of distillation takes into account that evaporation and condensation events will occur multiple times within a single distillation process. Furthermore, it takes into account that fermented products contain a multitude of compounds with inherently different boiling points, each with its own characteristic physiochemical and sensory properties.

The art of distillation, however, is the balance of having the clarity of the final product, whether it will be matured in wooden barrels for decades or have all its congeners and heads removed for vodka production. The art of distillation requires the distiller to understand not only the basics of distillation and fermentation but also the ability to see, feel, and sense how the product is to be developed after distillation. The art of distillation can only be taught through experience, nosing the product as it comes off the still, tasting consecutively until the desired quality is achieved—the soul of the spirit.

The most basic form of distillation can best be described as converting a liquid into a gaseous state and then back to a liquid state—for example, boiling water "Bear Grylls style": If you had seawater or dirty stagnant water that was undrinkable you could distil this water into a drinkable water through distillation. This involves filling the kettle with the stagnant water, placing it on a fire, and waiting for it to reach the boiling point (100°C). This point is when the water begins to steam through the kettle's spout. Then, the steam is cooled to below its boiling point for the water molecules to return back into a liquid state. This can be done by connecting a hose to the spout and dropping the hose into the stagnant water and cooling the steam back to water, where it can be collected in a drinking container. This process allows the water to be purified so that it is fit for human consumption.

2.1.2 Basics of Distilling

Distilling of alcohol begins with a fermented base obtained from fermentable sugar from cereals, fruit, tubers, or some sort of molasses. The role of distillation is to capture the essence of the raw material while purifying it fit for human consumption. The base fermented alcohol is a heterogeneous mixture of compounds of varying boiling points which is delivered to the distiller at a range of 5%–13% alcohol by volume (abv). Ideally a distiller would like to distil the wine/beer within three days of the completion of the fermentation in order to avoid spoilage and degrading the distiller's raw material. Alcohol evaporates at 78°C while water evaporates at 100°C, the other chemical compounds created during fermentation have different boiling points, which might be higher or lower than that of water or ethanol. The distiller takes advantage of this difference in the boiling point between water and ethanol. As the fermented base is warmed, the alcohol evaporates first, and at a much more rapid rate, before the water, allowing the alcohol to be separated from the water through evaporation. The evaporated fraction is collected following condensation, which is done continually until the distillate is of exceptional quality.

2.1.3 Evaporation and Condensation

2.1.3.1 Evaporation

Evaporation is the change in the physical state of a compound from a liquid to a gas; condensation is the exact opposite and is the change in state from a gas into a liquid. The physical state of a compound is governed by two different energies that act on the molecules of that compound. One energy facilitates the force that allows a molecule to move (kinetic energy). All molecules move. This movement can vary from a slight vibration, without actually changing location, to an extremely rapid change in location. The second energy that acts on molecules facilitates the attraction between individual molecules (van der Waals forces). The energy of attraction between individual molecules is fixed, meaning that it represents a property of a compound. The ability of a molecule to move is variable and is governed by the amount of energy that is introduced to the molecules. The physical state of a compound is therefore governed by the balance between the energy of attraction that "binds" molecules together (van der Waals forces), and the magnitude of energy that allows molecules to move around (kinetic energy).

When the van der Waals forces are many times greater than the kinetic energy that acts on a compound, that compound is very likely to be a solid. When the van der Waals forces are roughly equal to the kinetic energy that acts on a compound, that compound is very likely to be in a liquid state; and when the kinetic energy that acts on a compound becomes (much) greater than the van der Waals forces, it is very likely that a compound changes from a liquid state into a gas state. Hence, in order for molecules to change from one physical state to another, either energy has to be added to those molecules (i.e., melting or evaporation), or energy has to be removed (i.e., freezing or condensation).

All compounds have their own chemical properties, which include the magnitude of interaction between a quality of identical molecules. For instance, individual compounds have a unique density, boiling point, and latent heat of vaporization (Table 2.1). The density is the mass of a compound per unit volume (kilograms per liter), which relates to the atoms that make up the molecules and the manner in which the molecules pack together. The boiling point is the temperature at which a liquid changes into vapor, which coincides with the point at which the vapor pressure of the liquid equals the pressure surrounding the liquid. The latent heat of vaporization is the heat (or energy) required for a fixed quantity of a given compound to change from a liquid to a gas. The fermented broth contains a wide variety of volatile and non-volatile compounds, which varies within the same substrate range and across substrate ranges. Variations across different substrate ranges (i.e., grains versus tubers versus fruits versus molasses) is very much due to the core nature of the plants they are obtained from; variation within the same substrate range (i.e., different grape varieties for wine, different barley or malt varieties for beer, different apple varieties for cider, etc.) can be due to variations in raw material cultivars and raw material processing, yeast and bacterial strains, and fermentation conditions.

While there are many factors that influence the quantities of the various volatile compounds that can be found in the fermented broth, they are usually present within reasonably distinct ranges. Evaporation commences when a liquid mixture of alcohol, water, congeners, and oils is heated up. As the mixture is heated, components within the mixture start to evaporate, with those with a relatively low boiling point evaporating rapidly at their boiling point; other compounds with higher boiling points remain largely in the liquid state. During a pot still distillation, a mix of organic compounds begins to evaporate—at first, methanol, simple aldehydes, and some higher alcohols. These compounds evaporate relatively rapidly and coevaporate with a significant portion of ethanol.

TABLE 2.1
Various Volatiles Present in Fermented Wash and Some of Their Properties

Compound	Density	Boiling Point (°C)	Latent Heat of Vaporization (kJ/mol)
Water	1.000	100.0	40.6
Ethanol	0.798	78.4	38.5
Methanol	0.791	64.7	35.3
Propanol	0.804	97.0	47.5
Butanol	0.810	117.7	62.5
Acetaldehyde	0.788	20.2	26.1
Ethyl acetate	0.900	77.1	31.9
Isoamyl alcohol	0.823	131.6	53.1
Isoamyl acetate	0.876	142.0	43.9
Butyl acetate	0.882	126.0	36.8

2.1.3.2 Condensation

Condensation is the change in the physical state of a compound from a gas to a liquid. As mentioned before, the physical state of a compound is governed by two different energies that act on the molecules of that compound. The van der Waals forces are the attractive forces that exist between individual molecules and keep them in close proximity to each other, restricting the movement of individual molecules. The opposite forces that allow molecules to move around freely are represented by the (kinetic) energy that comes in the form of heat. By transferring the heat contained within the heated vapors to a much colder medium, the kinetic energy of the molecules in the vapor phase is reduced to such an extent that they undergo a change in physical state from gas to liquid.

2.1.3.3 Reflux

Reflux occurs in each still regardless of its design. It occurs when a volatile reaches a temperature at which it transforms into a gaseous state and then returns to a liquid state within the same space. In the first instance, the compound (alcohol) gains sufficient energy to change its physical state from liquid to gas; however, conditions within the space in which the compounds evaporated cause the removal of just enough energy within the environment to cause the same compounds to revert back to the liquid state. The energy-rich conditions that caused the compounds to evaporate in the first instance will promote the re-evaporation of the slightly more volatile fraction and will ultimately allow a significant portion of the volatiles to cross the threshold to either reach the condenser or move on to the next stage/plate of a fractional distillation column. Hence, during distillation the physical state of volatiles is in constant flux, moving from liquid state to gaseous state over and over again. During continuous distillation, this can occur thousands of times as compounds travel up the column, moving from plate to plate, and being forced to change states due to the design of the trays. In a pot still, this can be forced by using a dephlegmator, a partial condenser, forcing the vapor coming through to be cooled and returned to the pot for redistillation.

In a simple pot still (Figure 2.1) the fermented broth is heated by some external heating source. The input of energy causes the temperature to rise and both the pot

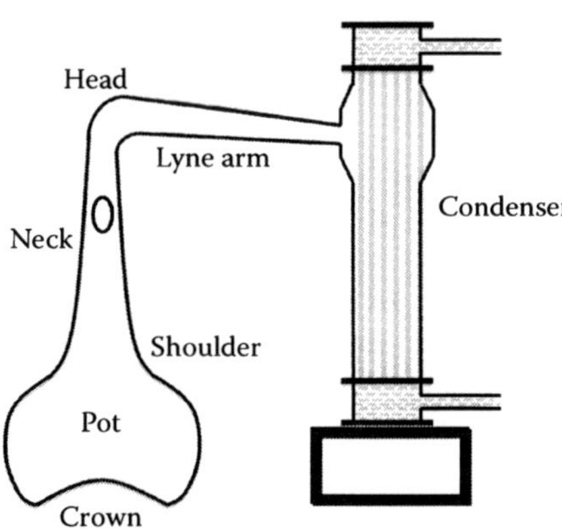

FIGURE 2.1 Pot still configuration.

still and the broth will be heated. The fermented broth will contain a mixture of compounds of varying boiling points. The more volatile compounds (those with a low boiling point) will evaporate rapidly and rise up the neck of the still. Initially, the vapors of the volatile compounds will come in contact with the surface that makes up the neck of the still and transfer some of their energy to the metal surface. When this occurs, the metal in the neck of the still heats up and the compounds cool down. The cooling down of these compounds can cause them to revert back to the liquid state and flow back into the heating broth under the influence of gravity. This occurs over and over again until the neck reaches a temperature that is equivalent to the vapor temperature; hence the reflux is high at the beginning of the distillation process. This has the effect of concentrating the compounds with similar boiling points while the vapors are being rectified.

Once the neck is at the vapor temperature, the reflux drops off as there is not much interpot cooling being facilitated, and the only way to create more reflux is to force the vapor back into a liquid state by introducing additional reflux opportunities, such as a doubler, dephlegmator, or purifier (see Sections 2.3.3, 2.3.4, and 2.3.5). The ability to use reflux allows the distiller to keep redistilling the heterogeneous vapor mixture until the distiller is happy with the product coming out of the condenser. It allows the alcohol to come out at higher concentrations, reducing the number of physical distillations required (i.e., double or triple distillations). Theoretically, distillation can take place using reflux in just one pass and 96% alcohol can be achieved; however, this depends very much on the reflux capacity built into the still design.

2.2 COMPONENTS OF A DISTILLATION UNIT

The most traditional method of distilling is through the use of a batch process. In a typical batch system materials are being processed in one fixed quantity at a time, from start to finish. When one batch has finished, the next batch can be started to repeat the process. Most batch distillations are carried out in copper pot stills. Pot stills come in a wide variety of designs, but they all have the same principal components. Figure 2.1 provides a common design for a pot still. Each pot still contains the following principal components:

- A heat source that provides the energy to facilitate the differential evaporation of the volatiles in the fermented broth
- A pot that contains the fermented broth that is to be distilled
- A shoulder, which represents the curved system that links the pot to the neck of the still
- The neck, which is usually the most upright component of the still
- The head, which is the highest point of the neck and leads to the lyne arm
- The lyne arm, which connects the neck of the still to the condenser
- The condenser, which facilitates a cooling process in which the evaporated compounds are forced back into the liquid phase to collection (Nicol, 2014)

The heat source can be any fuel. The most important input is whether the heating is facilitated through direct or indirect heating. When the heating is facilitated through direct heating, the crown of the pot has to be able to deal with the intense heat that is provided by the source. Indirect heating could be an electrical heating element or a steam or hot water coil. The pot can be of any shape; the most common ones are straight walled, plain, or onion shaped (Figure 2.2). It is usually made of copper and has openings that allow for filling, emptying, and cleaning. The volume of the pot represents the capacity of the still; however, not all pot stills are always filled to their capacity and the access surface area will play a role in reflux (see Section 2.1.3.3).

The pot is connected to the neck of the still via a curved component made up of the pot's shoulder and an ogee. The neck of the pot still is usually the tallest part of the still. However, within that arrangement it can come in short and very tall configurations. Furthermore, the neck can come in a variety of shapes (Figure 2.3). The plainest shape is a straight-shaped linking the shoulder all the way through to the head of the neck. However, various constrictions and widenings create shapes such as that resembling a lantern or a ball. Both shapes contribute to additional reflux capacity within the still. The head of the still is typically the highest point of the still's neck. The neck links the main evaporation vessel to the lyne arm. The most simplistic description of the lyne arm would be the connecting pipe that links the evaporation vessel to the condenser. The positioning of the lyne arm has a significant bearing on the spirit character that comes from the still. There are three principal alignments for a lyne arm (Figure 2.4). When the lyne arm is positioned in a near-perfect horizontal position, the still yields a "common spirit" that represents the middle-of-the-road spirit that the combination

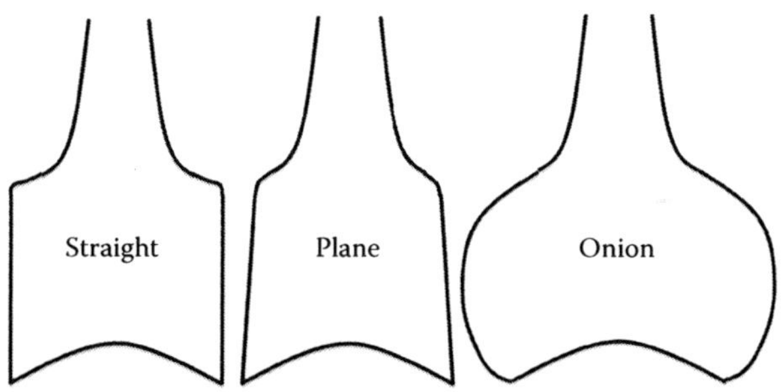

FIGURE 2.2 Common shapes of the pot of a pot still.

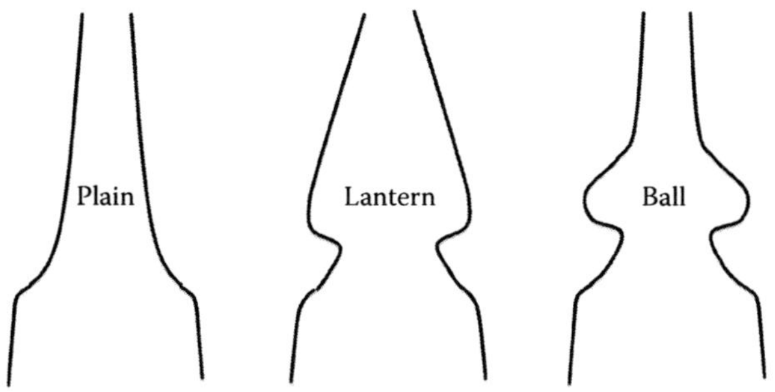

FIGURE 2.3 Common shapes of the necks of a pot still.

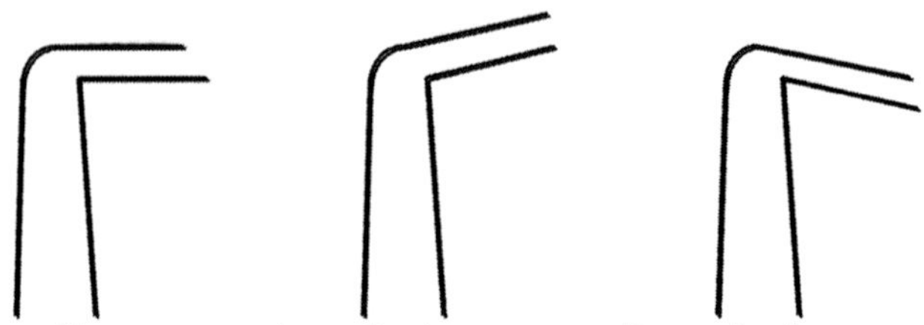

FIGURE 2.4 Various alignments of the lyne arm that connects the neck of the pot still with the condenser.

of the evaporation vessel and the condenser can yield. This means that the lyne arm plays no, or a limited, role in the character of the final spirit. However, if the lyne arm is positioned in an upwards (inclined) plane, it will essentially become a part of the evaporation vessel because the vapors can still reflux in the lyne arm and run back into the pot. The additional reflux will yield a lighter spirit with a lighter character. On the other hand the lyne arm can be positioned in a descending plane, which essentially makes it a part of the condenser because any vapor condensing in a descending lyne arm will run into the condenser of the distillation unit. This means that all vapors that cross the threshold at the top of the neck will be collected in the new spirit. As a rule of thumb, pot stills with an ascending lyne arm will be lighter in character; while those with a descending lyne arm will be heavier in character.

2.2.1 The Condenser

The condenser is a cooling apparatus designed to remove energy from the vapor that passed over the still's threshold and forced the compounds to convert from the vapor phase into the liquid phase. In order to make the condenser an efficient apparatus for the removal of heat from the vapor, it is set up to act as a heat changer in which cold water absorbs the heat from the hot vapors, which causes the vapors to become liquid. There are two principal types of condensers. The first is a worm tub where the hot vapor passes down a coiled tube (the worm) that is submerged in a tub of cold water. The heat energy will pass from the vapors into the water through the wall of the coiled tube. This means that the water will slowly warm up while the vapors are being cooled. In order for the worm tub to remain an effective apparatus to facilitate the cooling down of the spirit, it requires continuous replenishment with cold water. This usually occurs by cold water being pumped into the tub near the bottom of the tub. The (slightly) warmer water near the top of the tub will flow over the edge of the tub and maintain an appropriate cooling apparatus.

The second, and more common, condenser style is the tube-in-shell configuration. This condenser is designed as a heat exchanger in which the vapors enter into a vapor chamber in which many copper tubes are placed. The vapors are cooled by means of the heat exchange that occurs across the copper tubes in which a cool liquid runs in the opposite direction on the insides of the tubes. This forces the heat energy from the gaseous state vapors to be transferred into the cool liquid, allowing the vapors to return to the liquid state.

2.3 BATCH DISTILLATION

Batch distillation refers to the application of distillation in individual lots. In our context, this means that an alcoholic liquid is distilled to be separated into two or more fractions based on the volatility of the components in those fractions. The term "batch" applies because one individual charge at the time will be distilled. This starts with filling a pot in which the charge is heated, followed by the separation based on volatility where the volatile fractions are collected as condensate, and the nonvolatile fraction remains in the pot. When the separation is deemed to be complete, the pot is emptied and a new charge is introduced into the pot. This process can be repeated many times over. There are various design variations and iterations that apply to the equipment used in batch distillations. The following sections aim to provide a basic outline of their ability to facilitate the operations of evaporation, reflux, and concentration.

2.3.1 Pot Still Designs

Distillation units come in a wide array of styles, shapes, and modes of operation. In all instances their design will influence the characteristics of the spirit produced from them. Most pot stills can be seen as the same principal components, made up of a heating vessel in which the volatile components are encouraged to change their physical state from liquid to gas (evaporation), and a cooling vessel in which the volatiles that emerge from the heating vessel are returned from their gas state into a liquid state (condensation). The heating vessel (the pot) is connected to the cooling vessel (the condenser) via a lyne arm. Very early pot stills were made from clay or ceramics, which represented a readily available construction material that could be fashioned into a form that allowed a practical shape to be obtained. All modern stills are made from metal, with copper the most commonly used construction material. Metals are practical in that they allow for ease of heating and heat transfer; metal components are sturdy and are unlikely to crack upon a minor impact.

While the heating pot is exclusively made from metal, other components within the still could be made from alternative materials such as glass or wood. Copper is the metal of choice when it comes to still design. Copper is extremely malleable and ductile, allowing it to be shaped and stretched into varied and complex surfaces without yielding or breaking. Furthermore, copper is an extremely good conductor of heat, which allows for optimal heat transfer at the heating and cooling points within the still. A readily available alternative to copper would be stainless steel; however, stainless steel is a relatively poor conductor of heat, with the thermal conductivity of copper being approximately 30 times higher—14 W/(m•K) vs. 399 W/(m•K)—for stainless steel and copper respectively. This means that the use of copper in high-throughput condensers (see later discussion) is highly recommended. The high malleability and thermal conductivity were the initial reasons why copper became the metal of choice for still design. An unwanted side effect of the elevated exposure to copper can cause high levels of copper in the final spirit, which can cause serious health issues (Neves et al., 2007). Furthermore, copper catalyzes the formation of ethyl-carbamate from various nitrogen compounds found in raw fermentation substrates, or from the addition of urea as yeast food (Nóbrega et al., 2009).

Despite this, a positive consequence of the use of copper has been that the new spirit made in a copper still is significantly more refined and smoother. The influence of copper on the sensory profile of new spirit has been found to be due the fact that copper provides a highly reactive, catalytic surface on which sulfur compounds react with copper ions and create nonvolatile sulfur compounds. This leads to a marked increase in variety of sulfur compounds, but, more importantly, a decrease in volatile sulfur compounds (Harrison et al., 2011). The beneficial effects of copper on spirit quality are predominantly due to the contact that copper has with the compounds while in the vapor phase. This means that sufficient copper surface area is required in areas where reflux and condensation occur. Overcharging a still will reduce the available area of copper surface for vapor exposure; undercharging will enhance the available copper surface.

The most efficient part of the copper surface is the patina that forms on the vapor exposed components. While it will occasionally be necessary to clean the various distillation components, the removal of the copper patina during cleaning will compromise the sulfur reactivity of the still. Because of the reactive nature of copper with various compounds in the vapor phase, the copper will gradually dissolve during the distillation process so that stills and other copper parts need periodic replacement (every 10–15 years). Variations in copper–vapor interactions can be construed when comparing similar spirits, those produced at distilleries with a worm tub as condenser compared to those with a shell and tube condenser. The more rapid and evenly distributed flow of distillate through a shell and tube condenser is thought to keep the tubes free of detritus, whereas in a worm tub the flow is more gentle over less surface area and deposits buildup on the copper surfaces. When the copper surfaces in worm tubs become covered in detritus, the spirit emerging will be relatively high in sulfur compounds, which contribute to the so-called heaviness or meatiness or body of the spirit.

2.3.2 Pot Still Operation

If the traditional Scottish pot still (Figure 2.1) is taken as the base design for all batch distillations, then it shows that the desired purity of the new spirit requires at least two distillation runs. The first distillation run is carried out in a "wash" still, which is charged with the fermented wash/beer/wine (from now on referred to as "wash") in its crude form. Upon heating of the fermented wash, the alcoholic liquid will start to froth because of the inclusion of the carbon dioxide, yeast cells, and proteins from the fermented substrate. In order to manage the expanding volume of the frothing wash, one or two sight glasses are included in the neck of the still. This allows the still operator to observe the behavior of the wash, and avoids the froth from the wash being carried over into the condenser. The froth consists of a high concentration of proteins that will foul the condenser and require an immediate shutdown of the still to carry out a complete cleaning. In order to manage the inevitable frothing of the broth, the shape of the still neck can be designed to be bulbous, which can accommodate a large volume of material as well as acting as an additional amount of surface for reflux to occur. Another design feature that could control/subdue frothing could be inclusion of a cold finger near the top of the still head. The cold finger will be supplied with cooling water and will act to disrupt the froth, forcing it to collapse. All volatiles from the wash still will be collected as "low wines." The progress of the wash distillation can be followed by measuring the specific gravity of the spirit stream coming from the condenser.

The completion of the wash still run occurs when the spirit stream is less than 1% abv. The time and energy required to continue distilling and only collect a very weak spirit is uneconomical. The remaining liquid in the pot is referred to as "pot ale," which tends to be high in proteins and unfermentable sugars, as well as any other nonvolatile compounds originating from the wash. In a traditional Scottish double distillation, the wash still is typically followed by another pot still referred to as the "spirit still," which is operated to refine the new spirit by essentially redistilling the low wines and selecting the heart of the spirits by removing the foreshots (heads) and the feints (tails) (Figures 2.5 and 2.6). Part of the refining of the spirit is

FIGURE 2.5 Distillation flow chart for a traditional Scottish double distillation.

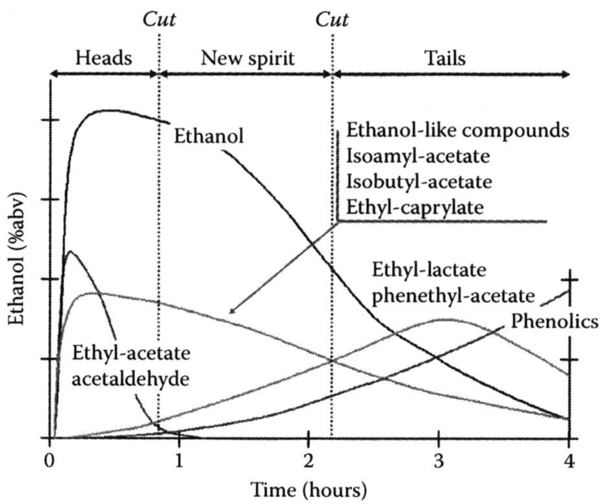

FIGURE 2.6 The evolution of various volatile fractions as they apply to a typical spirit still in a double or triple distillation. The cuts between the heads and the new spirit, and the new spirit and the tails, are approximate only, and will vary from product to product.

due to the application of another set of reflux that will occur within the spirit still. The foreshots are the first runnings of the spirit distillation run and contain many volatiles, such as acetaldehyde, ethyl-acetate, and methanol, including much ethanol. The principal aim for eliminating the foreshots from the new spirit is to remove the potentially toxic methanol and acetaldehyde from the spirit. While grain-based low wines yield relatively low levels of methanol, fruit-based low wines tend to contain substantial quantities of it.

Because the spirit still is used to process one batch after another, many compounds that were part of the tails from a previous run are often deposited on the inner surfaces of the still. The very high strength of the early runnings of the spirit still will displace much of the fatty material deposited in the still (Nicol, 2014). In order to judge the appropriate timing for the cut from foreshots to the hearts to be collected as new spirit, a demisting test can be carried out. This test involves mixing the foreshots with water in order to reduce the alcohol strength to 45.7% abv. Foreshots will be collected for as long as the foreshots/water mixture remains turbid in the demisting test. When the sample mixture of foreshots and water clears (at 45.7% abv), the spirit is deemed to be of sufficient quality to be collected as the hearts (Nicol, 2014).

While the demisting step is a helpful tool in managing the spirit still run, many experienced distillers will use either time or volume collected or spirit strength as the point at which the cut is made. They might only use the time-consuming demisting test when new recipes or significantly different raw materials are used. The heart of the spirit distillation run will contain much of the ethanol, plus ethanol-like compounds such as various higher alcohols and esters (Figure 2.6). It will also contain some of the compounds that were targeted for removal into the foreshots, but their presence will only be at trace levels. The end of the hearts run is difficult to describe as a one-size-fits-all definition, since the cut from heart to tails will inevitably be part of the principal character of the new spirit collected and depends on the brand, product, and/or application of the new spirit. Therefore, it is very much the call of the master distiller where to finish collecting spirit and let the remainder of the spirits be collected as tails. The tails contain heavier higher alcohols and esters, plus many fatty esters and phenolic compounds. Many of these heavier compounds will already have coevaporated during the earlier stages of the distillation run. In instances where a heavy phenolic presence might be desirable, the cut might be pushed back to allow a more heavy character to be included in the new spirit.

There are many variants of the typical Scottish double distillation method. For instance, many Irish whiskies are distilled thrice. One of the reasons for the triple pot distillation is the fact that most Irish pot stills are often much larger than most Scottish pot stills. The larger volume of the pot will result in a significantly smaller surface area within the pot and neck on which reflux can occur. Limited reflux can result in a very crude spirit. The application of triple distillation increases the overall reflux that the spirit is exposed to before the hearts are collected as new spirits. Furthermore, the triple distillation process also awards multiple opportunities to move forward various fractions closer to the final spirit outcome and return other fractions backwards for redistillation (Figure 2.7).

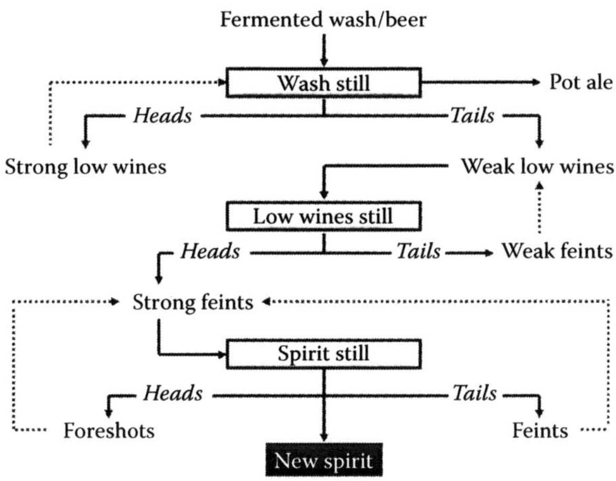

FIGURE 2.7 Distillation flow chart for a typical triple distillation.

The operation of the first pot still will entail a single cut and yield "strong" low wines and "weak" low wines. The strong low wines represent the highly volatile fraction of the spirit containing approximately the same mixture as the heads from a double distillation run. The remaining weak low wines are carried over as the charge for the second pot still (the low wine still). The low wine still operates very similarly to the wash still and will again entail a single cut, which will this time around yield strong feints and weak feints (Figure 2.7). Whereas the weak low wines from the wash still become the charge for the next still, in this instance, the strong feints from the low wine still will be carried over as the charge for the spirit still. The weak feints are returned to the low wine receiver. The spirit still in the triple distillation setup operates in very much the same manner as the spirit still in the double distillation setup. The new spirit from a triple distillation run will create a far more delicate spirit that will mature faster when matured in wooden casks, compared to the Scottish double distillation system.

2.3.3 The Dephlegmator

The dephlegmator is a partial condenser that is typically used as an internal, intermediate condensation unit in a multicomponent vapor stream. It is designed to selectively reflux vapor components in order to enhance the purity of the final spirit. Alternative terms to describe a dephlegmator are "partial condenser" or "reflux condenser;" both are correct and explain their role in the system they are employed in. A dephlegmator can be run as an air-cooled or a liquid-cooled dephlegmator. An air-cooled dephlegmator will provide only a limited opportunity to induce reflux; however, that might be all that is required. An air-cooled dephlegmator allows the vast majority of vapors to pass from the pot to the final condenser and can be operated as an active cooling system where air is

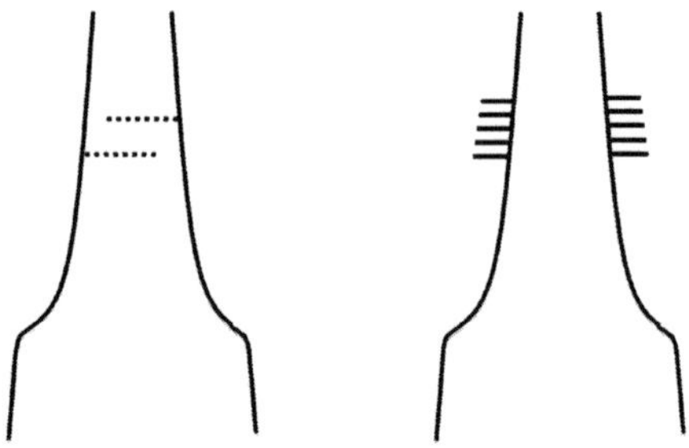

FIGURE 2.8 Placement of internal plates and external plates (fins) on the neck of a pot still to enhance reflux by providing: (a) more internal surface area for reflux to occur on and (b) external surface area for a more efficient heat exchange.

actively pumped towards the external surface of the pipes the vapor streams runs through. Or it can be operated in a more passive cooling manner where cooling fins are employed to enhance the surface area for heat exchange to occur on, not unlike a radiator. Operating the dephlegmator as liquid cooled allows much greater control of the reflux within a still, allowing tighter control of the quality of the spirit.

With a liquid-controlled dephlegmator, the still operator could achieve 100% reflux. In this instance the dephlegmator is actually operating as a full condenser. 100% reflux can be employed during the start-up procedures of a still, which allows the main components of the still to equilibrate prior to spirit collection. There are numerous alternative dephlegmators that will all promote additional reflux within the distillation unit prior to the volatile vapors entering the final condenser. Packed columns, purifiers, and internal plates or pipes (cooled or not) are not uncommon dephlegmators. Internal plates that are positioned within the neck of a pot still act as additional internal surfaces onto which refluxing condensation can occur. When internal plates are used, they are often perforated (Figure 2.8a), which allows much of the vapor to pass through to the condenser via the lyne arm. External plates can be added as external cooling fins on the outside of the still's neck (Figure 2.8b). When external fins are added to the neck of the still, a greater surface area for heat exchange on the outside of the still can be achieved, which will essentially act as a heat sink for the efficient removal of heat from the pot's surface to the surrounding air.

2.3.4 Doubler and Thumper

Both a doubler and a thumper are essentially an additional pot still. Both are almost always operated in a continuous mode, but can be part of a batch distillation as well as a continuous distillation. In a batch distillation, the doubler and the thumper will be employed as an intermediate vessel between the pot and the condenser. While the overall operation is in batch mode; the flow through the doubler or thumper vessel itself is continuous. The doubler or the thumper operates to increase the overall reflux of the still; hence they are both dephlegmators. The principal aim of their inclusion is to reduce the water content of the hot vapors that come off the first stage of the still. Both the input and the output of doubler and thumper are vapors. The principal difference between a doubler and a thumper is that the vapors entering the doubler will bubble through a hot aqueous liquid, which will encourage water to condense. A thumper is essentially a dry doubler with hot vapors entering into the gaseous space of the thumper vessel. The sheer presence of the thumper's inner surface is what induces the reflux. The continuous flow of vapors into the doubler or thumper will push the vapors through the vessel and allow volatile organic compounds to continue through to the next stage of the still (Figure 2.9). Because the doubler and thumper will continuously collect water, most are fitted with some sort of control to allow excess condensed water to be removed. The condensate from either doubler or thumper can be rerouted into the initial pot of the still to allow recovery of any co-condensed ethanol and other volatiles. The doubler or thumper can be aligned in such a manner that the collected condensate can return to the pot to optimize yield (Figure 2.10). In a sense this operates like a purifier incorporated into a lyne arm. Purifiers will only be effective in either horizontal or descending lyne arms, rather than in ascending

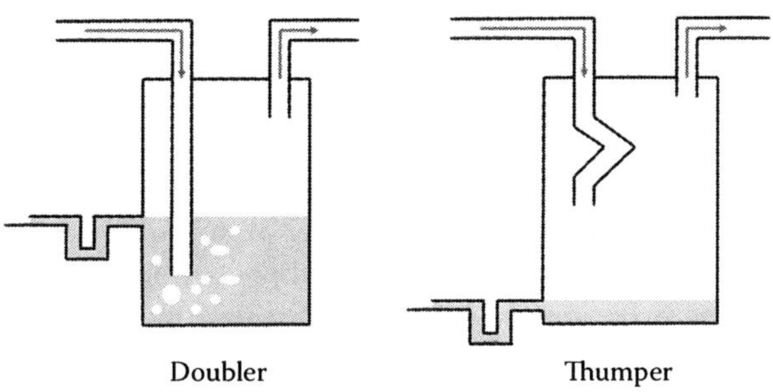

FIGURE 2.9 Overview of a "doubler" and a "thumper" set up as an alternative second distillation unit.

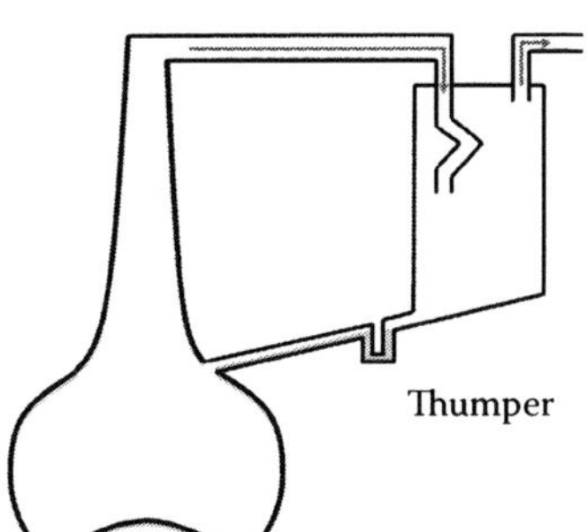

FIGURE 2.10 Inclusion of a thumper as a dephlegmator to promote addition reflux. Hot vapors enter the thumper from the top of the pot still. Easily condensable compounds can return to the pot, while the remaining vapors can continue to the next part of the distillation unit.

lyne arms, because ascending lyne arms will already provide ample opportunity for vapors to reflux back into the pot.

2.3.5 Fractional Distillation

Fractional distillation is often single-handedly relegated to the realm of continuous distillation. When fractional distillation is usually the main mode of distillation in continuous distillation, it often finds its application in batch distillations. In this section, the principal mode of operation of fractional distillation will be explained. This explanation will suit both the batch and the continuous application of fractional distillation.

Fractional distillation involves a column still with numerous plates where vapors pass from a lower chamber to an upper chamber through a plate with holes in it. Each plated chamber allows for additional reflux to occur. Because reflux entails the condensation of less volatile compounds within the chamber, it indicates that energy is lost to the surroundings. This loss of energy is represented by a slight decrease in temperature across multiple plated chambers within the column (Figure 2.11a). In the most simplistic terms, a fractional distillation column represents a stacked series of doublers. Hot vapors rise through the perforated plate where some of the vapors condense and form a liquid layer on the top of the plate. This liquid layer will constitute the liquid through which other hot vapors rise up. Volatile compounds that are still hot enough to remain in the gas phase will travel through to the next plated chamber in the stack. At each layer some condensation will occur and compounds that have their boiling point near the temperature of the liquid layer on the bottom of the plated chamber will remain in an equilibrium between evaporation and condensation within that chamber's environment. This is an equilibrium that is influenced by both the temperature of the liquid on the plate and the boiling point of the various volatiles; equilibrium is also influenced by the fact that vapors sweep across the plates and carry over some volatiles to higher plates regardless of the temperature and/or boiling points.

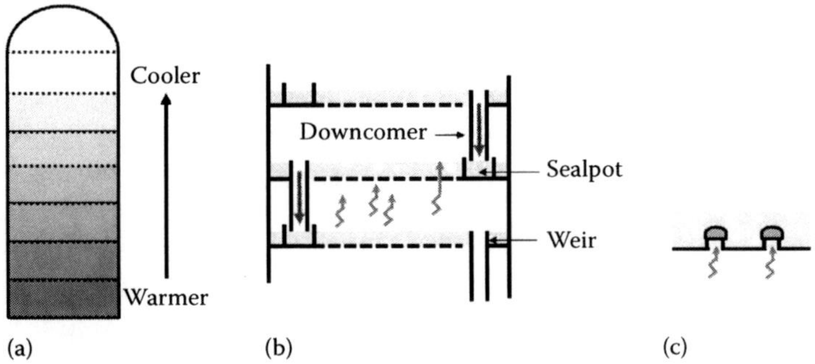

FIGURE 2.11 Details of fractional distillation column; (a) diagram indicating the layered temperatures across the various plated chambers within a fractionating column; (b) configuration of downcomers, weirs, and perforated plates; and (c) alignment of bubble caps.

In order to maintain a degree of management over the flow of liquids and vapors across a multitude of plates within the column, a number of internal components can be introduced. These come in the form of downcomers, weirs, and bubble caps and/or valve caps. A downcomer allows liquid to flow back from a plate to the one below it. The liquid in the liquid layer on a plate will keep accumulating throughout the time the column operates. In order to avoid any plates flooding the plated chamber, a downward position pipe (the downcomer) allows any excess liquid to be returned to the lower plated chamber (Figure 2.11b). The liquid level at the upper plate is controlled by the positioning of the top of the downcomer pipe, which usually protrudes up from the bottom of the plate by a couple of centimeters (a mini weir). At the other end of the downcomer pipe, the fluid-returning end, the downcomer pipe returns below the fluid level within a small seal pot. The combination of the mini weir and the seal pot ensures that vapors travel upwards through the column with very little opportunity to seep back down the column. The application of bubble caps (Figure 2.11c) ensures that the vapor–liquid contact at the liquid level on the plate is optimized for heat exchange and stripping out of volatiles.

2.4 CONTINUOUS DISTILLATION

Continuous distillation is not dissimilar to batch distillation in that it relies on differences in volatility between different compounds to achieve their separation. The principal difference between batch and continuous distillation is the need to continuously feed fresh wash into the still and be able to capture the various volatile fractions as they (continuously) emerge from the system. Most continuous distillation processes occur in a manner where reflux (Section 2.1.3.3) is optimized, which typically occurs by means of fraction distillation (Section 2.3.5) in a column design. As mentioned earlier, fraction distillation can be portrayed as a stacked series of pot stills (doublers) where the product of one chamber becomes the feed of the next, and so forth. Because continuous distillation is a series of repeating distillation chambers, it allows for a far more refined application of reflux and can achieve very pure runoff of volatile fractions. This can be further refined by employing column configurations that are optimized for the separation of specific compounds or groups of compounds.

A basic continuous still is made up of at least four components: a heat source (usually steam), a stripping section, a rectifying section, and a condenser. The stripping section of the continuous still allows for most of the volatiles to be stripped from the predominant aqueous base. This is achieved by continuously entering the fermented wash on the feed plate, which is positioned between the stripping section and the rectifying section (Figure 2.12) while steam is being introduced at/near the bottom of the column. The rising steam will provide enough heat to cause all volatiles to be stripped from the fermented wash while it cascades down the stripping plates. The stripping plates usually contain large openings for both the liquid wash to cascade down the various plates and the steam and entrained volatiles to rise up. The efficiency of the stripping section is a balance between the flow rate of the wash entering the feed plate, the amount of energy contained in the steam introduced at the bottom

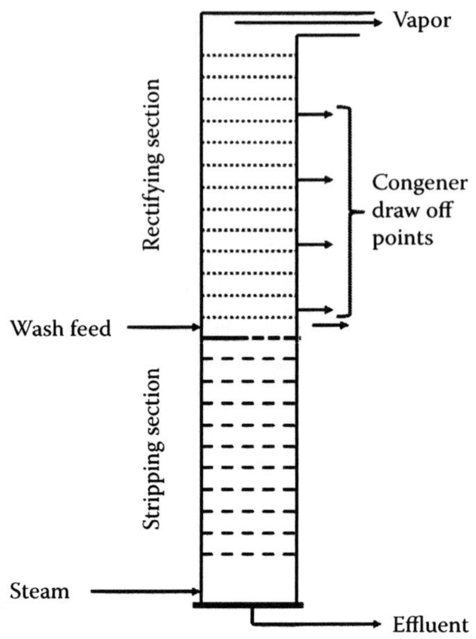

FIGURE 2.12 Basic schematic for a continuous still.

of the column, and the ability of the steam to achieve a sufficient level of contact with the cascading liquid running down the column. The rectifying section of the column allows for the vapors originating from the fermented wash to be purified into various streams. The heat held within the steam will carry the volatiles through the remainder of the rectifying column, which operates as a fractional column (see Section 2.3.5). Various congeners of different attributes (boiling points) will be able to be collected from the various plates along the rectifying section as liquids. The vapors leaving the top plate are condensed in a total condenser (all vapors are condensed). Most of these vapors are a high-grade ethanol with few impurities (congeners) and will be collected as neutral alcohol. However, some of the liquid is rerouted to the top plate of the column to maintain a sufficient liquid level and the downflow of reflux to the lower plates.

2.4.1 The Coffey Still

The Coffey still was not the first attempt to operate a distillation process for the production of potable alcohol in a continuous mode, but it was the first successful attempt to run a continuous distillation in a pragmatic and manageable manner. The existing continuous still as mentioned in Section 2.4 and Figure 2.12 required the still to be extremely tall, and the invention by Coffey separated out the two components of the still and placed the stripping column next to the rectifying column (Figure 2.13). This negated the need to have the continuous distillation plant housed in a tall edifice. Furthermore, great energy savings were obtained by allowing the wash to be preheated by utilizing the latent heat of vapor in the stripping column.

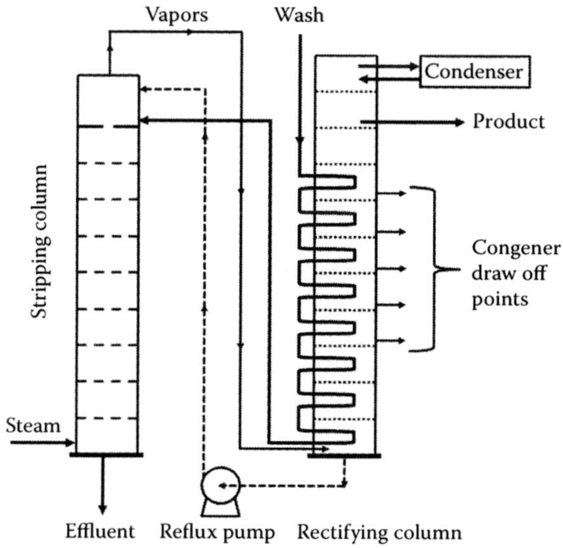

FIGURE 2.13 Basic schematic of a Coffey still.

The already warmed wash is introduced at the top of the stripping column, requiring less energy to strip the volatiles from the wash. The preheating of the wash by the vapors in the rectifying column also meant that much of the entrained heat was already removed by the time the vapors reached the condenser so that a much smaller condenser sufficed and less cooling water was required.

The Coffey still will produce a significantly lighter spirit compared to the spirit obtained from a pot still, which retains significant levels of both light and heavy congeners. The Coffey still is a commonly used continuous distillation system with applications in large-scale grain whisk(e)y, rum, and Armagnac production, and as the first two columns in multistage continuous distillation in the production of high-purity spirits suitable for the production of vodka, gin, etc. Products like grain whiskies, light bodied rums, and Armagnac will obtain most of their final character from the maturation process in wooden barrels, which means that minor amounts of congeners do not pose a serious obstacle to product quality. However, spirits such a vodka or gin require a virtual absence of congeners and other offensive substances, which represents a quality demand that cannot be obtained in a Coffey still. In order to achieve an even greater reduction in congeners, additional column designs can be employed.

2.4.2 Hydroselection Column

Hydroselection distillation is the more commonly used term for extractive distillation in the production of potable alcohol. Extractive distillation is defined as a distillation process that ensues in the presence of a miscible, relatively nonvolatile (high-boiling point) liquid (Gerbaud and Rodriguez-Donis, 2014). In the production of highly refined potable alcohol it is used to separate out the more volatile fractions (i.e., methanol, small esters, and acetaldehyde) from the already produced new spirit. Hydroselection distillation employs water as the separation solvent, with the remaining product being a mixture of water and predominantly ethanol. Water is the solvent of choice in extractive distillation for potable alcohol because it is nonreactive with the other components in the remaining mixture, causes no corrosion in the equipment, and can be easily separated from the remaining mixture. In the production of potable spirits, the hydroselection column is fed with the cooled spirit coming from the rectifying column, which is fed into the hydroselection column about halfway up. A very large quantity of hot water (up to 20 times in volume compared to cold spirit) is fed in near the top of the column. The addition of the hot water provides the energy required for the evaporation of the remaining aldehydes, esters, and methanol; at the same time, the heat provided will not be enough to drive off significant quantities of ethanol. The heads can be reintroduced into a preceding rectifying column in order to recover any coevaporated ethanol. The diluted spirit that emerges from the hydroselection column is then concentrated to strength in a subsequent rectifying column to recover water and further purify the spirit (Aylott, 2003).

2.4.3 Packed Bed Distillation

In some instances, packed bed columns can be used in the distillation process for the production of potable alcohol. There are two principal methods in which a packed bed column can be applied. The first is simply the introduction of additional reflux surfaces. If additional reflux is required, this can be achieved by employing any inert material that adds no flavor of its own to the spirit, nor should the inert material play any catalytic role. At the same time, the additional surfaces should not overly impede the flow of the vapors through the column, meaning that a highly structured packing material will be used. The second reason for employing a packed bed column in the production of potable alcohol is to provide additional opportunity of reactive distillation to occur. As mentioned earlier (Section 2.3.1), copper plays an essential role in the refinement of distilled spirit. Copper provides a highly reactive, catalytic surface on which sulfur compounds react with copper ions and create nonvolatile sulfur compounds, significantly improving the sensory profile of a new spirit. This means that reactive distillation is part and parcel of the distilling process in copper pot stills.

However, many distillation units are constructed with stainless steel, where there might not be enough opportunity for sufficient copper contact with the hot vapors in order to reduce the sulfur content of the new spirit. Distilled beverage products such as vodka should be devoid of any tastes or aromas, which means that great care has to be taken to remove all traces of sulfur from the spirit. The new spirit used in the production of botanically flavored beverages such as gin and akvavit also requires a base spirit that is devoid of sulfur. The high-purity alcohol required for vodka and gin production is usually achieved by means of continuous distillation, where stainless steel is the construction material of choice. In those instances, a column packed with (sacrificial) copper is often employed. The packed column containing copper can be a stand-alone unit, where hot spirit vapors are passed through the copper packing material en route to the next column. Or one or more packed sections are included as an integral part of, for instance, a rectifying column. The copper packing material can come in the form of copper pipes, saddles, or copper wool. Because the copper actually reacts with the sulfur compounds in the vapor stream, the copper will wear out and has to be replenished—hence the term "sacrificial copper."

2.5 DISTILLATION OPERATION AND MAINTENANCE

2.5.1 Distillation Operation

Once the temperature is up, the wash pump is started. Once alcohol vapor warms the base of the second column, the second pump is turned. And once that column is warmed up, then the pump for the third column is started. Most columns are automated these days; however, it is important to understand the operation so that solutions can be found during issues. It is important to balance the columns; this means the mass balance needs to be reached. Energy in equals energy out, which allows each column and each attempt to reach its target temperature, which allows separation to occur on each tray. This energy equilibrium is critical to the functioning of the columns as this will ensure that separation will occur at the critical points, which will then produce the ideal spirit required.

2.5.2 Distillation Maintenance

2.5.2.1 Daily/Weekly/Annual Checks

Checks are important when running a still. Listen for familiar sounds and for foreign sounds. Also check gauges regularly at 15-minute intervals; these can be quick glances at the gauges. This will give immediate feedback on what is happening within the still. When the still is shut down for maintenance, this will give indications of where to look for potential issues.

Keep a lookout for pumps and any leaks; some leaks occur during start-up and may pull up during distillation once steady-state temperature is reached. As a minimum, check plates annually with a visual inspection, looking for product buildup and scaling. Check sacrificial copper and add more if required. Bring a hose fitted to a pump and pump out all silt that has built up on the plates; this once again depends on column design because raw materials being processed have been taken from the farm, where a lot of sand can be deposited on plates over time. All pumps should be checked and well maintained. Temperature and pressure gauges should be checked and tested if possible. Check seals because leaks can occur over time and seals may need to be replaced, as the still heats and cools this expansion and contraction works the seals. Leaks are easily spotted visually against the copper and stainless steel. Remember that, if any welding is done around the columns, the column needs to be flushed three times and if possible flooded so that all alcohol and vapors are removed. If possible, place an air blower overnight to vent the internal area of the column prior to any welding. There have been many instances of a grinder or welder creating an explosion and destroying columns. Use a mobile sensor to check the air prior to welding after flushing and airing the columns.

2.5.2.2 Cleaning

Cleaning in a continuous still is undertaken every 3 to 6 months depending on usage. If there is a block in the still, then cleaning needs to be undertaken urgently and quickly; this does not occur regularly with new still designs. However, older stills can have blockages because of which an emergency shutdown would occur, and then, once cooled, the specific trays would be inspected.

Routine cleaning is normally done right after distillation is complete. Once the last of the wash is pumped into the column while the steam is still on, then water is pumped in for 9 minutes or longer depending on the still design. This allows the wash column to be stripped of all alcohol; once this occurs, then a caustic solution is pumped into the wash column and distilled. The majority of cleaning to be undertaken is in the wash column as this is where all the solids are removed from the distillate and it has the highest proportion of scaling and solids buildup. Once the caustic has gone through the system, follow by a water flush of about 15 minutes; then the still can be shut down or distilling fresh wash can begin.

2.6 DISTILLATES

There are numerous spirits on the market. Some are readily recognized because of their worldwide penetration across cultures; others have remained local in their orientation and are difficult to obtain outside the area where they evolved and are produced. While a product like tequila is unequivocally associated with Mexico, whisk(e)y has a much wider penetration into multiple countries and cultures, and can still be defined according to the variations that exist. Both tequila and whisk(e)y are readily available all over the world, while some regionally produced spirits rarely see the light of day outside the region in which they are produced. Examples of the latter are cashew and coconut feni and topinambur, which are rarely available outside their production locality.

Some of the differences between spirits are the raw materials used as fermentation substrates. Much of the choice of raw materials is influenced by local availability, which can impose a very large influence on the characteristics and quality of the final product. The development and differentiation of various (local) spirits is to a degree influenced by the distillation techniques employed in their production. The following sections contain a number of examples of various spirits, organized by raw material sources, with an explanation of the distillation techniques typically used in their production.

2.6.1 Fruit-Based Spirits

There are many fruit-based, distilled alcoholic beverages. Among the most common fruits used as the fermentation substrate are grapes, apples, apricots, and plums. Of course, fruit distillate can be produced from many different fruits and berries. Ideally, when working with fruit as the fermentation substrate, the fruit is best overripe, picked or harvested towards the end of the season of the fruit type. This ensures that the fermentable sugars are the highest possible. Many hard fruit (i.e., apples, pears, etc.) distilleries might use fruit that is picked and then let it sit for about a month before processing into juice; this allows the fruit to mature and increase in sugars as well as to soften so that it can be more easily processed.

2.6.1.1 Wine-Based Spirits (Brandy)

Wine-based distillates vary culturally and historically based on local traditions and local grape varietals. Collectively, all wine-based spirits are referred to as "brandy." Cognac and Armagnac are typical examples of French brandies from the Cognac and Gascony regions respectively. Stravecchio and brandy de Jerez are examples of Italian and Spanish brandies respectively, with pisco being a brandy produced in Peru and Chile. The previously mentioned brandies are distilled from wines made from intact grapes. As a complement to these brandies, there are also brandies made from the grape marc left over after the primary wine fermentation. These brandies include Italian grappa, Greek rakija, Spanish orujo, and Serbian komovica.

2.6.1.1.1 Cognac

Cognac is a brandy whose production is related to a fixed geographical area on the west coast of France, slightly north of the city of Bordeaux. The base wines are produced from a selected prescribed range of grape varieties (Ferrari et al., 2005). The base wine is a relatively weak wine with a typical alcohol content of only 7%–8% abv. The distillation of cognac is carried out in a two-stage pot distillation, facilitated in traditional alembic or Charentais stills (Figure 2.14) that have to be heated by an open flame. The first still is limited to a capacity of 10,000 L, while the second still (the *bonne chauffe*) is limited to a capacity of 3,000 L. The Charente distillation process is tightly regulated. The first distillation yields three fractions that are due to two cuts. The heads and the tails are returned to the wine-holding vessel, while the hearts from the first distillation run (the brouillis) become the charge for the second distillation run (Figure 2.15).

The second distillation run typically yields four fractions, governed by three cuts. The first and the last cuts again yield heads and tails, which again are returned to the wine-holding vessel. However, following the first cut, the principal hearts are collected. It is then up to the master distiller to determine at which point these hearts are collected and the seconds are collected. The seconds are returned to either the wine-holding vessel or the brouillis-holding vessel, or a split between the two. A greater

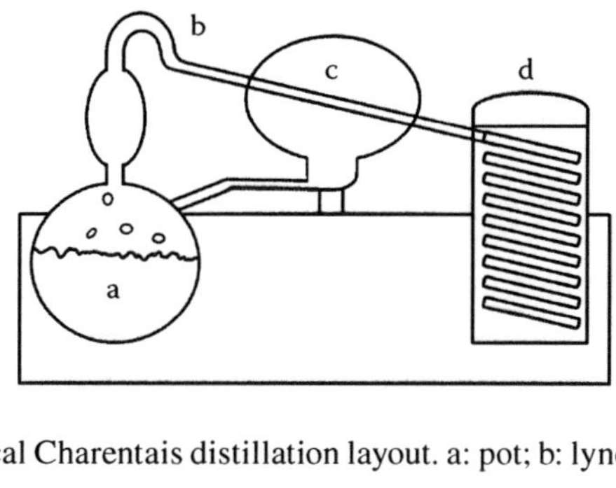

FIGURE 2.14 A typical Charentais distillation layout. a: pot; b: lyne arm; c: wine preheater; d: worntub condeser.

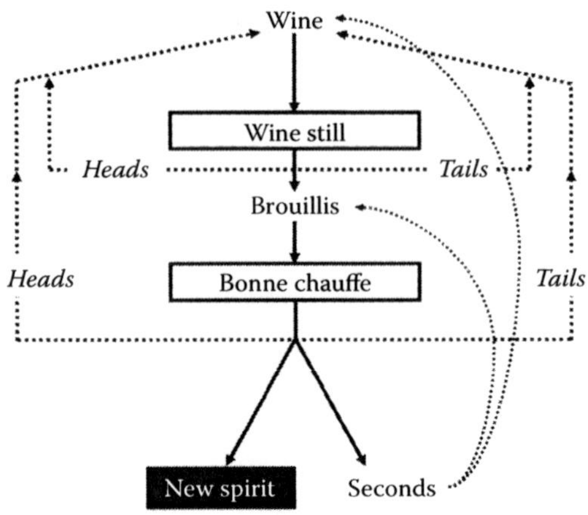

FIGURE 2.15 Distillation flow chart for a typical double Charentais cognac distillation.

inclusion of the spirits into the principal hearts will result in a greater proportion of congeners. Cycling the seconds back into either the wine or the brouillis will affect the proportions of congeners entering the two distillation stages and therefore the makeup of the principal hearts. The split between the principal hearts and the seconds, and the management of the return of the second back to either the wine-holding vessel of the brouillis, confer an authentic quality characteristic that underpins the differentiation between different cognac houses (Cantagrel and Galy, 2003). The principal hearts of the double cognac distillation cannot yield a spirit with an alcohol strength of greater than 72% abv. There is no prescribed alcohol strength at which new cognac spirit is matured; most cognac is put in barrels for maturation at ~70% abv. The current regulations that govern the production of cognac oblige the pot still to be switched off before March 31 the year following the harvest of the grapes (Ferrari et at., 2005)

2.6.1.1.2 Armagnac

Armagnac is another French brandy whose production is related to a fixed geographical area in the Gascony Province near the west coast of France, south of the city of Bordeaux. The base wines are produced from a selected prescribed range of grape varieties (Bertrand, 2003). Unlike cognac, modern Armagnac can be distilled in either a pot still or in copper continuous stills; however, by far the greatest proportion of Armagnac is produced in continuous Armagnac stills. The typical continuous still used in the production of Armagnac has five to six plates above the boiler and runs continuously (Figure 2.16). Wine is fed into the wine-cooled, coiled condenser from a feed tank positioned above the condenser. The cold wine is introduced at the bottom part of the wine-cooled condenser and flows upwards into the top section. The separate sections allow for optimized cooling across both sets of coils. The warm wine is fed into the column at the top and falls over the plates in the column where the alcohol boils off. The heated wine collects in the bottom, where it boils, generating vapors that eventually collect in the condenser. The alcohol-stripped wine at the bottom of the pot boiler gets emptied every now and again, and the base wine in the

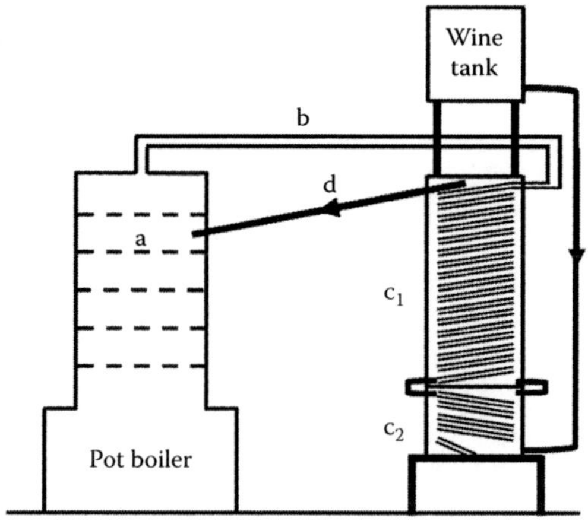

FIGURE 2.16 Continuous Armagnac still. (a) Fractionating column; (b) lyne arm; (c_1) top wine-cooled coiled condenser; (c_2) bottom wine-cooled coiled condenser; and (d) warm wine entry pipe into fractionating column.

wine tank is replenished in order to keep the still operating continuously. The new Armagnac spirit is produced at a strength that can range from 52% abv to about 72% abv, and it is matured in 400 L new oak barrels at that strength. Most new Armagnac is kept in the new oak barrels for up to 1 year, after which the juvenile spirit is transferred to old, reused casks. In line with the regulations that govern the production of cognac, the distillation of Armagnac ceases on March 31 the year following the harvest of the grapes. Furthermore, the age calculation of Armagnac is based on the end of August in the year of its vintage; therefore, Armagnac is deemed to be 1 year old on September 1 in the following year.

2.6.1.1.3 Grape Pomace Spirit

Almost all traditional wine-growing regions in the world also yield a secondary alcoholic product associated with the "waste streams" of the principal wine they produce. Pomace spirit is made from *eau de vie de marc*, or simply, marc. The production of red wines in particular yields large amounts of grape marc that accompanies the primary fermentation. The skins of the grapes are responsible for the color of the resulting wine, especially in red wines, where the dark pigments are abundantly present in the skin. In order to extract sufficient quantities of these pigments into the wine, the skins are kept in contact with the fermenting must. The increasing alcohol concentration aids in the dissolution of the pigments into the wine. At the completion of the primary fermentation, the skins and the pips (the pomace or marc) are removed from the wine. The grape marc will contain the same ethanol concentration as the wine itself and is therefore a valuable source of alcohol. Examples of grape marc–based distillates are Italian grappa, Greek rakija, Spanish orujo, and Serbian komovica.

Most of the grape pomace–based spirits have traditionally been a home-produced, artisan product; however, in many locations the spirits are now produced at an industrial scale. For instance, orujo comes from the Galicia region in the northwest of Spain. Traditionally, orujo has been produced and consumed by the "common" people. Distillation of the marc was typically carried out by roving specialists who traveled the countryside with their distillation equipment (Cortés et al., 2005). However, this practice was prohibited in 1987. Since then, fixed installations have been required, causing a gradual loss of the tradition associated with the production of orujo. Nevertheless, this did not lead to an abandonment of orujo production. Most winemakers in the Galicia region produce their own brand of orujo, in their own distilleries. Almost all fixed distillation equipment for the production of orujo consists of a stainless steel pot in which the extract from the pressed marc is heated by low-pressure steam, which provides the energy to facilitate the evaporation of the volatiles. Above the pot there will be a short copper fractional distillation column from which three fractions are received (Cortés et al., 2005). The heads and tails are discarded, leaving the hearts with an alcohol content of approximately 60% abv.

Grappa is probably the best known of the grape pomace spirits. It originates from Italy, including the Italian part of Switzerland, and San Marino. In order to be called "grappa," the fermentation and distillation must occur on the pomace without the addition of water to rinse out more solubles. Furthermore, EU regulations with regard to grappa prohibit the use of pure grape juice for the production of grappa. In contrast to orujo, where the pressed marc is subjected to the distillation treatment, in the production of grappa, the distillation must occur in the presence of the marc solids. Because of this, the distillation is carried out with a bain-marie or steam distillation instead of employing a direct fired heat source. On a smaller artisan scale, grappa distillation will be carried out using a pot still setup with either a substantial neck or a short fractionating column; on an industrial scale, the distillation tends to be carried out using continuous distillation. In most batch distillation, the fermented marc is placed into special copper baskets that are placed one on the other to avoid excessive compression of the marc during the distillation (da Porto and Longo, 2005). During the continuous distillation, steam is passed through a column of marc from which the vapors are collected at a relatively low strength (15%–20% abv). These condensed vapors are then run through a rectification column.

Grape marc will contain high levels of pectin, which correlates to elevated levels of methanol. Care needs to be taken to remove most (if not all) methanol from the final spirit. While grappa can be distilled to an alcohol level of 86% abv, the more commonly attained alcohol content in the final spirit is 72%–76% abv. A higher alcohol content might well be beneficial for the production of rectified base spirit for the production of vodka or gin; a significant part of the unique character of grappa is the very high levels of congeners. Hence, a less refined product will produce a better quality grappa.

2.6.1.2 Apple Brandies

Apple brandies are produced from fermented cider, including calvados, from the Normandy region of France, cider brandy from the Somerset region in the United Kingdom, and perhaps even the American applejack.

2.6.1.2.1 Calvados

Calvados is distilled cider made from selected apples that are characterized as sweet, tart, or bitter from over 200 named apple varieties. Pears can also be used in the production of calvados. In fact, the AOC calvados Domfrontais requires a minimum of 30% of the fruit to be pears from the designated geographical area. The selection and proportion of the various apple and pear varieties will influence the final product. When the fermentation is finished, the resulting cider is distilled by either a batch or a continuous distillation method. However, the distillation mode is influenced by the appellations of calvados. The appellation "AOC calvados Pays d'Auge" demands that the distillation be carried out in batch mode (Mattsson, 2005). The batch method utilizes the exact same Charentais still design as is used in the production of cognac. However, it appears that the distillation and its cuts in the calvados Charentais still is more like that used in the Scotch double distillation, instead of the more complex cognac cuts across the two distillation runs. The remaining appellations tend to use continuous column stills, similarly to the system used in the distillation of Armagnac in that only a single fractionation column is employed.

2.6.1.2.2 Applejack

Traditionally applejack was not a distilled spirit. The original applejack was a fermented beverage that used apples as the fermentation substrate; it was essentially a cider. The term "jacking" is derived from the process of freezing the cider and continually removing the ice from the liquid. This process could be referred to as freeze distilling, even though the traditional definition of distilling as a combination of evaporation and condensation to remove the volatile fractions does not apply. A more appropriate term would be "freeze concentrating." During freeze concentrating of alcoholic beverages, the water freezes as pure water crystals, leaving behind a concentrated alcoholic beverage. If the temperatures are low enough and the process is continued, the remaining concentrated applejack could contain 25%–35% abv. The traditional process of applejack production by means of freeze concentrating is very rarely practiced these days. Apart from the process being very time consuming and energy demanding, the fact that methanol and fusel alcohols also concentrate in this process causes a significant health risk associated with methanol consumption. Modern varieties of applejack are still being produced. However, these tend to be produced in more conventional pot and fractionating stills.

2.6.2 Grain-Based Spirit

Grains are commonly used for the production of potable alcohol. Most grains contain a high proportion of carbohydrates in the form of starch, which is a polymer of glucose. Typical grain sources used in the production of grain-based spirits are wheat, barley, rye, corn, and sorghum. The glucose contained within the starch is not readily available as a fermentable substrate for the yeast, which typically carries out the alcoholic fermentation. Most grains lend themselves well to the process of malting, in which the grain is hoodwinked into producing the innate amylolytic enzymes that are produced when grains are germinating. By carefully controlling the germinating conditions, a maltster can produce a malt that contains most of its original starch but includes the enzymes for its controlled hydrolysis.

2.6.2.1 Whisk(e)y

Whisky, or whiskey, is a type of distilled alcoholic beverage produced from a fermented grain mash. A variety of grains (malted or unmalted) are used for different varieties of whisk(e)y, including barley, corn, rye, and wheat. The distilled spirit arising from the grain-based fermentation is typically aged in wooden casks, generally made of charred white oak.

2.6.2.1.1 Scotch Whisky

There are numerous types of scotch whiskies that, from a distillation point of view, can be characterized into three main groups. Single malt whiskies are typically distilled twice in a pot still in a batch operation. The first distillation is referred to as the wash still and the second distillation as the spirit still. Grain whiskies, on the other hand, are typically produced from a large proportion of unmalted grains (wheat or corn), and only a small proportion of malted barley for the supply of hydrolytic enzymes. The fermented grain mash is distilled using a Coffey still (see Section 2.4.1). A very large proportion of the scotch whiskies on the market are blended scotch whiskies, which are blends of carefully selected single malts and grain whiskies. The regulations for Scotch whisky stipulate that the definition of "Scotch whisky" mean a whisky produced in Scotland that has been distilled at a distillery in Scotland from water and malted barley (to which only whole grains of other cereals may be added) all of which have been: (i) processed at that distillery into a mash, (ii) converted at that distillery into a fermentable substrate only by endogenous enzyme systems, and (iii) fermented at that distillery only by the addition of yeast. Scotch whisky has been distilled at an alcoholic strength by volume of less than 94.8% so that the distillate has an aroma and taste derived from the raw materials used in, and the method of, its production. Scotch whisky has been matured only in oak casks of a capacity not exceeding 700 L, has been matured only in Scotland, has been matured for a period of not less than 3 years, has been matured only in an excise warehouse or a permitted place, and retains the color, aroma, and taste derived from the raw materials used in, and the method of, its production and maturation. Furthermore, no substance can have been added except water and/or plain caramel coloring. Scotch whisky must have a minimum alcoholic strength of 40% abv.

2.6.2.1.2 Irish Whiskey

Irish whiskies include all whiskies produced in the Republic of Ireland and in Northern Ireland. Many Irish whiskies are made from a blend of unmalted barley and malted barley and are either distilled in large pot stills in a triple distillation manner (see Figure 2.8) or by distillation in a Coffey still. The Irish regulations state that the spirits shall have been distilled in the state or in Northern Ireland from a mash of cereals that has been saccharified by the diastase of malt contained therein, with or without other natural diastases; fermented by the action of yeast; and distilled at an alcoholic strength of less than 94.8% by volume in such a way that the distillate has an aroma and flavor derived from the materials used. Furthermore, the regulations state that the spirits shall have been matured in wooden casks, in a warehouse in the state for a period of not less than 3 years, or in a warehouse in Northern Ireland for such a period, or in any warehouse in the state and in Northern Ireland for periods the aggregate of which is not less than 3 years.

2.6.2.1.3 Bourbon

Bourbon is a whiskey produced in the United States made from a grain mixture that is at least 51% corn with wheat and/or rye added to it, and malted barley as the source of hydrolytic enzymes. However, commercial enzyme preparation can be used too. Those bourbons containing a significant proportion of wheat will produce a soft, sweet flavor of berries, sometimes somewhat nutty. Bourbons containing a significant proportion of rye will possess a bold, spicy flavor, which becomes increasingly bolder with an increase in rye content. The fermented wash is typically distilled in a continuous fractionating still followed by a doubler. The wash is distilled to no more than 80% abv after which it is diluted to no more than 62.5% abv to be matured in new, virgin charred oak (*Quercus alba*) barrels. Bourbon is bottled at an alcohol strength no less than 40% abv. It can be sold at any age, but when aged for less than 4 years it must contain a statement to that effect, and in order to be sold as "straight bourbon" it must be aged for at least 2 years.

2.6.2.2 Neutral Grain Spirit

Neutral grain spirit (NGS) is typically used as the base spirit for the production of vodka and botanically flavored spirits such as gin and jenevers. The base grains are typically wheat or corn, depending on availability in the location of the producer. The grains are usually cooked in a cereal cooker in order to gelatinize the starch, which is then degraded by an exogenously added enzyme preparation for the enzymatic production of glucose. The fermented wash is typically distilled in a continuous, multi-column system comprising a stripping column, followed by a rectifying column (see Section 2.4), a hydroselection column to remove most of the light boiling impurities (see Section 2.4.2), a reactive, and packed column containing sacrificial copper (see Section 2.4.3), followed by a second rectifying column to reconcentrate the ethanol. This will yield a raw spirit (NGS) of high ethanol concentration at a relatively high purity. In regions where other fermentation substrates are more readily available, those fermentation substrates could be used to produce their equivalent of NGS, such as neutral grape spirit, neutral cane spirit, and neutral beet spirit. Regardless of the fermentation substrate, the distilling method will be similar, with the aim to produce a relatively clear base spirit that can be used in multiple applications

2.6.2.3 Miscellaneous

Sorghum is an alternative grain that is in use in the production of potable spirits. While some artisan distillers have started using sorghum as a direct substitution to traditional grains in the production of whiskies, a more traditional use of sorghum as a principal fermentation substrate for the production of a distilled spirit applies to baijiu

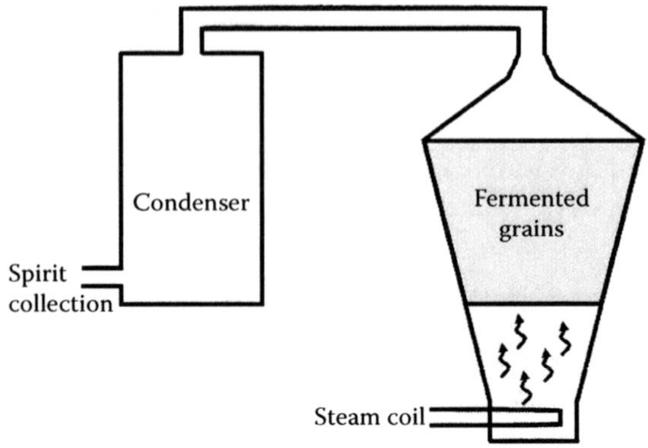

FIGURE 2.17 Typical baijiu distillation equipment. (After Lin, D. In *Distilled Spirits: Science and Sustainability* (3rd World Wide Distilled Spirit Conference, Edinburgh), Eds. Walker, G.M., Goodall, I., Fotheringham, R., and Murray D., pp. 103–108, Nottingham University Press, 2011.)

(Lin, 2011; Zheng and Han, 2016). However, baijiu can also be produced with the inclusion of other grains including rice. The hydrolysis of the starch contained within the grain is facilitated by a mold (mainly *Rhizopus* spp.) in a solid state, simultaneous saccharification cofermentation with a range of yeasts (including *Saccharomyces, Pichia, Kluyveromyces*, and others) in earthen pots (Lin, 2011; Sakaguchi et al., 1992). The distillation of modern baijiu can be carried out in a range of traditional stills. However, the traditional distillation approach for baijiu is carried out by using large steamers in which the fermented grains with liquid and all are deposited. Steam is purged through the grain–liquid mixture and the volatiles collected from the condenser (Figure 2.17). The efficiency of the baijiu distillation depends on the steam flow rate, retained water content, distillation speed, and packing porosity of the fermented grains. A low steam flow rate might not be sufficient to generate the thermal conditions for the thorough evaporation of ethanol, while a higher steam flow rate may cause the fermented grains to stick together and increase the diffusion resistance.

2.6.3 Tuber-Based Spirits

Tubers like potatoes are a very suitable carbohydrate sources for the production of potable alcohol. Potatoes are high in starch, but also very high in protein. The high protein content can give rise to elevated levels of congeners in the final wash, but can cause issues during the distillation process such as excessive foaming and rapid fouling of the equipment. Other tubers that have the potential to be used in the production of potable alcohol are cassava and Jerusalem artichokes. The former, however, is typically used in the production of a beer-like beverage rather than a distilled spirit.

2.6.3.1 Potato

While potatoes are abundantly used in the production of potable alcohol, other than in potato vodka, there are no common spirits where potatoes are the unique fermentation substrate. The production of vodka as an extremely neutral spirit is described in Section 2.6.6.

2.6.3.2 Topinambur

Topinambur is a distilled beverage produced from Jerusalem artichokes, commonly encountered in the German Baden-Württemberg region. The principal carbohydrate associated with Jerusalem artichokes is inulin, a polymer of fructose. The tubers are usually covered in soil and will need to be cleaned. This will limit the contamination of the fermentation by soil bacteria and carryover of unwanted earthy flavors. Following a period of cold weather, the endogenous inulase enzymes will initiate the hydrolysis of the tuber's inulin. However, commercial enzyme preparations are available that also include a pectinase that aids in the degradation of cellular material, simultaneously releasing the knoll's own inulin. The fermented artichoke wash is distilled immediately following the completion of the fermentation. The production of topinambur is almost exclusively small scale and distillation occurs in batch mode only. A standard double pot still setup is able to produce a quality topinambur, as long as sufficient congeners are carried over into the spirit run. Alternative batch distillation systems are also used. In most of these instances, a gentle but extensive reflux is promoted to maintain the characteristic artichoke aroma. The distillation should yield a spirit of less than 86% abv. The spirit is not allowed to be flavored and is rarely matured. It retails at no less than 38% abv.

2.6.4 Sugarcane- or Molasses-Based Spirits

Sugarcane belongs to the grass family, which includes many grains including sorghum. Both sugarcane and sorghum accumulate large quantities of sucrose in the stalk internodes, which then attract moisture (sucrose is relatively hydroscopic), yielding the sweet syrup. The best known cane crop is sugarcane; however, sweet sorghum is also harvested for its sugary liquid, which is normally boiled down to produce sorghum molasses. Occasionally, the sorghum liquid is used in the production of ethanol to produce sorghum whiskey or sorghum rum.

2.6.4.1 Rum

Rum is a distilled beverage made from the sugarcane by-product molasses. Molasses is the residual syrup that is left over after crystallized sugar is produced from the cane sugar juice. The molasses is diluted, fermented, distilled, and then aged in oak barrels. Sugarcane is cut and crushed to extract the liquid held in the inner tubes of the cane and then boiled to produce sugar, bagasse, molasses, and mill mud. When molasses is blended with water, yeast is added to start the fermentation, yielding a wash at ~5%–10% abv. Rum can be distilled by employing a batch process in a pot still, or it can be produced by means of a continuous distillation process. On many occasions, rums are blended from spirit obtained from both pot stills and continuous still distillates. The variation in pot stills used for the production is great. While some pot stills are not dissimilar to the pot stills encountered during the production of Scotch malt whisky, many rum pot stills have additional reflux built into them, such as single thumpers or doubles. Multiple doublers in different configuration are often employed (Figure 2.18).

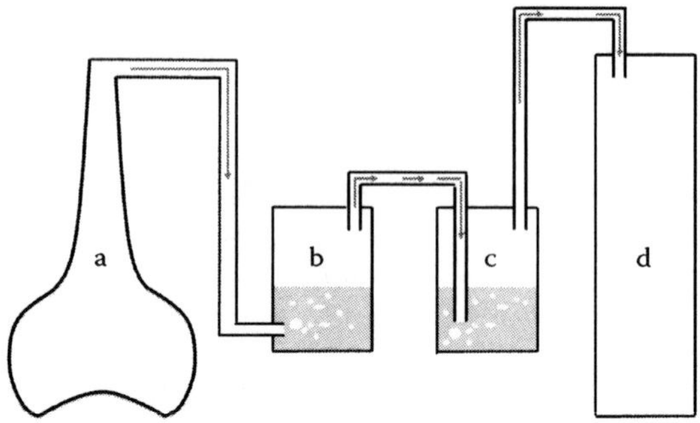

FIGURE 2.18 Pot still configuration including two doublers as used in the production of rum; (a) pot; (b) doubler with vapor inlet from the side; (c) doubler with vapor inlet from the top; (d) condenser.

The doubler as used in rum distilling is also referred to as a retort (Nicol, 2003), which is often charged with the low wines collected from the previous distillation run. This provides a distinctively full-bodied rum. Another method for producing a full-bodied, high-ester rum is to utilize fermented dunder as the feedstock. Dunder is pot ale that has undergone a spontaneous refermentation by wild yeast and various bacteria. The high levels of amino acids yield high levels of higher alcohol and, subsequently, esters. Most residual sugars will also be fermented and yield some ethanol. The resultant distillate yields extremely high levels of esters and is not commonly used as rum, but instead for blending purposes. Large rum houses will produce the bulk of their rums by means of continuous distillation. The Coffey still is the most common configuration, which can occasionally be supplemented by a hydroselection column (see Section 2.4.2) in order to make lighter rums.

2.6.4.2 Rum Agricole and Cachaça

Rum agricole and cachaça are also spirits that are derived from sugarcane. The principal difference is the fact that the fermentation substrate is sugarcane juice rather than molasses. Rum agricole is a variant of rum that came about when the French sugar market dried up and the French elected to utilize sugar from beet instead. The various countries that were relying on sugarcane as a significant part of their income started producing rum from sugarcane juice instead of molasses. In order to differentiate between the rum produced from molasses and the rum from sugarcane juice, the term "rum agricole" was adapted. Cachaça is very much associated with Brazil, where it is the national beverage. Similarly to cognac and Armagnac, both rum agricole and cachaça are very much seasonally produced products, whose production relies on the availability of locally available raw materials. For instance, in Brazil production of cachaça occurs during the sugarcane harvest period, between May and November or October and March (depending of the region). The distillation processes that are applied to produce agricole or cachaça are nearly identical to the distillation process that is applied to rum.

2.6.5 Agave-Based Spirits

Agave-based spirits include tequila, mescal, Bacanora, sotol, and Cocuy, which all use different agave varieties as their raw material. The best known agave-based spirit is tequila, which is produced from *Agave tequilana* Weber (aka blue agave). The products marketed as "100% agave tequila" must be produced from only *Agave tequilana* Weber. Products marketed as "tequila" are required to be produced from at least 51% sugars derived from the *Agave tequilana* Weber, whereas the remainder of the sugars can be obtained from any other source. The agave plant can take 8 to 12 years to mature before it is harvested. At harvest only its core is utilized; it is cooked in an oven that converts the carbohydrate polymer inulin into fermentable sugars (fructose). Once cooked, the piñas are pressed to remove the sugary liquid for fermentation. The fructose-rich agave juice is fermented either by a spontaneous fermentation or by using a specific *Saccharomyces cerevisiae* strain. This is fermented over 3–5 days and, once fermented, distilled immediately. Because the distillate will taste bready if the agave wash sits for longer than 3 days, distillation occurs immediately following the completion of the fermentation. The distillation is done in pot stills in combination with a rectifying column or a doubler to enhance the reflux and fine-tune the final spirit. However, larger distilleries do utilize continuous stills.

Once distilled, tequila is sold as a blanco or aged oak in barrels, mostly in former bourbon barrels. There are, in principle, two maturation processes: resposado or añejo. Resposado tequila is matured for at least 2 months, either in oak barrels or in large vats (often 10,000+ L) to impart subtle wood hints without compromising the delicate agave-driven tequila flavors. Añejo tequila must be matured in oak barrels no greater than 600 L and for no less than 12 months. Occasionally, tequila is matured as "extra añejo" in oak barrels no greater than 600 L and for at least three years (López-Ramírez et al., 2013). Prolonged maturation emphasizes the wood character, but often to the detriment of the delicate agave-driven tequila flavors.

2.6.6 Neutral Spirits

Neutral spirits are utilized as a base product in many commercial bottling facilities to produce an assortment of beverages. The spirit is produced above 96% abv and is blended down with clean filtered reverse osmosis water. Once watered down, flavors and colorings can be added to create the desired product. For a spirit to become neutral, the desired alcohol is to be at 96%–96.6% and have less than 5ppm of methanol as well as very low levels of other congers. To achieve this, the spirit is distilled by continuous distillation through three to five columns, in a manner identical to that described for neutral grain spirits in Section 2.6.2.2.

2.6.6.1 Vodka

Vodka is derived from a neutral spirit, watered down and bottled. The neutral spirit for the production of vodka can be produced from a wide range of fermentation substrates, including potatoes, wheat, rye, milk, grapes, and apples. Because of the extensive rectification process, supplemented by additional specialty distillation columns, the raw materials used in the production have only limited influence on the final product. In some instances, the final spirit used in production of vodka will have been processed to eight or more columns. The application of multiple columns in a continuous distillation system has been yielding very high-quality vodkas and has, in many instances, negated the need to use carbon filtration in order to remove specific impurities. The use of carbon filtration has gradually been made redundant, and the quality of alcohol produced has increased significantly to the point that placing the neutral alcohol through a carbon filter will actually dirty the alcohol and add impurities.

2.6.7 Distilled Botanicals

Many spirits undergo a redistillation in order to extract specific flavor and aroma compounds from botanicals after the principal distillation that yields the new spirit has been completed. In this instance the distillation process can take two routes. The first one is very similar to the common spirit still operation in a pot still, while the second route is more akin to the steam distillation process as seen in the extraction of essential oils from plant materials. Additionally, there is a third route for the extraction of flavor and aroma compounds from botanicals, but that method does not involve a distillation step and relies entirely on extraction by maceration.

A common method for the extraction of the volatiles from botanicals is using a combination of maceration and distillation, also referred to as a "Bennett still" operation (Figure 2.19). In this method, the botanicals are typically mixed and soaked overnight in a base sprit at an appropriate ethanol strength, following which distillation takes place in the same manner as in a spirit still. The extraction of the botanicals in this method is influenced by the direct contact with water and ethanol and the direct application of heat through contact with liquid. The process of reflux in the pot will apply in the exact same manner as it does in a normal pot still operation as described before (Sections 2.2 and 2.3.1).

An alternative method of extraction of the botanical is by suspending the botanicals in the ethanol vapors above the liquid level in the pot, also referred to as the "Barnett still" operation (Figure 2.20). In this particular process, the volatiles from

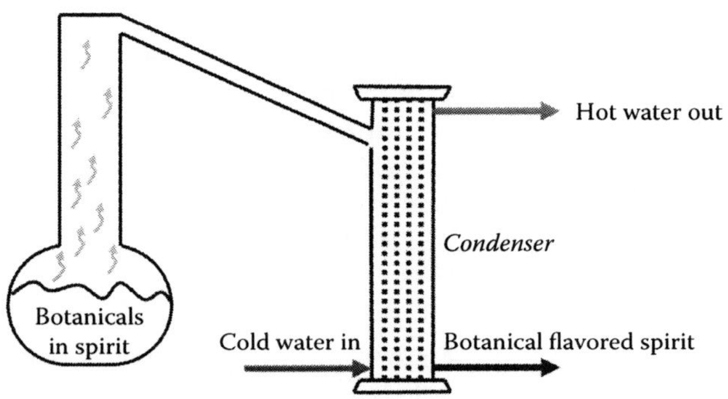

FIGURE 2.19 Batch operation of a Bennett still, in which botanicals are often macerated in the raw spirit before distillation commenced. In a Bennett still configuration, the botanicals are submerged in the raw spirit in the pot.

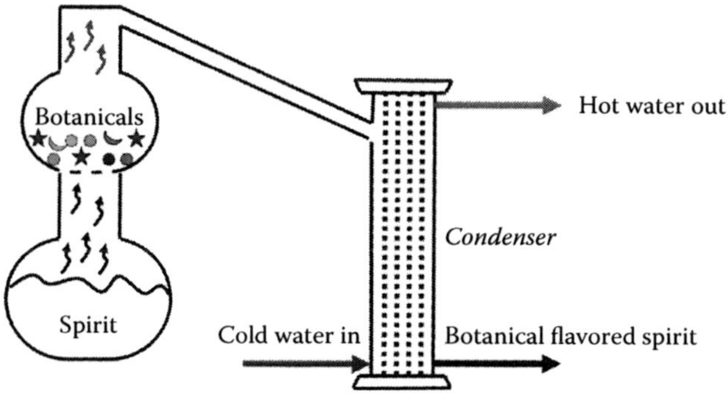

FIGURE 2.20 Batch operation of a Carterhead still, in which botanicals are placed in a mesh compartment that is located within the warm vapor stream emerging from the raw spirit being distilled in the pot.

the botanicals are extracted almost directly into the ethanol vapor phase with only limited opportunity for reflux. At the start-up of the still there will be substantial reflux and, while that is common to all stills, until the botanicals have the same temperature as the ethanol vapor, the botanicals themselves will provide a surface for reflux to occur on. However, as soon as the botanically induced reflux is minimized, the volatiles within the botanicals are simultaneously extracted and distilled and carried through to the condenser. The maceration process that precedes the distillation step in a Bennett still operation will also aid in dissolving much of the volatiles and the resin material into the water:ethanol mixture.

This means that many more compounds will be carried over with the ethanol vapors during a distillation run. This might well favor a greater extractability, but it might also coextract some unwanted compounds. For instance, the macerate from juniper berries will have a bittersweet taste that typifies juniper, but it also has a harsh pine-like resin character (Coumou, 2007). When distilled in a Bennett still, the typical juniper bouquet will come through in the earlier fractions, while the resin-like aromas are carried over in the later fractions. These resin-like aromas are not always desirable. The macerate of cardamom is yellowish and the typical cardamom aromas are immediately evident in the earlier fractions, increasing in potency through to the mid fractions. The later fractions during the Bennett still run will reveal a more bitter, camphor-like taste. The macerate of coriander has a yellow-brown hue with the early- to mid distillation fractions containing all the typical coriander flavors; the mid- to late fractions reveal increasingly stronger aniseed aromas. In contrast to the changing flavor intensity of juniper, cardamom, and coriander, the distillation of a cocoa macerate will provide an intense cocoa aroma throughout all distillation fractions, with the later fractions somewhat sweeter. However, the bitterness or color of cocoa does not carry over into the condensate (Coumou, 2007).

The overall character of the volatile character of the resulting spirit will be very dependent on the management of the cuts of the collected distillate. Avoiding the extreme and more delicate flavors that dissipate rapidly from the final spirit requires tight management of the foreshots, while the harsher and less volatile compounds can be avoided by careful management of the heads, selecting only the hearts of the distillation run to be included in the final product. In some instances, distillers may choose to distil some botanicals using the Bennett still while others use the Carterhead still. The resultant distillates can then be carefully blended to fine-tune the delicate flavors and aromas from each to create a characteristic brand. The most common spirits that rely on botanicals for the flavoring of the spirit are jenever, gin, and akvavit. However, many more spirits rely on the distillation of botanicals for the flavoring of spirits.

2.6.7.1 Spirits Flavored with Juniper

Both jenever and gin rely on similar botanicals to flavor the base spirit. In modern terms, a gin is based on a fully neutral spirit that has no real character of its own. This allows the gin botanicals to dominate the final spirit. Jenever, on the other hand, has a proportion of grain- and/or malt-based spirit as its base, which provides for a base character through which the botanical volatiles are intertwined. The principal botanicals used in both gin and jenever are juniper berries, which yield substantial quantities of pinene and γ-muurolene; coriander seeds are also often used botanicals in both spirits, yielding linalool and γ-terpinene. A wide variety of other botanicals are used, including angelica root, orris root, licorice root, cinnamon, dill, cardamom, cloves, lavender, almond, hops, orange peel, lemon peel, pink grapefruit peel, and a variety of teas.

2.6.7.2 Spirits Flavored with Caraway

Akvavit is a botanically driven spirit that is typical of the Scandinavian countries. Akvavit is typically produced from potato or grain as the source of fermentable carbohydrates; the alcohol from this is distilled by a multistage continuous distillation, followed by a charcoal filtration. This yields a neutral spirit that has no flavor or aroma of its own, but provides an excellent base for the botanicals that characterize akvavit. The quintessential botanical for akvavit is caraway. Caraway seeds yield substantial quantities of S-carvone and limonene. The latter is the metabolic precursor to S-carvone and provides a citrus aroma; the S-carvone provides the typical caraway aroma. The limonene will distil in the earlier fractions, while the S-carvone is much less volatile and will only distil in the mid fractions. The later distillation fractions yield a crude, oily taste.

2.6.7.3 Spirits Flavored with Aniseed

Numerous spirits obtain a significant part of their aroma and flavor from a redistillation of a new spirit with aniseed as the dominant botanical. Examples of these redistilled aniseed spirits are ouzo, raki, pastis, and absinthe—in contrast to other aniseed flavored spirits where the aniseed aroma is obtained by maceration only, such as sambuca and Jägermeister. The most characteristic compound from aniseed is anethole, which is a compound also readily noticeable in fennel, licorice, camphor, and star anise. During the distillation of the macerated aniseed in a fractionation column, the anethole-related flavors will elute in the mid fraction of the distillate; while in a pot still the anethole-related flavor will be noticeable somewhat earlier (Coumou, 2007).

For tightly refined extracts, multiple distillations are recommended. The macerate itself (without distillation) is also useful in the production of strong aromatic bitters, but it will retain its acidic and astringent mouthfeel. The anethole in the aniseed-derived spirits is responsible for the louche effect that is noticeable when water is added to any of these spirits. The louche effect is caused by the decrease of the ethanol proportion below 30% abv, which leads to a clouding of the beverage and a colloidal separation of the solution, but also liberates the herbal bouquet within the drinks.

2.6.7.3.1 Ouzo

The base spirit for ouzo can be derived from any agriculturally derived alcohol source. The botanicals for ouzo vary from brand to brand, but always include aniseed, often complemented by star anise, fennel, mastic, cardamom, coriander, cloves, or cinnamon. The botanicals are always macerated for 24 hours in an appropriate mixture of rectified ethanol and water. Ouzo comes in two varieties: a 100% distillate, and a 20% distillate (aka cold/crude ouzo). The distillation process for the production of 100% distillate takes place in a double pot still system (Varvagiannis et al., 2005). In the first pot still, a slow rate of distillation is encouraged, with the typical distillation run taking 10–12 hours. The heads and tails are rejected and only the heart of the distillation run collected. The concentrated hearts from the first distillation run are distilled again in the same manner; rejection of the heads and tails, however, takes place at a higher commencing ethanol concentration. The resulting final spirit will have an ethanol content of approximately 75%–80% abv. The rejected heads and tails from the first and second pot still runs are often distilled again to produce a second quality ouzo. The 100% ouzo is diluted to ~42%–48% abv for retail. The distillation process for the production of crude/cold ouzo often involves only a single distillation from which the heads and tails are rejected. The hearts are diluted with anethole-flavored potable alcohol so that the final product contains no less than 20% of the original ouzo (100% distillate) spirit (as stipulated by Greek legislation). The crude ouzo is then diluted to no less than 35% abv for retail (Varvagiannis et al., 2005).

2.6.7.3.2 Raki

Raki is produced from the rectified distillate (94%–96% abv) from marc pomace called suma (Cabaroglu and Yilmaztekin, 2011). The suma is typically produced by batch fractional distillation, which is then reduced in strength to about 30%–40% abv in which aniseed is macerated for 24+ hours. The subsequent distillation is carried out in a pot still at very low energy input. From the batch distillation, the heads are collected at a temperature below 80°C over a slow 6-hour period. The hearts are collected over a 30+ hour period during which the temperature is racked up to 91°C and yields the main raki spirit at 80% abv. The tails are collected while the temperature is further raised to 101°C over another 6- to 7-hour period. Any volatiles that emerge from the still when the temperature in the pot rises above 101°C are discarded and the distillation is discontinued; the collected heads and tails are recycled into the next batch of raki. Raki is rarely matured, but it is stored in oak casks for 1–3 months to allow the anise volatiles to marry with the ethanol. Finished raki must contain at least 800 ppm of the principal aniseed compound, anethole.

2.6.7.3.3 Absinthe

Absinthe is a highly alcoholic spirit that contains the distilled macerated extracts from aniseed, grand wormwood (*Artemisia absinthium*), and sweet fennel. When distilled, the macerate of wormwood concentrates the vermouth-like aroma (thujone) in the earlier distilled fractions, which coincides (as mentioned earlier) with the elution of the anethole from the aniseed. While the macerate has a slight green tinge, the distillate is completely colorless. The original distillate containing the botanical volatiles from aniseed, grand wormwood, and sweet fennel are diluted to 85% abv and additional botanicals are added only to be macerated (Lachenmeier et al., 2006). These botanicals are Roman wormwood (*Artemisia pontica*), hyssop, and lemon balm. The Roman wormwood is less bitter than the grand wormwood, but still contains thujone and also camphor. The hyssop macerate provides a hint of ginger and also camphor; however, all of the botanicals used in the second maceration provide a slight green tinge to the absinthe. An alternative method for the production of absinthe is by 100% maceration of the principal botanical with the addition of star anise. Thujone is, in principle, a toxic, mind-altering substance that was the cause of absinthe's initial success as a distilled beverage, but also became its downfall when absinthe became banned around the turn of the nineteenth century in many countries. That ban has now been lifted in almost all countries.

2.6.7.3.4 Pastis

Pastis is often described as the replacement for absinthe, in that it maintains a strong aromatic flavor but does not contain the mind-altering substance thujone. Pastis does not contain grand wormwood and obtains its anise flavor from a distillation of star anise. Furthermore, pastis has a distinct flavor of licorice root, which is not a part of a traditional absinthe. Finally, unlike a traditional absinthe, pastis is a "liqueur," which means it is always bottled with sugar.

REFERENCES

Aylott, R.I., 2003, Vodka, gin and other flavoured spirits, In *Fermented Beverage Production*. Eds. Lea A.G.H., and Piggott, J.R., pp. 289–308, Kluwer Academic, New York.

Bertrand, A., 2003, Armagnac and wine-spirits, In *Fermented Beverage Production*. 2nd. ed. Eds. Lea, A.G.H., and Piggott, J.R., pp. 213–238, Kluwer Academic, New York.

Cabaroglu, T., and Yilmaztekin, M., 2011, Methanol and major volatile compounds of Turkish raki and effect of distillate source. *Journal of the Institute of Brewing*, 117(1), pp. 98–105.

Cantagrel, R., and Galy, B., 2003, From vine to cognac, In *Fermented Beverage Production*. 2nd. ed. Eds. Lea, A.G.H., and Piggott, J.R., pp. 195–212, Kluwer Academic, New York.

Cortés, S., Gil, M.L., and Fernández, E., 2005, Volatile composition of traditional and industrial Orujo spirits. *Food Control*, 16(4), pp. 383–388.

Coumou, A., 2007, Kleinschalig destilleren and likeurmaken. Brouwland, Beverlo, Belgium.

Da Porto, C., and Longo M., 2005, Grappa: The Italian distillate, In Distilled Spirits: Production, Technology and Innovation (2nd World Wide Distilled Spirit Conference, Edinburgh), Eds. Bryce, J.H., Piggott J.R., and Stewart G.G., pp. 315–322, Nottingham University Press, Nottingham.

Ferrari, G., Galy, B., Dumot, V. and Canagrel, R., 2005, The production of Cognac, In *Distilled Spirits: Production, Technology and Innovation* (2nd World Wide Distilled Spirit Conference, Edinburgh), Eds. Bryce, J.H., Piggott, J.R., and Stewart, G.G., pp. 309–314. Nottingham University Press, Nottingham.

Gerbaud, V., and Rodriguez-Donis I., 2014, Extractive distillation, In *Distillation: Equipment and Processes*. Eds. Górak, A. and Olujic, Z., pp. 201–245, Academic Press, London.

Harrison, B., Fagnen, O., Jack, F. and Brosnan, J., 2011, The impact of copper in different parts of malt whisky pot stills on new make spirit composition and aroma. *Journal of the Institute of Brewing*, 117(1), pp. 106–112.

Lachenmeier, D.W., Walch, S.G., Padosch, S.A. and Kröner, L.U., 2006, Absinthe—A review. *Critical Reviews in Food Science and Nutrition*, 46(5), pp. 365–377.

Lin, D., 2011, Chinese white spirit, In *Distilled Spirits: Science and Sustainability* (3rd World Wide Distilled Spirit Conference, Edinburgh), Eds. Walker, G.M., Goodall, I., Fotheringham, R., and Murray D., pp. 103–108, Nottingham University Press, Nottingham.

López-Ramírez, J.E., Martín-del-Campo, S.T., Escalona-Buendía, H., García-Fajardo, J.A. and Estarrón-Espinosa, M., 2013, Physicochemical quality of tequila during barrel maturation. A preliminary study. *CyTA-Journal of Food*, 11(3), pp. 223–233.

Mattsson, H., 2005, *Calvados*, ISBN 91-631-5545-1, Flavourrider, Malmo, Sweden.

Neves, E.A., Oliveira, A., Fernandes, A.P. and Nobrega, J.A., 2007, Simple and efficient elimination of copper (II) in sugar-cane spirits. *Food Chemistry*, 101(1), pp. 33–36.

Nicol, D.A., 2014, Batch distillation, in *Whisky, Technology, Production and Marketing*, Ed. Russell, I., Academic Press, London, UK.

Nicol D.A. 2003, Rum, In *Fermented Beverage Production*. Eds. Lea A.G.H., and Piggott J.R., pp. 263–288, Kluwer Academic, New York.

Nóbrega, I.C., Pereira, J.A., Paiva, J.E. and Lachenmeier, D.W., 2009, Ethyl carbamate in pot still cachaças (Brazilian sugar cane spirits): Influence of distillation and storage conditions. *Food Chemistry*, 117(4), pp. 693–697.

Sakaguchi, K., Takagi, M., Horiuchi H. and Gomi K., 1992, Fungal enzymes used in oriental food and beverage industries. In *Applied Molecular Genetics of Filamentous Fungi*. Eds. Kinghorn, J.R., and Turner, G., pp. 54–63, Blackie Academic & Professional, London.

Varvagiannis, A. Margomenou, L. and Zabetakis, I. 2005, Ouzo and raki, In *Distilled Spirits: Production, Technology and Innovation* (2nd World Wide Distilled Spirit Conference, Edinburgh), Eds. Bryce, J.H., Piggott J.R., and Stewart G.G., pp. 323–325, Nottingham University Press, Nottingham.

Zheng, X.W. and Han, B.Z., 2016, Baijiu, Chinese liquor: History, classification and manufacture. *Journal of Ethnic Foods*, 3(1), pp. 19–25.

3 Proteins and Bentonite in Winemaking: Chemical and Practical Aspects and Sensory Consequences

Federico Casassa

CONTENTS

3.1 INTRODUCTION

The postfermentation stage of white, red, and rosé wines encompasses a collection of treatments and procedures that winemakers apply to wines after the completion of alcoholic fermentation. It is usually during the postfermentation stage that fermentable sugars are reduced to appropriate levels (defined by style, fermentation conditions, etc.) and CO_2 concentration has dropped significantly enough so that its physical (i.e., stirring), chemical (blanketing effect and thus protection against oxidation), and sensorial (increase in perceived acidity) effects are of lesser importance than during active alcoholic fermentation. Also, during postfermentation, the main sensory aspects of wines can start being effectively gauged and, in the same way that positive sensory aspects emerge, other potentially negative aspects may surface as well. In the same way in which some of these positive sensory aspects may subside and even disappear downstream during winemaking (e.g., certain fermentation esters), some negative sensory aspects may do so too (e.g., certain volatile sulfur compounds or the lessening of astringency in red wines along with barrel and bottle aging).

However, some of these potentially negative sensory features, together with specific physical and chemical instabilities, may remain, and still others may even manifest due to variations of the wine matrix conditions during winemaking, including changes in wine ethanol, pH, ionic strength, redox potential, or external conditions (e.g., temperature). In the case of the former, the winemaker should be able to predict future instabilities based on, for instance, fruit condition at harvest, fermentation and/or maceration conditions (Note 3.1) historical records, etc.; in the case of variations of the wine matrix, the winemaker may be forced to apply preventive treatments to comply for downstream variations on the chemical composition of the wine that may result in instabilities.

Note 3.1: Maceration is a crucial step during red winemaking whereby the fermentation solids (skins, seeds, lees, and eventually stems) are kept in contact with the fermenting wine and during which phenolic compounds, free aroma, and aroma precursors are extracted into wine. Maceration is also referred to sometimes as "skin contact time" or "maceration time."

The general goals of postfermentation processing in winemaking include (1) achieve clarity, (2) achieve stability, (3) adjust the chemical composition of the wine when required, (4) comply with stylistic and/or production goals, and (5) prepare the wine for bottling and packaging. This chapter reviews current advances in bentonite usage, chemistry, and sensory consequences, recognizing bentonite as a fining agent routinely added to wines to achieve long-term clarity (goal 1) and protein stability (goal 2) and to prepare the wines for bottling (goal 5). An overview of the chemical aspects that underpin the interaction between bentonite and wine proteins in winemaking is presented, together with certain practical aspects and expected sensory consequences on the finished wines.

3.2 FINING TO ACHIEVE LONG-TERM CLARITY

The process of fining in winemaking is in reality a two-sided story that includes (a) fining agents, and (b) targeted compounds. The chemical and/or physical interactions between fining agents and targeted compounds is specific and matrix dependent, meaning that certain additives may interact favorably with targeted compounds in a given wine but may not necessarily yield the same positive outcome in a different wine. Fining agents are chemicals deliberately added to wine that react with specific compounds within the wine matrix, thereby altering their chemistry, reactivity, and sensory properties. Fining agents are usually not soluble in wine and, as such, they often need to be eliminated post-treatment by physical means, including decanting, racking, centrifugation, or filtration.

Contrastingly, targeted compounds are usually fairly soluble in wine but their presence may not be desirable or it may lead to instability. The pair protein–bentonite exemplifies this concept nicely, as bentonite, arguably the most commonly used fining agent in winemaking, is an insoluble clay, whereas proteins in finished wines are usually soluble but very prone to become unstable if left untreated. The interaction between wine proteins and bentonite is electrostatic, although the nature of this interaction is an adsorptive phenomenon (Blade and Boulton, 1988). This means that the targeted compounds (wine proteins) and the fining agent (bentonite) are partially or totally removed from the wine due to opposite charges (attractive electrostatic interaction) coming together to form insoluble complexes, whereby proteins adsorb into bentonite platelets (Section 3.4). Cloudiness and/or haziness is caused by suspended particles or colloids that scatter light (Horne et al., 2001). Insoluble, partially soluble, and colloidal particles in musts and wines include yeast and bacteria; phenolic compounds such as tannins; protein and proteinaceous-like structures; complex carbohydrates such as polysaccharides, mannoproteins, and gums; and exogenous wine additives (such as bentonite, CMC, PVPP, etc.). Yeast and bacteria are negatively charged at wine pH (Amory et al., 1988; Fisher, 1975; James, 1982) and, once CO_2 resealed has ceased after alcoholic fermentation, yeast cells are usually decanted (i.e., flocculated) by natural settling, with ethanol enhancing the process (Amory et al., 1988).

Haze due to tannins is limited to (and very uncommon in) white and rosé wines and arises from tannin–protein interaction that forms light-scattering aggregates (tannin concentrations in white wines below 5% w/w have been found to cause haze) (Waters et al., 1995). In white and rosé wines, cloudiness due to unstable proteins is common, particularly after heat exposure (Figure 3.1). This is problematic because

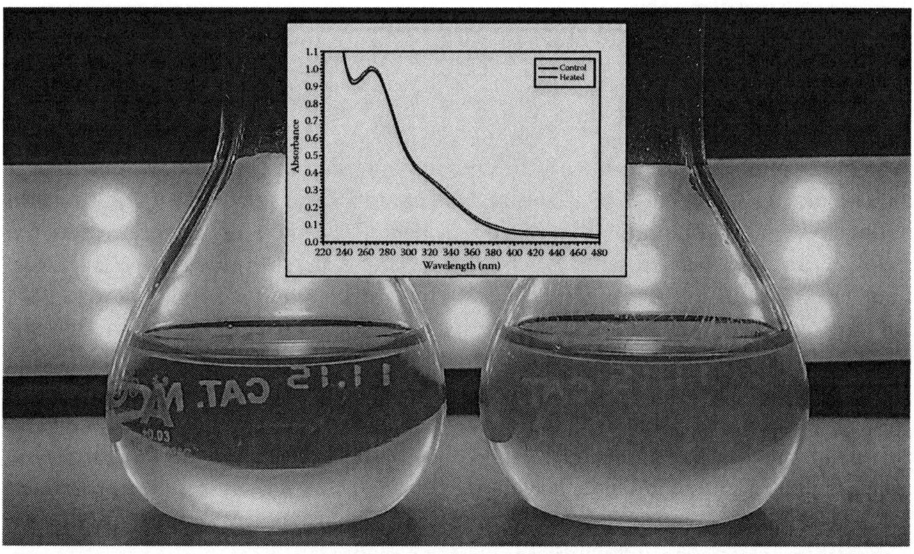

FIGURE 3.1 Example of protein instability in a Chardonnay wine. The wine on the left is the original wine (control); the wine on the right was submitted to the fast heat test (6 h at 80°C, Section 3.8), thereby resulting in protein haze. The central inset shows the full absorbance spectrum of the control sample (solid line) and that of the heated sample (dotted line).

cloudy or hazy wines typically cause consumer rejection, sometimes regardless of other aspects of the wine such as aroma and flavor (which may be otherwise unrelated to the haze). Under controlled conditions, detection thresholds for cloudiness in yellow and red liquids are reportedly as low as 0.38 nephelometric turbidity unit (Note 3.2) (Carrasco and Siebert, 1999) and, under more realistic, "store"-like conditions, as high as 3.55 NTU in apple juice (Horne et al., 2001). For wines, however, protein haze manifested in a Riesling wine sample at 3.1 NTU (Pocock and Waters, 2006).

Note 3.2: The determination of nephelometric turbidity units, or NTU, is a common measurement in winemaking that uses a nephelometer (essentially, a spectrophotometer that measures the amount of scattered light). A nephelometer measures the amount of scattered light impinging in a 90° angle from the incident light beam. Light might be scattered (i.e., reflected) by suspended solids (including yeast, bacteria, tartrates) and colloids. For example, unsettled white juice prior to fermentation may have turbidity levels in the vicinity of 500–600 NTU, whereas the same finished wine after membrane filtering (0.45 μm nominal pore size) prior to bottling may have turbidity levels below 0.3 NTU (looking essentially transparent to the unaided eye).

In this context, white and rosé wines need to be rendered protein stable before bottling. On the other hand, red wines typically contain sufficient amounts of native grape-derived tannins that will readily interact with and subsequently precipitate native wine proteins. Indeed, proteins readily react with tannins, which are phenolic compounds normally present in red wines in mean concentrations of 544 mg/L (as protein precipitable tannins) (Harbertson et al., 2008). As such, with the sole exception of low tannin Pinot Noir wines (as well as some low-tannin wines from hybrid European–American varieties), haze-forming proteins are naturally fined during red wine production and are not known to cause problems in red wines.

3.3 WINE PROTEINS

Proteins in must and wines from *Vitis vinifera* L. represent between 5% and 10% of the total nitrogen content of musts/juices and about 35% of the total nitrogen content of finished wines. Total protein in freshly fermented red and white wines varies from as low as 9 mg/L to as much as 500 mg/L (Batista et al., 2009; Ferreira et al., 2001; Lambri et al., 2012b; Marangon et al., 2011b; Murphey et al., 1989; Okuda et al., 2006; Pocock and Waters, 2006; Sarmento et al., 2000). Variations are due to vintage (Lambri et al., 2012b), variety (Murphey et al., 1989), fruit maturity (Murphey et al., 1989), growing conditions (Sommer et al., 2015), and processing conditions (Ramey et al., 1986). Varieties that are notorious for their intrinsically high protein content, particularly those susceptible to haze, and potentially susceptible to protein instability issues include Sauvignon blanc, Riesling, Sylvaner, Muscat Gordo, Gewürztraminer, and Muscat of Alexandria (Bayly and Berg, 1967; Pocock and Waters, 2006; Pocock et al., 2007; Somers and Ziemelis, 1973; Zoecklein et al., 1995).

Proteins in finished wines include those derived from yeast (Bayly and Berg, 1967; Dambrouck et al., 2003; Rowe et al., 2010; Waters et al., 1994a; Wigand et al., 2009) and those originally present on the grapes, including glycoproteins, pathogenesis-related proteins (e.g., thaumatin-like proteins and chitinases), osmotin-like proteins, lipid-transfer proteins, and many enzymes (including pectinolytic enzymes, invertases, and β-glucanases) (Dufrechou et al., 2011; Okuda et al., 2006; Tian et al., 2015). Total protein content, however, is not correlated with protein instability and haze formation (Bayly and Berg, 1967) and as such, protein instability is difficult to predict by referring only to the total wine protein content.

Among all wine proteins, proteins that cause haze belong to the group known as pathogenesis-related proteins (PR), which includes thaumatin-like proteins (Marangon et al., 2014; Tian et al., 2015; Waters et al., 1996) and chitinases (Marangon et al., 2011b). Interestingly, pathogenesis-related proteins and chitinases likely confer antifungal activity against pathogens such as *Phomopsis viticola* and *Botrytis cinerea*, which makes them desirable candidates as biocontrol agents. Both chitinases and thaumatin-like proteins have the ability to "survive" the fermentation process, are stable at wine pH, and are resistant to proteolytic degradation (Waters et al., 1992, 1996). These proteins display molecular weights ranging from 21,252 to 32,000 Da (Note 3.3) (Waters et al., 1992, 1996; Van Sluyter et al., 2009) and isolectric points (pI) ranging from 4.1 to 5.8 (Dufrechou et al., 2011), which means they are positively charged at wine pH.

Note 3.3: Molecular weight (MW) of haze casing proteins is an important feature in determining molecular weight cutoff (MWCO) for partial separation of these proteins by a technique known as ultrafiltration. Ultrafiltration with MWCO of 10,000 to 40,000 Da is typically limited to treatment of white wines because its use in red wines may remove desirable phenolics with MW of 20,000 Da.

The concentration of PR proteins on a per-berry basis is much higher in the pulp than in the skins (Tian et al., 2015), but prefermentative skin contact time increases their content in wines. Thaumatin-like proteins, in turn, exist in different isoforms, which explain why certain wines, having similar amounts of these proteins display vastly different propensity to haze formation (Gazzola et al., 2012; Jaeckels et al., 2017) (e.g., Sauvignon blanc, highly susceptible to haze, versus Chardonnay, generally much less prone to protein haze than Sauvignon blanc). Additionally, chitinases are in general much more prone to form haze than thaumatin-like proteins and typically form larger protein aggregates (Gazzola et al., 2012; Marangon et al., 2011a). These proteins known to cause instabilities in wines are often informally referred as "heat-sensitive" proteins as they readily unfold when submitted to temperatures in the range of 56°C to 62°C (Marangon et al., 2014), thereby causing haze (Figure 3.1). However, they can also unfold by, for instance, exposure to rather mild temperatures (30°C) for as little as 22 h (Falconer et al., 2010; Marangon et al., 2011b). For instance, the unfolding of chitinase (half-life) will spontaneously occur after 9 years at 15°C, 4.7 days at 30°C, and 17 min at 45°C (Falconer et al. 2010). Protein haze can also develop in properly stored wines because unfolding is also regulated by pH and ionic strength, and such conditions that may change during wine storage (Dufrechou et al., 2011; Moretti and Berg, 1965). Haze formation in white wines is not only regulated by the intrinsic protein profile but also by the presence of nonproteinaceous materials. It has been recently demonstrated that heat unstable proteins do not produce haze when added to model wines (Note 3.4), but the opposite occurs when these proteins are added to actual commercial wines (Pocock et al., 2007). This suggests the presence of a factor or factors, other than proteins, within the matrix of the wines as triggering the haze. This factor has been dubbed the "X" factor, and evidence for its existence in wines has been provided (Batista et al., 2009; Pocock et al., 2007). This factor may include, among other potential candidates, sulfate ions. Other compounds, such as phenolics, may act as "haze modulators" as well, possibly by aiding in the cross linking of denatured proteins (Pocock et al., 2007; Waters and Colby, 2009).

Note 3.4: Model wine solutions are simple wine-like solutions typically made up by DI water, ethanol (between 12% and 13.5% (v/v)), and adjusted to pH 3.5 or 3.6 using tartaric acid.

The mechanism of haze formation is not fully understood but a general overview of both the protein unfolding and haze formation processes has been provided by Pocock and coworkers. Unstable proteins remain soluble in wine in the form of compact coiled structures held together by disulfide bridges. Protein unfolding may occur due to heating of the matrix containing these proteins, subsequently followed by aggregation of the unfolded proteins, thereby resulting in hazy particles. Sulfate ions result in the "salting out" of previously denatured proteins. Protein "salting-out" is a protein precipitation process whereby anions (such as sulfate) compete with proteins for solvation water molecules (those water molecules surrounding the protein and allowing proteins to remain in solution). Sulfate ions deplete water molecules solvating the surface of the protein, ultimately resulting in a decrease in solubility (and precipitation) of unfolded proteins (Pocock et al., 2007). It has also been suggested that protein haze may also occur by the ability of sulfate ions to cross link heat unfolded proteins to promote their aggregation or precipitation (Marangon et al., 2011b; Pocock et al., 2007). Preharvest factors that affect protein instability include the variety and growing conditions, as mentioned before, as well as fungal attack by powdery mildew (*Erysiphe necator*). This is because the powdery mildew infection increases the levels of grape thaumatin-like proteins (Note 3.5) (Girbau et al., 2004).

Note 3.5: Total pathogenesis-related (PR) protein content in healthy Chardonnay grapes was reported to be 67 mg/L as compared with 184 mg/L in the same Chardonnay grapes in which >30% of the clusters were affected by powdery mildew.

Counter to previous assumptions, grape-derived polysaccharides in compromised grapes (e.g., those making their way into wine from grapes affected by powdery mildew or *Botrytis cinerea*) have been associated with preventing protein haze (Waters et al., 1994a,b). However, soluble fungal-derived polysaccharides are negatively charged at wine pH (Vernhet et al., 1996) and may lead to clarification problems and haze formation. This dichotomy between polysaccharides being stabilizing agents against protein haze or haze-promoting compounds may be due to differences in the polysaccharide/protein ratio, but it has been suggested that the type of protein is more relevant than that of the polysaccharide in haze formation (Gazzola et al., 2012). In this regard, the enhanced production and presence of chitinases and thaumathin-like proteins, also known as pathogenesis-related proteins, in must and wines obtained from grapes affected by *Botryis cinerea* and/or powdery mildew may be the determining factor leading to protein instability and haze formation in such wines, as is addressed later in this chapter.

Postharvest factors that affect protein instability include prefermentative skin contact time, blending, tartrate stabilization, and shipping and storage conditions

(Waters et al., 2005). For example, skin contact time during the prefermentative phase during processing of white wines increased protein content (Dizy and Bisson, 1999), thereby increasing bentonite needs.

As mentioned before, white and rosé wines need to be rendered protein stable before bottling. This means that the haze-causing proteins need to be partially or totally removed and/or reduced to lower molecular weight fractions that will not cause haze. However, achieving protein stability is a relative concept. As such, this parameter needs to be determined according to the production goals (Toland et al., 1996). For example, winemaking style, projected shelf-life, and shipping conditions of the finished wine will determine the appropriate level of protein instability. If a winery has control over storage conditions and market shelf-life is estimated to be short (i.e., high rotation)—for example, 2–3 days on the shelf followed by quick consumption after purchase—then a higher protein level on the wine can be risked. Conversely, if temperature abuse is suspected to occur during shipping, then a more complete protein stability should be achieved.

As mentioned in the introduction, in white wines, haze-forming proteins are typically removed with bentonite, a negatively charged clay (Section 3.4). However, a plethora of must or wine pretreatments as well as fining agents as alternatives to bentonite fining have been proposed as winemaking aids for protein fining since the early 1950s, including nylon powder (Moretti and Berg, 1965), mannoproteins derived from *Saccharomyces cerevisiae* (which may act as protective colloids) (Dupin et al., 2000), the addition of hydrocolloids such as carrageenan (Marangon et al., 2013) and polymers such as chitin (Vincenzi et al., 2005), the use of yeast-derived or immobilized proteolictic enzymes (Bakalinsky and Boulton, 1985; Lagace and Bisson, 1990), and technologies such as removal of proteins by immobilized tannic acid on a silica carrier (Weetall et al., 1984), grape-derived tannins immobilized on agarose (Powers et al., 1988), ultrafiltration (Hsu et al., 1987), and the combined use of flash-pasteurization and proteases (Marangon et al., 2012). Very recently, magnetic nanoparticles coated with acrylic acid plasma were used to successfully remove phatogenesis-related proteins from white wines—for example, reducing the concentration of chitinases in a Verdejo wine to 10% of its original content (Mierczynska-Vasilev et al., 2017). Even though the latter technology seems not to affect the phenolic composition of the treated white wines, thereby potentially preserving desirable sensory aspects of the treated wines, it does suffer from a potential high cost per liter of treated wine. Thus, an equally efficient, cheap, disposable, and fast alternative to bentonite fining has yet to be discovered.

3.4 CHEMISTRY OF BENTONITE FINING

Proteins known to cause haze, having low molecular weight and pI between 4.1 and 5.8 (Hsu and Heatherbell, 1987; Vincenzi et al., 2005; Waters et al., 1992) are positively charged at wine pH. This positive charge of wine proteins at wine pH is at the core of the bentonite–protein interaction established during fining. Bentonite is perhaps the most ubiquitous fining agent of the wine industry. In the early 2000s, for example, an average of 0.66 kg of bentonite per ton of crushed white grapes were used in Australia, representing about 8,000 tons of bentonite per year (Bacic, 2003).

Bentonite is a montmorillonite clay that works as an ion-exchange system (Note 3.6) by virtue of its chemical composition based on a layered structure of sequential platelets. Each platelet, in turn, consists of three layers arranged like a sandwich: a central octahedral alumina (Al_2O_3) layer, and two tetrahedral silica (SiO_2) layers (Figure 3.2). The elemental composition of a Na-based bentonite may look like this: SiO_2 = 63%; Al_2O_3 = 22%; Fe_2O_3 = 3.25%; FeO = 0.35%; MgO = 2.67%; Na_2O = 2.57%; CaO = 0.65%; and trace compounds = 0.72%. The central alumina layer is not entirely made up of Al, but also of Mg, which turns the layer surface to a negative charge. This surface negative charge is neutralized by a narrow interlayer of cations, including Na, K, Ca, and Mg, which render the dry structure of bentonite very stable and account for bentonite's ion exchange capacity. Different bentonite types are the result of variations in the proportion of the cations in the interlayer. Because of its platelet and layered structure, bentonite has a large surface area (as high as 750 m^2/g) and due to the capacity of the surface and interlayer to form hydrogen bonds with nearby oxygen or hydrogen groups, bentonite can adsorb polar molecules (but also nonpolar ones, albeit with lower affinity). In fact, third-party labs and some wineries typically use bentonite "traps" placed on key areas within the winery or cellar layout to monitor certain environmental odors such as 2,4,6-trichloroanisole (TCA) or ethyl-phenols (produced by the yeast *Brettanomyces* sp.). This hopefully gives the reader an idea of the adsorption capacity of bentonite if environmental off-odors and/or taints are present in the winery environment. When bentonite is dispersed in water, the interlayer cations readily form hydrogen bonds with water, resulting in the "house of cards" configuration depicted in Figure 3.2. The end result is the swelling of bentonite due to the widening of the interlayer; the extent of the swelling is dependent on the main cation in the interlayer. Small monovalent cations (predominantly Na in bentonites from Wyoming [see Note 3.7], but also Li and K) largely form hydrogen bonds with water molecules transforming the system into a sol state, almost resulting in the full separation of the bentonite platelets.

Note 3.6: An ion exchange system is established when ions from one electrolyte are swapped by ions of another electrolyte, thereby changing the composition of both compounds after the exchange. In fact, different types of bentonites are classified as a function of the maximum amount of exchangeable cations, which is indicated by their specific cation exchange capacity (CEC), measured in milliequivalents per gram (meq/g) or per 100 g (meq/100 g).

Note 3.7: Volclay is a common trade name that refers to Na-based bentonites mined in Wyoming (USA).

Conversely, in bentonites based on divalent ions such as Ca or Mg or mixtures of Na and Ca (as in the so-called activated bentonites, Section 3.5), the separation of the platelets and thus the swelling is limited to a maximum interlayer distance of a few nanometers (Brindley and Brown, 1980).

The (variable) swelling capacity of bentonite is at its core of its interaction with wine proteins. Once bentonite is properly hydrated, the widening of the interlayer increases the exposed negative surface of the platelets, allowing positively charged proteins to engage in ion-exchange reactions with Na and/or Ca cations. Additionally, hydrophobic interactions and hydrogen bonding cooperatively participate in this

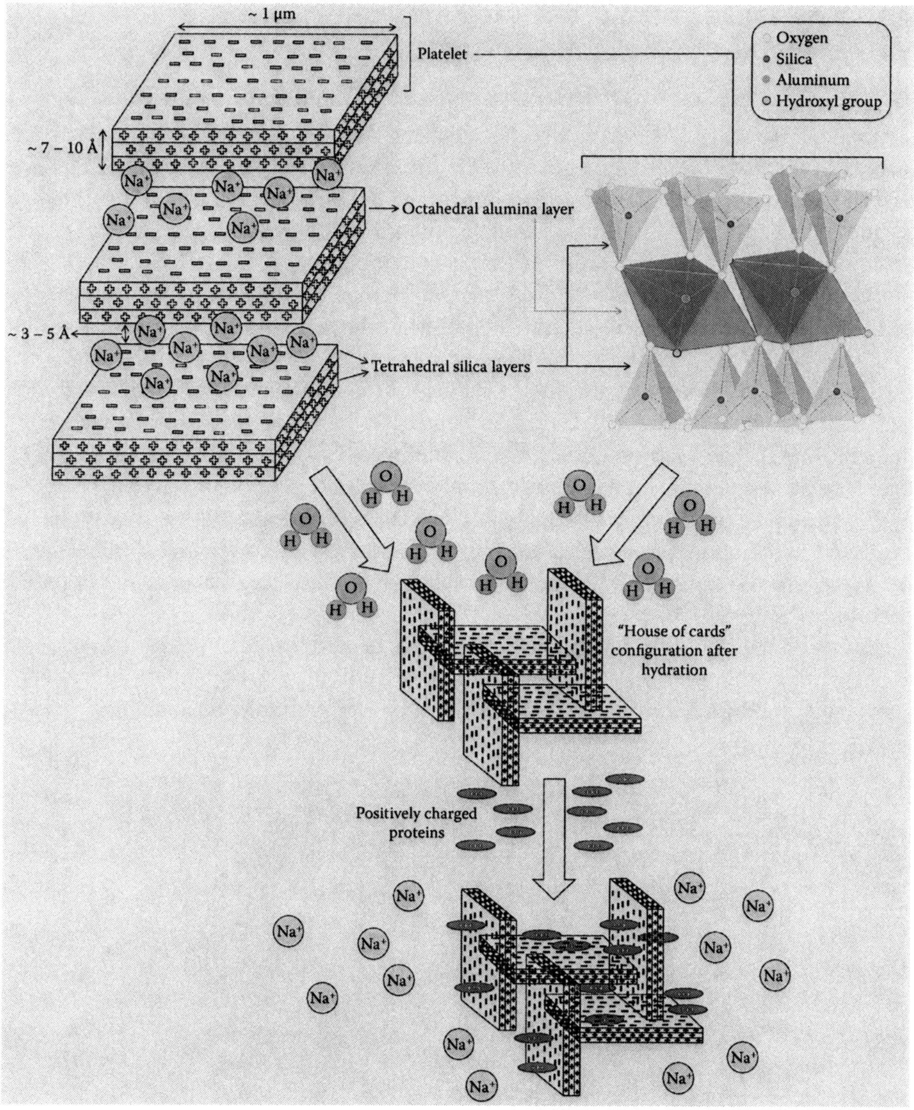

FIGURE 3.2 Oversimplified and idealized structure of a Na-based bentonite also showing the "house of cards" configuration after hydration and the electrostatic binding of positively charged wine proteins.

ion-exchange process. As a result, the interlayer acts by electrostatically "sequestering" wine proteins while releasing equal molar amounts of Na and or Ca cations to the surrounding wine. Thus, the mechanism of protein removal by bentonite is complex and modulated by both electrostatic and noncovalent interactions. It seems that proteins are mostly adsorbed in the edges of the platelets via interactions between protein side chains and the silicate tetrahedral layers, and there may be limited actual penetration of the protein backbone into the interlayer space (Gougeon et al., 2003b).

3.5 NA- AND CA-BASED BENTONITES AND "ACTIVATED" BENTONITES

The ion-exchange capacity of Na-based bentonites is comparable with that of Ca-based bentonites. However, by virtue of the superior swelling capacity and concurrent increase of negatively charged surfaces in the platelets, Na-based bentonites have adsorptive protein capacities per unit of agent that doubles that of Ca-based bentonites (Blade and Boulton, 1988). As a result, Na-based bentonites produce less compact and more voluminous lees (Figure 3.3) but can add between 1.7 and 3.5 g/100 g of Na to the wine (Zoecklein et al., 1995). Na-based bentonites are also easier to dissolve and result in less irritation of respiratory pathways during handling. Empirical results exist that indicate that Na-based bentonites are more abrasive when treated wines are put through cross-flow filtration and are thus not recommended for wines undergoing this treatment. Conversely, Ca-based bentonites have comparatively lower swelling capacity and as such produce lower amounts of less—and more compact—ones than Na-based bentonites (and, concomitantly, result in higher wine recovery) (Figure 3.3); however, rate additions of Ca-based bentonites tend to be higher than those of Na-based bentonites to compensate for their comparatively lower swelling capacity. As a result of the latter, as well, Ca-based bentonites form lees faster than Na-based bentonites.

Ca-based bentonites are not as popular in the United States and Australia (due to widespread availability of Wyoming Na-based bentonites in the United States) and, conversely, are more commonly used in Europe—for example, as a riddling aid for

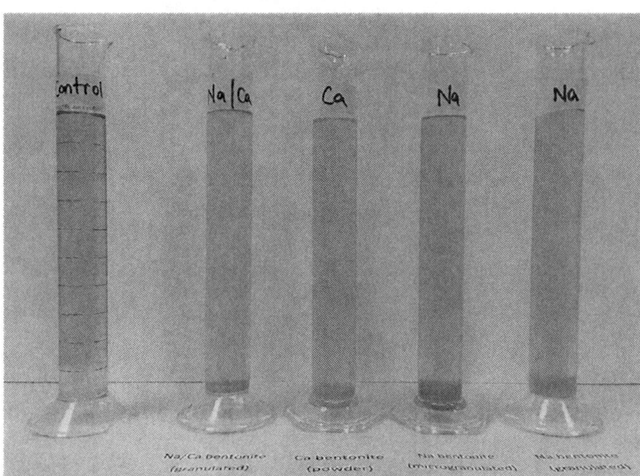

FIGURE 3.3 From left to right: a control wine (untreated) and detail of lees generated in a rosé wine treated with a standard rate addition (60 g/hL) of a Na/Ca (activated) bentonite, Ca bentonite, Na bentonite (microgranulated), and Na bentonite (granulated). The volume of lees generated was, respectively, 1.5%, 2.5%, 4.9%, and 6%.

production of sparkling wines (Marchal and Jeandet, 2009); Ca-based bentonites may also enrich the wine in Ca, leading to tartrate instabilities. Furthermore, specific Ca-based bentonites have been developed (e.g., Microcol® FT) that are specifically marketed as compatible with cross-flow filtration. Finally, the use of Ca-based bentonites is not recommended for wine pH > 3.4. Overall, Na-based bentonites are preferred based on their superior swelling capacity and higher range of functional pH values.

Activated bentonites are Ca-based bentonites treated with CO_3Na_2 at 80°C, resulting in the partial replacement of Ca with Na cations at the edge of the interlayer (Gougeon et al., 2003a); as such, these bentonites typically exhibit "the best of both worlds," showing the superior adsorption efficiency of Na-based bentonites relative to Ca-based bentonites but offering better compaction (and thus higher wine recovery) than pure Na-based bentonites (Salazar et al., 2017). Empirical reports indicate that activated bentonites produce lees volumes as low as 1.6% relative to 6.5% for Na-based bentonites and 3.4% for Ca-based bentonites (Figure 3.3). Activated bentonites also result in higher retention of volatile compounds (Salazar et al., 2017). Activated bentonites are also recommended for fermentation on bentonite lees (Section 3.11).

3.6 RATE ADDITIONS AND USES OF COMMERCIAL BENTONITE PRODUCTS

Bentonite is considered a generally recognized as safe (GRAS) additive and some commercial brands are kosher certified, so its use is widespread in winemaking of white and rosé wines. An overview of general bentonite rate additions for juice fermentation, juice fining/clarification, and wine fining is presented in Table 3.1. Must

TABLE 3.1
Rate Additions of Bentonite in Different States of the Winemaking Process

Type of bentonite	Fermentation	Grape juice	Wine protein content: Low	Medium	High	Tirage (sparkling wine)
Na-based	–	50–100	6–48	12–60	24–120	–
Ca-based	–	200–400 (30–80)	50–150 (10–70)	100–200	150–300	3.5–5.5 (1–4)
Na/Ca-based	–	150–300 (20–100)	30–100	80–180	150–250	–
Activated	100–180	150–300	30–100	80–180	150–200	–

Note: Rate additions are given in grams per hectoliter. Ranges given from different manufacturers are provided in parentheses.

additions are generally higher than additions to finished wines, and, overall, risk of overfining (Note 3.8) is also lower in unfermented juice relative to finished wine.

Note 3.8: Over-fining refers to the situation in which a portion of the fining agent (in this case, bentonite) remains in the wine after treatment. Over-fining can be prevented by the use of counter-fining agents and/or settling aids. In the case of bentonite, alginates can be used to produce more compact lees; the use of silica gel (Kieselsol) is also popular as a counter-fining agent in bentonite additions.

Bentonite additions to juices prior to the onset of alcoholic fermentation, however, can and will strip a portion of the yeast available nitrogen (Note 3.9) on the must because basic amino acids, such as arginine, are positively charged at wine pH and are thus readily fined by bentonite (Bisson and Butzke, 2000).

Note 3.9: Yeast available nitrogen (YAN) is a parameter regularly measured in the winery that encompasses free amino-nitrogen (basically, aminoacids), NH_3, and NH_4^+. Nitrogen is an important macronutrient for yeast (*Saccharomyces cerevisiae* sp.) metabolism during alcoholic fermentation.

Bentonite additions to juices may also remove essential fatty acids and sterols necessary for *Saccharomyces cerevisiae*, which explains why resupplementation with a nitrogen source (e.g., diammonium phosphate, or DAP) does not prevent stuck fermentations in fined musts (Weiss and Bisson, 2002). Overall, excessive bentonite rate additions in musts can lead to decreases in fermentation rate (Zoecklein et al., 1995) and, eventually, to stuck or sluggish fermentations. Bentonite is also used as a riddling aid during sparkling wine production (Marchal and Jeandet, 2009; Pérez-Magariño et al., 2013) but it needs to be used judiciously because it removes proteins and polysaccharides that favorably affect the foaming height and foaming stability of the wines (Lira et al., 2014; Martinez-Rodriguez and Polo, 2003; Vanrell et al., 2007). For this application, Ca-based bentonites are preferred due to the tendency of these bentonites to form more compact lees than Na-based bentonites. The use of bentonite is not recommended in red wines due to the stripping of wine color (Figure 3.3), as bentonite electrostatically adsorbs the (positively charged) flavilum form of anthocyanins (Donovan et al., 1998; Gomez-Plaza et al., 2000; Gonzalez-Neves et al., 2014). In red wine production, however, very low rate additions of bentonite (6 to 12 g/hL) are reported to improve filterability due to reduction in colloidal suspended particles (Zoecklein et al., 1995). In Australia, a technical survey concluded that normal rate additions in white wines ranged from 74 to 84 g/hL (Bacic, 2003). For all intents and purposes, therefore, and for most must/wine applications, bentonite additions may vary from as low as 6 g/hL to as high as 300 g/hL.

3.7 HYDRATION PROTOCOLS

The protein removal capacity of bentonite is not only related to the type of bentonite or the wine chemistry. Contact time and hydration protocols also affect the efficiency of protein fining by bentonite. As shown in Figure 3.2, upon dispersion in water, bentonite platelets repel each other, thereby exposing more surface area. Once hydrated, 1 g of Na bentonite may have an adsorption surface of 5 m^2 (Peynaud,

1984). Expectedly, water composition affects bentonite swelling. Water used for bentonite hydration should be chlorine-free and have low mineral and metal content, as these can replace Na or Ca ions on the surface of the platelets, affecting the binding capacity of bentonite (Zoecklein et al., 1995). Hydration using must or wine should be avoided for similar reasons. For smaller rate additions, the desired amount of bentonite should be dispersed in water (Note 3.10) following an approximate 16:1 ratio [equivalent to a 6% slurry (bentonite to water, w/v)] (water (L):bentonite (kg)), ideally under slow but continuous agitation and preferably using an industrial mixer or blender (Figure 3.4a and b).

Note 3.10: And NOT the other way around! Adding water on top of bentonite (versus correctly dispersing bentonite into water) will result in clump formation.

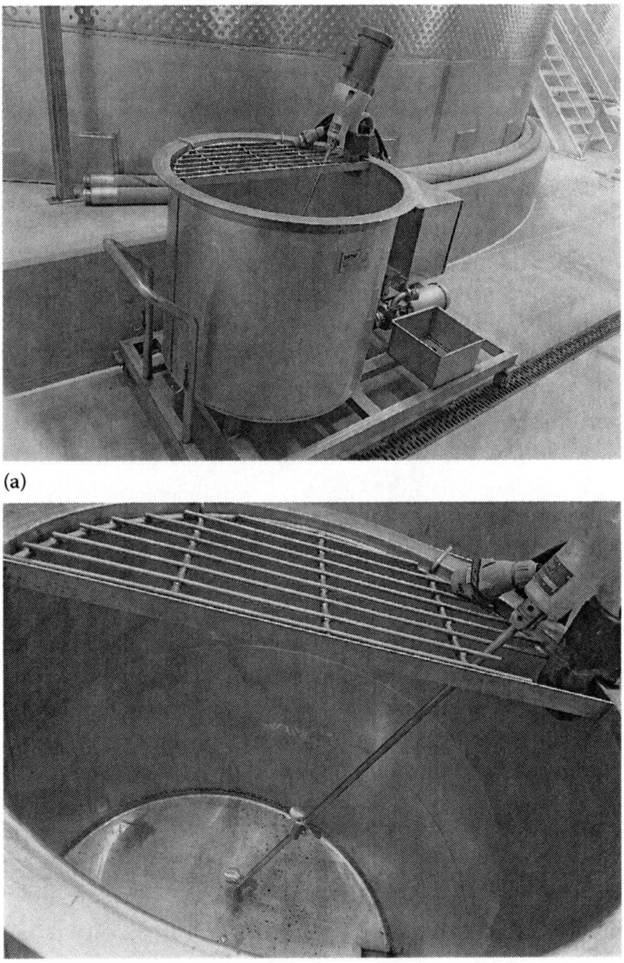

(a)

(b)

FIGURE 3.4 (a) An industrial mixer for bentonite hydration. (b) Details of an industrial mixer.

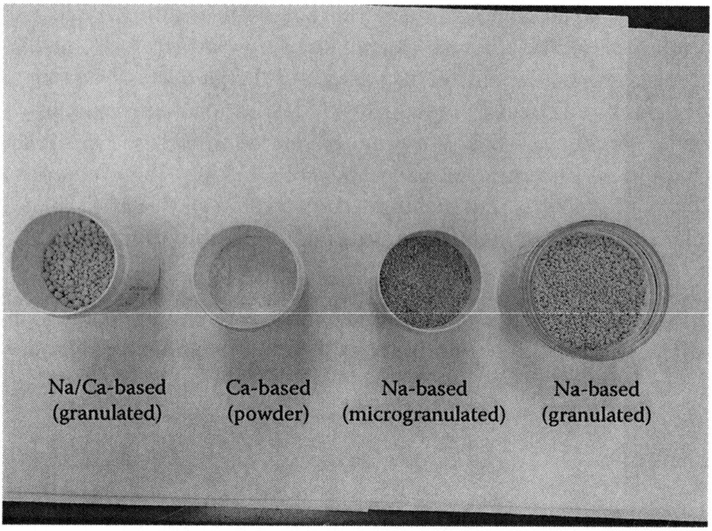

FIGURE 3.5 Detail of different types of bentonites, showing also different granulometries.

The temperature of the hydration water is also important. Hydrating bentonite with water at 50°C is preferred; however, products like certain Na-based bentonites (such as Vitiben®) exist that, according to manufacturer's indication, can be dispersed in cold water. In the United States, the total amount of water added should not exceed 1% of the volume of wine treated (Zoecklein et al., 1995). In practice, this indicates that no more than 16.7 L of water can be used to dissolve 1 kg of bentonite under practical winemaking conditions (Butzke, 2009). When large additions are required (Table 3.1) and those exceed 120 g/hL, an 8:1 ratio is recommended (Butzke, 2009). Some more modern bentonite preparations allow for even smaller water-to-bentonite ratios (3:1 or 5:1) with the potential drawback of more tendency to clump. Also available are the so-called agglomerated bentonites (e.g., Volclay® KWK Agglomerated or Bentogran®), which, based on their specific granulometry, can be directly hydrated in cool water (Figure 3.5). After initial hydration, it is recommended to leave the bentonite suspension to rest overnight (~16 to 18 h) at room temperature and resuspend prior to addition. Hydrated bentonite should not sit for longer than 18 h as this may result in microbial growth. Bentonite is also notorious for absorbing environmental odors, which can be subsequently transmitted into the wine. The bentonite slurry can be added to the tank with a closed pump-over and the addition mixed by a full volume (closed) pump-over or by the use of a device such as the Guth portable agitator (Figure 3.6a and b).

3.8 TESTS TO DETECT PROTEIN INSTABILITIES

Methods to predict protein haze are critical for practical winemaking decisions. Measurement of total protein content is not used for technical decisions or to calculate bentonite additions. This is because total protein content does not correlate with protein instability (Bayly and Berg, 1967). Protein instability, instead, should be tested by forcing

(a)

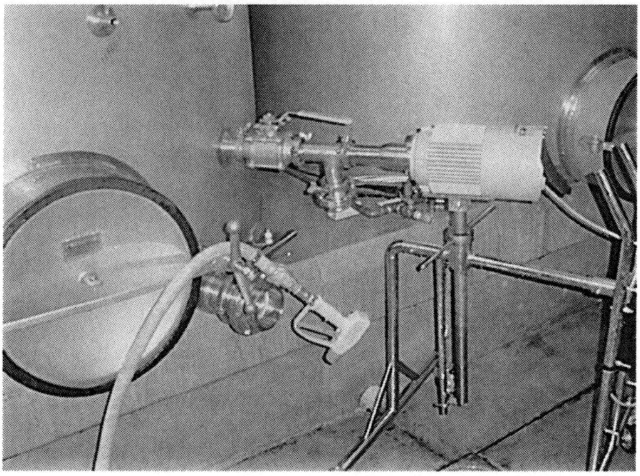

(b)

FIGURE 3.6 (a) Details of several Guth agitators in a commercial winery. (b) A Guth agitator at work for homogenization of bentonite additions in tanks.

the selective denaturation of the protein fractions susceptible to causing haze. Available methods differ in the approach they follow to trigger the denaturation of haze forming proteins. Of note is the fact that each test precipitate specific protein fractions, as demonstrated by the variable chemistry of the obtained protein precipitates after applying various available tests (Esteruelas et al., 2009). Common to all methods, samples to be tested for protein instabilities must be first filtered (preferably through a 0.45 μm membrane filter) and/or centrifuged at 15,000 g for at least 10 min (Figure 3.7). After denaturation and/or a decrease in solubility has occurred, the treated sample is typically reassessed for either NTU or, alternatively, by recording the absorbance at 540 nm (Batista et al., 2009) or 560 nm (Sarmento et al., 2000) with a spectrophotometer (Figure 3.1).

In Australia, hazes produced by unstable proteins are recommended to be assessed at 520 nm, with samples deemed unstable when the difference in absorbance between

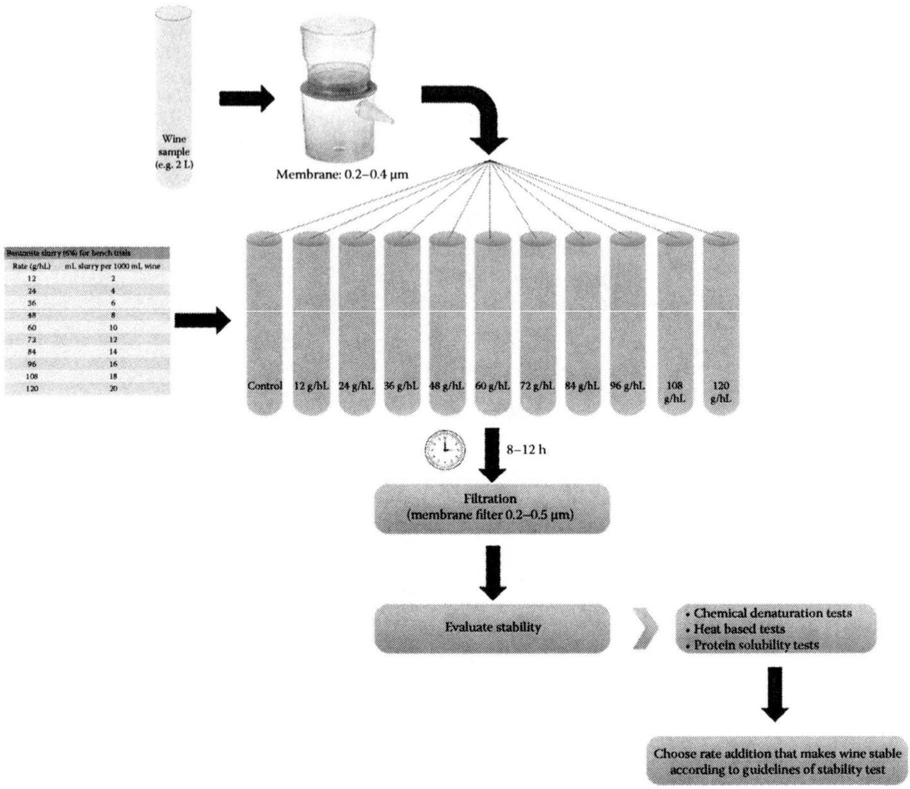

FIGURE 3.7 Diagram to set up bench trials to evaluate bentonite rate additions.

the control and the treated sample is greater than 0.02 absorbance units (Pocock and Waters, 2006). Tests to detect protein instabilities should only be carried out in finished wines and not in wines that are going to be subsequently blended and/or will experience pH or alcohol content shifts. If a haze stable wine is blended with another wine (whether this wine may be heat stable or not), retesting of the final blend must be performed (which may in turn lead to further bentonite additions). Furthermore, different tests applied to the same wine result in variable bentonite rate additions required to render the wine heat stable (Esteruelas et al., 2009; Toland et al., 1996). Protein stability tests include those based on denaturing heat unstable proteins by the addition of chemicals, or by the use of heat; additionally, some tests are based on inducing changes in protein solubility by shifting the ethanol level of the sample.

3.8.1 Chemical Denaturation Tests

These tests include the so-called trichloroacetic acid test and the so-called Bentotest. The former, originally termed "Nephelos analysis," is based on treating a 10 mL wine sample with trichloroacetic acid and heating the mix for 2 min on a test tube placed in boiling water (Berg and Akiyoshi, 1961). NTU are measured after heating with a cutoff criterion of 19 NTU or fewer for stable wines, a minimum of 40 NTU for wines deemed unstable, and a range between 19 and 40 NTU in which the wines are classified as "undetermined stability" and, as such, recommended to undergo further tests. The wide range of "undetermined stability" this test entails is perhaps an indication of a poor relationship between this chemical denaturation mechanism and the solubility of unstable protein fractions.

The Bentotest has been around since the early 1970s (Somers and Ziemelis, 1973) and is based on reacting the wine sample with a solution of phosphomolybdic acid in hydrochloric acid (the reagent is commercially available). The solution is then inverted three times and maintained at room temperature 20 min, followed by measurement of turbidity with a nephelometer. Stability is estimated to be in the 13–14 NTU range. Both the trichloroacetic acid test and the Bentotest overestimate bentonite rate additions, particularly relative to heat-based tests (Toland et al., 1996), probably because these tests cause denaturation of higher molecular weight proteins in the range of 67,000 to 68,000 Da (Esteruelas et al., 2009). In addition, the Bentotest fails to react with heat-sensitive proteins, resulting in bentonite rate additions about 28% higher than the fast heat test (Ribéreau-Gayon et al., 2006). Overall, chemical denaturation tests behave like crude protein assays with little relationship with the solubility of unstable protein fractions (Boulton et al., 1999), and as such they are not recommended.

3.8.2 Tests Based on Changes in Protein Solubility

Two main tests exist whose principles rest on altering the solubility of unstable proteins, based on the addition of an excess molar amount of ethanol or the addition of tannins. The former test, known as the ethanol test, rests on the fact that a decrease in the dielectric constant of a solution occurs with increasing additions of ethanol, thus resulting in a decrease of protein solubility. However, rather than specifically targeting protein unstable fractions, the ethanol test precipitates the least soluble fractions at wine pH. After recording the initial NTU of the wine sample, the test involves the addition of 10 mL of absolute ethanol to a 10 mL sample of degassed wine, giving a final ethanol content of approximately 56% (Boulton et al., 1999). The sample is kept for 2 h at room temperature and then NTU are reassessed in the treated samples. The ethanol test suffers from various drawbacks, including the fact that it also precipitates mannoproteins and pectins, resulting in overestimation of bentonite rate additions. The use of certain yeast strains during alcoholic fermentation that are known to produce and release mannoproteins (e.g., BM 45, D-254) or winemaking techniques (such as *sur-lies* aging in Chardonnay) may also overestimate required doses of bentonite. The ethanol test is also sensitive to interferences due to pH variations. Lastly, Ca is known to interfere with the ethanol test (Sarmento et al., 2000), making the test of little use if Ca-based bentonites are used.

The tannin addition test, originally described by Mesrob (Sarmento et al., 2000), entails mixing 5 mL of wine with 0.4 mL of a 1% tannin solution (enological tannin) in 40% ethanol. The wine sample is then held for 15 min at room temperature and then heated for 3 min in a boiling water bath, after which turbidity (or absorbance at 560 nm) is measured and compared against the turbidity (or absorbance) of the untreated wine. The method is fast and easy to execute, but suffers from many interferences, including the formation of a black precipitate in the presence of Fe (Sarmento et al., 2000). The tannin test also suffers from variations in pH, K, and Cu concentration. Like the chemical denaturation tests, the tannin test seems to give an estimation of the total crude protein content rather than that of the unstable protein fractions (Esteruelas et al., 2009).

3.8.3 Heat-Based Tests

Heat tests are by far the most popular tests in the wine industry to predict protein instability (Esteruelas et al., 2009). This makes practical sense, as, for one, these tests are easy to perform and require equipment that is readily available. Furthermore, the application of heat during testing mimics shelf-life conditions, as proteins known to cause haze become unstable typically after bottled wines have experienced temperature changes and, more specifically, temperature increases (Pocock and Waters, 2006). There is also experimental evidence of a good correlation between protein affinity for bentonite and sensitivity of those protein fractions to heat denaturation (Sauvage et al., 2010).

Heat tests are divided into those in which heating of the sample occurs slowly, typically over a period of a few days, as opposed to fast heating tests, in which heating typically occurs in a short period of time and often under high temperatures. For the slow heating test, 40 mL of wine is stored on an oven during 4 days at 60°C, subsequently followed by the chilling of the wine sample at 4°C for 6 h in a refrigerator (Esteruelas et al., 2009). The slow heat tests have been abandoned not only because they are time consuming (and usually immediate results are required for winemaking decisions) but also because slow heating fails to precipitate haze-forming thaumatin-like proteins (Esteruelas et al., 2009). Furthermore, slow heating may cause progressive oxidation of phenolics followed by condensation with proteins, thus blurring the results of the test.

Fast heat tests, on the other hand, are based on submitting 40 mL of wine to 90°C for 1 h (Sarmento et al., 2000), or to 80°C for 2 h (Pocock and Waters, 2006), in both cases in a thermostatic water bath, followed by chilling of the sample at 4°C for 6 h. It is important to ensure that during heating in the water bath, no water is entering the tubes or vials, which should always be kept above water level. A more stringent version of the fast heat test calls for heating the sample at 80°C for 6 h (Pocock and Waters, 2006) (Figure 3.1). In both cases, developed turbidity is measured against a blank of each untreated (but filtered) wine (Sarmento et al., 2000). If the difference between final NTU (after heating the wine sample) and initial NTU (untreated wine sample) is below two NTU, the wine is considered stable (Ribéreau-Gayon et al., 2006); tough, more stringent criteria exist that call for this difference to be below one NTU. This seems to be the criterion followed by most third-party labs that offer analytical services to check for protein instability in the United States. If a nephelometer is not available, the absorbance at 540 nm of the heated and unheated sample can be measured instead. Samples are considered to be protein unstable when the difference in absorbance between heated and unheated samples is >0.02 absorbance units (Mierczynska-Vasilev et al., 2017). Fast heat tests readily precipitate proteins in the range of 18,000 to 25,000 Da (Esteruelas et al., 2009), giving protein precipitates that chemically resemble spontaneous protein precipitates.

In addition, fast heat tests are relatively unaffected by wine pH and the presence of metal cations (Sarmento et al., 2000). The fast heat test also result in comparatively lower bentonite additions (up to 20 g/hL lower) than, for instance, the Bentotest (Esteruelas et al., 2009). In turn, within the two versions of the fast heat test, the less severe one (80°C for 2 h) generally yields lower dosage rates of bentonite, thus reasonably predicting short- to medium-term protein stability (Pocock and Waters, 2006). When long-term stability of the bottled wine must be assured, then the severe heat test (80°C for 6 h) is recommended, although, as an expected drawback, experience shows that the risk of overfining is higher in the severe heat test than in the slow heat test. Concerns have been raised about the fact that reported temperatures of the heat tests refer to the temperature of the heating apparatus rather than that reached in the wine. It has been established that when using heating blocks, for example, the temperature should be set 14°C above the required temperature and that an extra 30 min might be necessary for a 5 mL wine sample (Pocock and Waters, 2006). The reader is referred to the study of Pocock and Waters (2006) for more details on proper working temperatures during both slow and fast heat tests. Overall, both the ease of application and correlation with the wine's unstable protein fractions validate the election of the fast heat tests as the methods of choice to test for protein stability and thus the current industry standard.

3.9 BENCH TRIALS TO DETERMINE BENTONITE RATE ADDITIONS

Bench trials to determine bentonite additions are routinely carried out by wineries. Bentonite demands vary widely as a function of the vintage, even for varieties grown under the same viticultural conditions. For example, a 10-year study in the Palatinate region of Germany reported that, in the 2006 vintage, overall bentonite demand to achieve heat stability amounted to 71 g/hL, contrasting sharply with the 2009 vintage, which demanded on average 168 g/hL (Sommer et al., 2015). This explains why and supports the notion that bench trials should be carried out, even in a given wine from the same vineyard block, on a vintage-to-vintage basis.

Bentonite additions to freshly pressed musts and juices are typically predetermined and range from 30 to 300 g/hL (Table 3.1). These juice additions are higher than those performed in finished wines due to the fact that protein content generally decreases during fermentation (Murphey et al., 1989). Conversely, additions to finished wines require bench trials. The goal of bench trials is to determine the lowest possible bentonite rate addition that will make the problem wine protein stable, according to the test being carried out (Section 3.8). Minimizing bentonite rate additions is not only important from the perspective of wine quality but also to minimize costs associated with bentonite in itself, wine losses, and the environmental impact of bentonite lees (Pocock et al., 2011). Informal reports also indicate that bentonite lees can form a sludge at the bottom of wastewater treatment ponds, compromising algae growth and thus the biological functioning of the pond. On the other hand, bentonite may be beneficial in wastewater management systems as, by reducing protein content of wastewater, it causes reduction of the BOD (biochemical oxygen demand). Regardless, winemakers should strive for the lowest possible bentonite addition that would make the wine stable. A typical example of a bench trial to determine bentonite additions to a finished wine is shown in Figure 3.7. Test tubes (or better, a graduated cylinder, or even 750 mL wine bottles; see Note 3.11) can be used for the bench trials.

Note 3.11: Very small volumes of sampled wine (<100 mL of total reaction volume) are not recommended due to the difficulty of accurately pipetting the necessary milliliters of bentonite slurry and thus the error associated with this.

The analyst must first make a 6% slurry (w/v) of the bentonite to be used. This slurry must be made using the same bentonite (and, preferably, the same lot) to be used to fine the problem wine (Zoecklein et al., 1995). The slurry must be made at laboratory scale, by slowly dissolving the bentonite into chlorine-free (and preferably DI) water at 50°C–60°C and slowly dissolving the addition with the aid of a magnetic stirrer. The slurry must be free of clumps and clots, and preferably allow it to swell overnight before use. Once the problem wine is filtered, it should be distributed into a number of cylinders or bottles, each of them assigned with a specific rate addition. For maximum accuracy and to reduce bentonite rate additions, testing rate additions at 12 g/hL increments is recommended. The slurry must be gently added to each wine sample while the wine is being agitated, followed by a settling time of 8 to 12 h at room temperature. Subsequently, an aliquot of the wines is filtered again and submitted to one of the available stability tests (Section 3.8). The lowest bentonite rate that renders the wine heat stable is then chosen (Figure 3.8). If more refined bentonite additions are needed, a follow-up bench trial can be run using as a baseline the lowest rate addition that turned the wine stable, and progress, for example, at 2 g/hL increments. Rosé wines, in which color defines style, should also be checked for color changes after bentonite additions, contrasting the control (untreated) with

FIGURE 3.8 Samples of the same Chardonnay wine submitted to fast heat tests (6 h at 80°C) after: no treatment with bentonite (far left) and treated with a standard rate addition (60 g/hL) of a Na/Ca (activated) bentonite, Ca bentonite, Na bentonite (microgranulated), and Na bentonite (granulated), respectively. Note that, as assessed by the naked eye, all four bentonite-treated wines show no haze after heating.

the treated wine, preferably using analytical measurements, such as the Cie-Lab system (Ayala et al., 1997).

After bench trials are completed, it is also recommended to calculate the volume of lees as a percentage basis of the total volume of treated wine (Figure 3.3). In well-conducted bentonite additions at industrial scale, particularly in those in which the addition is homogenized with the help of a Guth agitator (Figure 3.6), lees can represent as little as 2% of the total volume of wine treated. However, the volume of bentonite lees volumes after treatment ranges from 5% to 10% under industrial scale conditions (Zoecklein et al., 1995). Treatment of bentonite lees to partially recover occluded wine may include the use of a rotary drum vacuum filtration system (Note 3.12), or centrifugation of the bentonite sludge, as it is typically done during the process known as in-line dosing of bentonite (Section 3.10).

Note 3.12: In the United States, mobile rotary drum vacuum filtration systems are available for rent if wineries do not own this technology.

Calculation of bentonite lees on a volume basis can provide an estimation of the volume of lees a given bentonite treatment will generate when applied at an industrial scale.

3.10 IN-LINE DOSING OF BENTONITE

The interaction of bentonite with haze-forming proteins occurs fast, with some models indicating as little time as 30 seconds for the fining process to occur (Blade and Boulton, 1988), but not more than 2 min in industrial conditions (Muhlack et al., 2006). This time frame (between 30 seconds and 2 min) contrasts sharply with the current industry standard of bentonite finning in which bentonite settling occurs by gravity, taking several days or even weeks for the process to be completed. Subsequent to settling of bentonite lees by gravity, a substantial amount of wine, ranging from 5% to 10% end up occluded in the lees. In-line dosing of bentonite addresses these two issues (time and amount of bentonite lees) with a simple setup (Figure 3.9). The wine to be treated is pumped and, throughout the pipework, meets a bentonite slurry, which is in turn being pumped from another tank. Upon injection, the mixture of bentonite–wine reaches a static mixer that optimizes contact between the flowing wine and the bentonite slurry, which are allowed to interact for 5 min through appropriate pipework. Pipework to provide contact time can be accomplished, for example, by using a shell-and-tube heat exchanger unit (used solely for the purpose of allowing contact time between wine and bentonite and not for cooling).

After contact time, the bentonite–wine mixture is centrifuged to separate bentonite lees from juice or wine (Muhlack et al., 2006). Using this setup, a Sultana wine and a Muscat Gordo juice were made stable within 2 min of contact time, without significantly affecting the sensory properties of the resulting wines. In Chardonnay and Semillon wines treated with in-line dosing of bentonite at a flow rate of 12,000 L/h, heat stability was achieved with a contact time of 3 min (Nordestgaard et al., 2007). Furthermore, cost analysis of this setup by Dr. Richard Muhlack with the Australian Wine Research Institute (AWRI) indicated that, in Sauvignon blanc wine, in-line

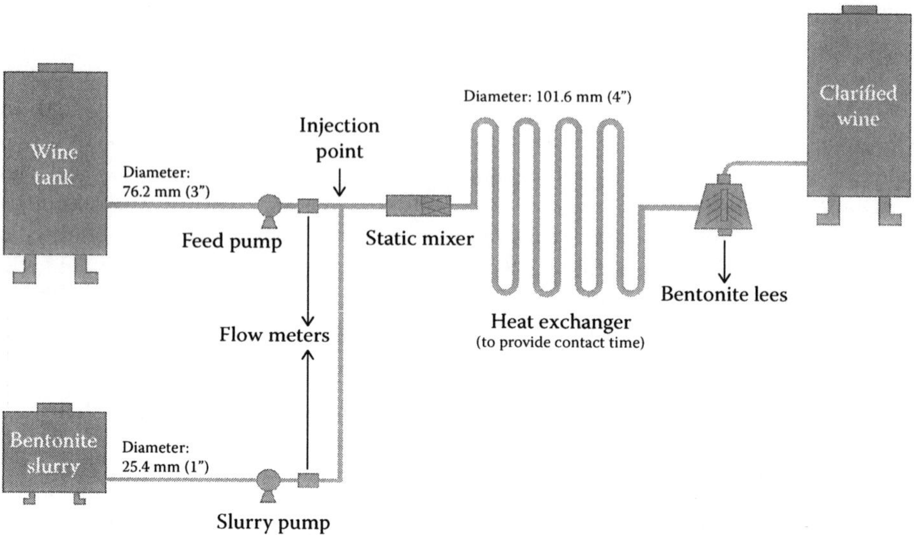

FIGURE 3.9 Schematic diagram of the setup for the in-line dosing of bentonite into juices and/or wines. (Modified from Muhlack, R. et al., 2006. *Australian Journal of Grape and Wine Research*, 12, 221–234, and Muhlack, 2006, Doctor of Philosophy thesis, University of Adelaide, Australia.)

dosing of bentonite performed at commercial scale cost \$0.0032/L versus traditional batch fining, which cost \$0.0194/L (representing about 104% savings).

While in-line dosing of bentonite has proven to have no negative sensory outcomes relative to the traditional batch fining process (Muhlack et al., 2006; Nordestgaard et al., 2007), a problem associated with in-line dosing of bentonite is that up to 30% of bentonite carryover into the clarified wine can occur (Muhlack et al., 2006). Bentonite carryover into clarified juice or wine, for example, can be reduced from 26% to 19% by reducing operating flow rate of the centrifuge from 17,700 L/h to 12,500 L/h (Nordestgaard et al., 2007). Bentonite carryover can also be reduced by the use of counterfining agents during the in-line process to improve bentonite separation (Waters and Colby, 2009). Counterfining agents are typically added to help in settling the original fining agent—in this case, bentonite. Commonly used counterfining agents for bentonite include sparkalloid, silicon dioxide, or gelatin. Carefully chosen counterfining agents hasten settling and decrease lees volume. During in-line dosing of bentonite it also appears that Na–Ca bentonites are separated more easily by centrifugation than Na-based bentonites at equivalent flow rates (Nordestgaard et al., 2007) and thus the former are usually preferred for this application.

3.11 FERMENTATION ON BENTONITE LEES

As mentioned before, bentonite can be added both to finished wines (after bench trials) and to unfermented juice. Early studies suggested the inefficiency of bentonite used as a prefermentation additive, with the argument that juices, relative to finished wines, contain higher protein content and N-derived materials than their respective finished wines do (Somers and Ziemelis, 1973). Indeed, larger bentonite additions are required for fermenting on bentonite lees relative to fining finished wines (Pocock et al., 2011). Furthermore, the adsorption capacity of bentonite towards aromas was reported to be enhanced in the presence of sugars in model wine solutions due to the ability of sugars to decrease the solubility of aromas (by competing with them for solvation water), thus increasing volatility and interaction with bentonite platelets (Lubbers et al., 1996). However, this effect may be of lesser importance due to the low proportion of free volatiles in freshly crushed must (Note 3.13).

Note 3.13: Notable exceptions include Muscat varieties or other terpenic varieties such as Gewürztraminer, Riesling or Malvasia juices, in which a variable but potentially important proportion of terpenes can be present as free volatiles in the freshly pressed or crushed must.

On the other hand, fermenting on bentonite lees may provide some advantages. Pocock and coworkers reported that in Sauvignon blanc and Semillon, fermentation rate increased when these juices were fermented on bentonite lees. These additions, which amounted to 170 g/hL in Sauvignon blanc and 150 g/hL in Semillon, were sufficient to turn the finished wines' protein stable in a single-step addition (Pocock et al., 2011). As an added benefit to the technique, the authors also implied less impact on wine sensory properties, as juice (which is more forgiving) is being treated instead of wine (Pocock et al., 2011).

In another report, the presence of bentonite during fermentation of Muscat of Alexandria, Sauvignon blanc, and Chardonnay juices (at rates of 12, 24, and 36 g/hL) showed no effect on fermentation rate and lowered protein concentration more than traditional juice fining (Weiss and Bisson, 2002). In a variation of this technique, addition of bentonite at 40 g/hL in Albariño at midfermentation resulted in reduced bentonite additions required to achieve protein stability. Furthermore, a preference test with untrained panelists indicated that must added with bentonite midfermentation was preferred over bentonite added prior to alcoholic fermentation, at the beginning or towards the end of alcoholic fermentation (Lira et al., 2015).

Conversely, addition of a Na bentonite to a Sauvignon blanc must during fermentation at rates ranging from 50 to 100 g/hL significantly decreased the concentration of the so-called "polyfunctional mercaptans," which are key varietal aroma compounds in Sauvignon blanc wines (Vela et al., 2016); however, these authors go on to suggest that aroma modification upon bentonite fining depends upon the chemistry of the aroma family, the vintage, and the timing of bentonite addition. Overall, although fermentation on bentonite lees appears as a promising technique to lower or even eliminate subsequent bentonite requirement of finished wines, there are no reports on the chemical effects (volatile aroma compounds) and the corresponding sensory consequences of this approach compared with that of traditional juice or wine fining. Furthermore, the preceding research indicates that the technique may be detrimental for Sauvignon blanc, suggesting that the final outcome of the application of fermentation on bentonite lees may be contingent upon the specific aromatic composition of the grape variety.

3.12 MUST AND WINE CHEMISTRY AND BENTONITE

The interaction between bentonite and wine proteins does not only rest on the specificities of wine proteins and bentonite chemistries. Expectedly, other factors within the wine matrix modulate the outcomes of this interaction. Likewise, the sensory consequences of the fining of wines with bentonite depend not only on these chemistries but also on other wine compositional and technical factors. Generally, wines of lower pH require lower bentonite rates to become heat stable (Zoecklein et al., 1995). This is because the pI of heat-sensitive proteins is typically above the normal pH for white wines (2.9 to 3.5) and thus most of these proteins are positively charged at those pH, resulting in enhancing exchange capacity with cations located at the bentonite's structure interlayer.

Haze by unstable proteins tends to occur when wine pH approaches the pI of the unstable proteins, which can be a problem in the case of high pH wines, as these are common in warm or hot viticulture areas. In this case, bentonite alone may not be able to turn these wines heat stable (Marchal and Jeandet, 2009) and other fining agents may have to be used in conjunction with bentonite. Ionic strength also plays a critical role in protein unfolding and aggregation (Dufrechou et al., 2011) and, as such, it determines bentonite rate additions as well. Increasing ionic strength affects the rate of protein unfolding by compacting protein conformation and minimizing accessible surface areas within the protein; however, increasing ionic strength also decreases the electrostatic repulsive contribution of colloidal protein–protein interactions, thereby increasing aggregation and resulting in more evident visual haze (Dufrechou et al., 2011; Marangon et al., 2011a).

From a practical standpoint, drought conditions in the vineyard can lead to high salt concentration on the resulting wines, ultimately increasing protein haze potential and even decreasing the efficiency of bentonite fining by decreasing the electrostatic interaction between bentonite platelets and wine proteins (Marangon et al., 2011a; Sommer et al., 2015). Growing season and overall climate patterns also play a role. In Chardonnay, for instance warm growing conditions increased protein content relative to cooler growing conditions, without qualitatively affecting the protein profile (Dizy and Bisson, 1999); however, as mentioned before, protein content is not related to protein haze. This ultimately suggests that even though heat-sensitive proteins may not be affected, other contributing nonprotein factors might, thereby resulting in higher bentonite rate additions required to turn the wines protein stable.

The content of wine-soluble protein is also positively correlated with increasing Brix during maturation (Murphey et al., 1989). Research suggests that this leads to higher bentonite needs. For Riesling and Gewürztraminer, bentonite rate additions of 48 g/hL were necessary to make a Gewürztraminer wine (Brix at harvest 24.5) heat stable, versus 24 g/hL necessary to achieve heat stability in the same wine made from fruit picked at 21.2 Brix (Murphey et al., 1989). Juice or wine fining with bentonite usually has little impact on wine's basic composition (alcohol, volatile acidity, titratable acidity, pH, dry extract, free SO_2) (Lambri et al., 2012a,b; Lira et al., 2015; Salazar et al., 2017; Vela et al., 2016; Weiss and Bisson, 2002) with only residual fructose being reportedly higher in wines made by juice treated with bentonite (Weiss and Bisson, 2002).

A long-term concern within winemakers has been the effect of bentonite fining on the volatile composition of the fined wines. Indeed, bentonite may form stable linkages with aroma compounds of sensory relevance (hexanol, ethyl hexanoate, isoamyl acetate, β-ionone, γ-decalactone), resulting in a loss of these volatiles as they settle from solution with bentonite (Moio et al., 2004; Voilley et al., 1990). Adsorption of aroma compounds onto bentonite platelets has been suggested to be the result of both hydrogen bonds and hydrophobic interactions (Lubbers et al., 1996; Vincenzi et al., 2015). Fixation rates of about 2.3 mg of volatiles per gram of bentonite added have been reported (Lubbers et al., 1996). Additionally, the volatile composition of treated wines could be indirectly affected as a result of the interactions between volatile compounds (especially esters) and macromolecules (e.g., proteins), the latter being removed by bentonite (Note 3.14) (Di Gaspero et al., 2017; Lambri et al., 2010; Lubbers et al., 1996; Sanborn et al., 2010; Vincenzi et al., 2015).

Note 3.14: In fact, it has been experimentally demonstrated that a specific thaumatin-like protein (VVTL1), is able to bind to ethyl esters of different chain lengths. These protein-esters complexes are subsequently removed by bentonite. For more information, the reader is referred to Di Gaspero et al. (2017).

Under this indirect mechanism, the hydrogen bond of aroma compounds with the protein's hydrophilic surface, together with hydrophobic interactions between aroma compounds and the internal hydrophobic protein structure, results in this so-called indirect "deproteinization" effect (Lambri et al., 2010). Ethyl dodecanoate, an ester having leafy and soapy notes, decreased in Chardonnay wines treated with Na–bentonite (rate addition 100 g/hL) from 0.031 mg/L (control wine) to 0.017 mg/L (treated wine). However, the remaining aroma profile of this wine, as determined by gas chromatography, was not affected by bentonite fining (Sanborn et al., 2010). In the same study, bentonite fining of Gewürztraminer wines significantly decreased the levels of linalool (a terpene having an intense floral note). However important this decrease in linalool in the bentonite-treated Gewürztraminer wines was (~33%), it was not sufficient to cause an overall sensory difference or to affect any aromatic, taste, or flavor descriptor in these wines as assessed by a sensory trained panel (Sanborn et al., 2010). In model wines reconstituted with Moscato (i.e., Muscat) aroma compounds, with and without proteins added, bentonite fining with a fairly low rate addition (10 g/hL) decreased the concentration of several ethyl-esters, leading the authors to suggest higher affinity of bentonite with esters of increasing carbon chain length, with a synergist effect played by the presence of proteins (Vincenzi et al., 2015). However, in this study, terpenes, which are prominent aroma compounds in Muscat wines, were unaffected by bentonite. Similarly, in Falanghina (a white variety) must, fining with bentonite (80 g/hL) decreased the concentration of free 2-phenyl ethanol (a volatile compound reminiscent of rose petal and/or honey), but had no effect on the concentration of terpenes such as free linalool, α-terpineol, and geraniol (Moio et al., 2004). Interestingly, the previous study found that must fining with bentonite preferentially decreased the concentration of the glycosylated precursors (Note 3.15) of linalool, geraniol, 2-phenyl ethanol, and eugenol; however, sensory analysis of the wines was not reported and it is thus unclear if this decrease on glycosylated aroma precursors was of sensory relevance.

Note 3.15: Glycosidically bound aroma precursors are aroma compounds covalently bound to a sugar moiety, and they represent a source of varietal flavor that may be released as free aromas via enzymatic and/or acid hydrolysis during winemaking and aging.

The adsorption of glycosylated terpenes on bentonite has also been suggested in the case of Albariño musts (Lira et al., 2015). An Albariño must fined with bentonite (60 g/hL) resulted in a 13% decrease of terpenes (including linalool, α-terpineol, nerol, and geraniol) and norisoprenoids (including α- and β-ionone, 1,8-cineol) relative to a control wine; however, this treatment concomitantly reduced the concentration of C6-alcohols (having herbaceous and/or grassy/fatty notes) by 33% relative to an untreated control, suggesting a positive effect of bentonite treatments on removing these undesirable compounds (Armada and Falqué, 2007). Overall, the previous studies suggest that treated musts of terpenic varieties could potentially have a lower aromatic potential relative to an untreated control, as bentonite may preferentially remove glycosidycally bound terpenes. As in the previous works, the sensory consequences of these otherwise significant chemical changes on aroma concentration as a function of bentonite fining were not explored. Therefore, it remains unclear how the sensory profile of the resulting wines may be modified.

3.13 CONCLUSIONS

Instabilities due to proteins resulting in hazy wines remain a perennial problem in the wine industry that, for certain wine styles and price segments, will always have to be addressed by winemakers. Consumers reject and penalize the presence of haze in white and rosé wines. In a global wine market of price squeeze and wine surplus, wines need to reach the consumer ready to withstand potential abuses (e.g., temperature) while maintaining sensory properties that are consistent, expectable, reproducible, and appropriate to style, variety, and price point.

Although emerging technologies for selective protein removal exist, bentonite remains the fining agent of choice for partial removal of haze-forming proteins in white and rosé winemaking. Its popularity in the wine industry rests on its many positive features, including affordability, availability, ease of manipulation and application, and its specific chemistry that allows this fining agent to preferentially remove thaumatin-like proteins and chitinases known to cause haze. Negative aspects of bentonite usage include potential environmental costs as well as wine loses on lees, coupled with the process of bentonite batch fining in itself, which is inherently time consuming. With regard to this last issue, emerging setups such as in-line fining, which rests on the fact that protein–bentonite interaction occurs in a 30 second to 2 min span, will probably become more mainstream in the future.

Different tests are available that will allow the winemaker to diagnose a potential protein instability issue downstream during winemaking, of which rapid heat tests remain the choice within the wine industry. Coupled with the incorporation of these protein stability tests in the winery laboratory SOP manual, bench trials to determine the rate additions that will minimize bentonite usage are a mandatory step in the process of bentonite fining. Several studies have attempted to link growing conditions or fruit maturity patterns with bentonite needs so as to propose models that may, ultimately, be able to predict bentonite needs, circumventing the need for time-consuming bench trials. In spite of the well-known poor correlation between total protein content and bentonite needs, most of these studies have targeted both total protein content and that of thaumathin-like proteins and chitinases, but no clear trends have emerged yet.

This points out the need of redirecting research towards study of the nonprotein composition of must and wines that prove susceptible to protein haze. Growing conditions, including viticultural practices, maturity, disease pressure, and the grape variety in itself, as well as prefermentative and postfermentative factors are among some of many factors responsible for the sometimes observed three- to fourfold variation in bentonite needs. This, once again, points out the need to establish protein stability, which although a relative concept in itself, needs to be determined on a wine-by-wine basis. Finally, the bulk of research reviewed in this chapter suggests that, in well-conducted bentonite additions previously determined by bench trials, the chemical impact of this addition is fairly selective on proteins. While an impact of bentonite additions has been reported on esters, norisoprenoids, and glycosylated terpenes, there is no evidence to suggest that decreases in aromatics may actually negatively impact the wines' overall sensory features, thereby possibly escaping the consumer notice.

REFERENCES

Amory, D. E., Rouxhet, P. G. and Dufour, J. P. 1988. Flocculence of brewery yeasts and their surface properties: Chemical composition, electrostatic charge and hydrophobicity. *Journal of the Institute of Brewing*, 94, 79–84.

Armada, L. and Falqué, E. 2007. Repercussion of the clarification treatment agents before the alcoholic fermentation on volatile composition of white wines. *European Food Research and Technology*, 225, 553–558.

Ayala, F., Echávarri, J. F. and Negueruela, A. I. 1997. A new simplified method for measuring the color of wines. I. Red and rosé wines. *American Journal of Enology and Viticulture*, 48, 357–363.

Bacic, T. 2003. Recovery of valuable products from lees and integrated approach to minimize waste and add value to wine production. Australia: The University of Melbourne.

Bakalinsky, A. T. and Boulton, R. 1985. The study of an immobilized acid protease for the treatment of wine proteins. *American Journal of Enology and Viticulture*, 36, 23–29.

Batista, L., Monteiro, S., Loureiro, V. B., Teixeira, A. R. and Ferreira, R. B. 2009. The complexity of protein haze formation in wines. *Food Chemistry*, 112, 169–177.

Bayly, F. C. and Berg, H. W. 1967. Grape and wine proteins of white wine varietals. *American Journal of Enology and Viticulture*, 18, 18–32.

Berg, H. W. and Akiyoshi, M. 1961. Determination of protein stability in wine. *American Journal of Enology and Viticulture*, 12, 107–110.

Bisson, L. F. and Butzke, C. E. 2000. Diagnosis and rectification of stuck and sluggish fermentations. *American Journal of Enology and Viticulture*, 51, 168–177.

Blade, W. H. and Boulton, R. 1988. Adsorption of protein by bentonite in a model wine solution. *American Journal of Enology and Viticulture*, 39, 193–199.

Boulton, R., Singleton, V. L., Bisson, L. F. and Kunkee, R. E. 1999. *Principles and Practices of Winemaking*, New York: Kluwer.

Brindley, G. W. and Brown, G. 1980. *Crystal Structures of Clay Minerals and Their X-Ray Identification*. Mineralogical Society of Great Britain and Ireland.

Butzke, C. 2009. Fining with bentonite. Purdue University Extension (https://www.extension.purdue.edu/extmedia/fs/fs-53-w.pdf).

Carrasco, A. and Siebert, K. J. 1999. Human visual perception of haze and relationships with instrumental measurements of turbidity. Thresholds, magnitude estimation and sensory descriptive analysis of haze in model systems. *Food Quality and Preference*, 10, 421–436.

Dambrouck, T., Marchal, R., Marchal-Delahaut, L., Parmentier, M., Maujean, A. and Jeandet, P. 2003. Immunodetection of proteins from grapes and yeast in a white wine. *Journal of Agricultural and Food Chemistry*, 51, 2727–2732.

Di Gaspero, M., Ruzza, P., Hussain, R., Vincenzi, S., Biondi, B., Gazzola, D., Siligardi, G. and Curioni, A. 2017. Spectroscopy reveals that ethyl esters interact with proteins in wine. *Food Chemistry*, 217, 373–378.

Dizy, M. and Bisson, L. F. 1999. White wine protein analysis by capillary zone electrophoresis. *American Journal of Enology and Viticulture*, 50, 120–127.

Donovan, J. L., Mccauley, J. C., Nieto, N. T. and Waterhouse, A. L. 1998. Effects of small-scale fining on the phenolic composition and antioxidant activity of Merlot wine. Chemistry of wine flavor. *American Chemical Society Symposium*, 714, 42–155.

Dufrechou, M., Poncet-Legrand, C., Sauvage, F.-X. and Vernhet, A. 2011. Stability of white wine proteins: Combined effect of pH, ionic strength, and temperature on their aggregation. *Journal of Agricultural and Food Chemistry*, 60, 1308–1319.

Dupin, I. V. S., Stockdale, V. J., Williams, P. J., Jones, G. P., Markides, A. J. and Waters, E. J. 2000. *Saccharomyces cerevisiae* mannoproteins that protect wine from protein haze: Evaluation of extraction methods and immunolocalization. *Journal of Agricultural and Food Chemistry*, 48, 1086–1095.

Esteruelas, M., Poinsaut, P., Sieczkowski, N., Manteau, S., Fort, M. F., Canals, J. M. and Zamora, F. 2009. Comparison of methods for estimating protein stability in white wines. *American Journal of Enology and Viticulture*, 60, 302–311.

Falconer, R. J., Marangon, M., Van Sluyter, S. C., Neilson, K. A., Chan, C. and Waters, E. J. 2010. Thermal stability of thaumatin–like protein, chitinase, and invertase isolated from Sauvignon blanc and semillon juice and their role in haze formation in wine. *Journal of Agricultural and Food Chemistry*, 58, 975–980.

Ferreira, R. B., Piçarra-Pereira, M. A., Monteiro, S., Loureiro, V. L. B. and Teixeira, A. R. 2001. The wine proteins. *Trends in Food Science and Technology*, 12, 230–239.

Fisher, D. J. 1975. Flocculation—Some observations on the surface charges of yeast cells. *Journal of the Institute of Brewing*, 81, 107–110.

Gazzola, D., Van Sluyter, S. C., Curioni, A., Waters, E. J. and Marangon, M. 2012. Roles of proteins, polysaccharides, and phenolics in haze formation in white wine via reconstitution experiments. *Journal of Agricultural and Food Chemistry*, 60, 10666–10673.

Girbau, T., Stummer, B. E., Pocock, K. F., Baldock, G. A., Scott, E. S. and Waters, E. J. 2004. The effect of *Uncinula necator* (powdery mildew) and *Botrytis cinerea* infection of grapes on the levels of haze-forming pathogenesis-related proteins in grape juice and wine. *Australian Journal of Grape and Wine Research*, 10, 125–133.

Gómez-Plaza, E., Gil-Muñoz, R., López-Roca, J. M. and Martínez, A. 2000. Color and phenolic compounds of a young red wine. Influence of wine-making techniques, storage temperature, and length of storage time. *Journal of Agricultural and Food Chemistry*, 48, 736–741.

González-Neves, G., Favre, G. and Gil, G. 2014. Effect of fining on the color and pigment composition of young red wines. *Food Chemistry*, 157, 385–392.

Gougeon, R. D., Soulard, M., Miehé-Brendlé, J., Chézeau, J.-M., Le Dred, R., Jeandet, P. and Marchal, R. 2003a. Analysis of two bentonites of enological interest before and after commercial activation by solid Na_2CO_3. *Journal of Agricultural and Food Chemistry*, 51, 4096–4100.

Gougeon, R. D., Soulard, M., Reinholdt, M., Miehé-Brendlé, J., Chézeau, J.-M., Dred, R. L., Marchal, R. and Jeandet, P. 2003b. Polypeptide adsorption on a synthetic montmorillonite: A combined solid-state NMR spectroscopy, x-ray diffraction, thermal analysis and N_2 adsorption study. *European Journal of Inorganic Chemistry*, 2003, 1366–1372.
Harbertson, J. F., Hodgins, R. E., Thurston, L. N., Schaffer, L. J., Reid, M. S., Landon, J. L., Ross, C. F. and Adams, D. O. 2008. Variability of tannin concentration in red wines. *American Journal of Enology and Viticulture*, 59, 210–214.
Horne, J., Olabi, A., Greenwalt, C. and Lawless, H. T. 2001. Visual haze detection threshold assessment by ascending method of limits and a transformed staircase procedure in apple juice simulations under "store like" conditions. *Journal of Sensory Studies*, 16, 447–460.
Hsu, J.-C. and Heatherbell, D. A. 1987. Heat-unstable proteins in wine. I. Characterization and removal by bentonite fining and heat treatment. *American Journal of Enology and Viticulture*, 38, 11–16.
Hsu, J. C., Heatherbell, D. A., Flores, J. H. and Watson, B. T. 1987. Heat-unstable proteins in grape juice and wine. II. Characterization and removal by ultrafiltration. *American Journal of Enology and Viticulture*, 38, 17–22.
Jaeckels, N., Tenzer, S., Meier, M., Will, F., Dietrich, H., Decker, H. and Fronk, P. 2017. Influence of bentonite fining on protein composition in wine. *LWT—Food Science and Technology*, 75, 335–343.
James, A. M. 1982. The electrical properties and topochemistry of bacterial cells. *Advances in Colloid and Interface Science*, 15, 171–221.
Lagace, L. S. and Bisson, L. F. 1990. Survey of yeast acid proteases for effectiveness of wine haze reduction. *American Journal of Enology and Viticulture*, 41, 147–155.
Lambri, M., Dordoni, R., Silva, A. and De Faveri, D. M. 2010. Effect of bentonite fining on odor—Active compounds in two different white wine styles. *American Journal of Enology and Viticulture*, 61, 225–233.
Lambri, M., Dordoni, R., Giribaldi, M., Violetta, M. R. and Giuffrida, M. G. 2012a. Heat-unstable protein removal by different bentonite labels in white wines. LWT—*Food Science and Technology*, 46, 460–467.
Lambri, M., Dordoni, R., Silva, A. and De Faveri, D. M. 2012b. Comparing the impact of bentonite addition for both must clarification and wine fining on the chemical profile of wine from Chambave Muscat grapes. *International Journal of Food Science and Technology*, 47, 1–12.
Lira, E., Rodríguez-Bencomo, J. J., Salazar, F. N., Orriols, I., Fornos, D. and López, F. 2015. Impact of bentonite additions during vinification on protein stability and volatile compounds of Albariño wines. *Journal of Agricultural and Food Chemistry*, 63, 3004–3011.
Lira, E., Salazar, F. N., Rodríguez-Bencomo, J. J., Vincenzi, S., Curioni, A. and López, F. 2014. Effect of using bentonite during fermentation on protein stabilization and sensory properties of white wine. *International Journal of Food Science and Technology*, 49, 1070–1078.
Lubbers, S., Charpentier, C. and Feuillat, M. 1996. Etude de la rétention de composés d'arôme par les bentonites en moût, vin et milieux modèles. *Vitis* 35, 59–62.
Marangon, M., Sauvage, F.-X., Waters, E. J. and Vernhet, A. 2011a. Effects of ionic strength and sulfate upon thermal aggregation of grape chitinases and thaumatin-like proteins in a model system. *Journal of Agricultural and Food Chemistry*, 59, 2652–2662.
Marangon, M., Van Sluyter, S. C., Neilson, K. A., Chan, C., Haynes, P. A., Waters, E. J. and Falconer, R. J. 2011b. Roles of grape thaumatin-like protein and chitinase in white wine haze formation. *Journal of Agricultural and Food Chemistry*, 59, 733–740.
Marangon, M., Van Sluyter, S. C., Robinson, E. M. C., Muhlack, R. A., Holt, H. E., Haynes, P. A., Godden, P. W., Smith, P. A. and Waters, E. J. 2012. Degradation of white wine haze proteins by Aspergillopepsin I and II during juice flash pasteurization. *Food Chemistry*, 135, 1157–1165.

Marangon, M., Stockdale, V. J., Munro, P., Trethewey, T., Schulkin, A., Holt, H. E. and Smith, P. A. 2013. Addition of carrageenan at different stages of winemaking for white wine protein stabilization. *Journal of Agricultural and Food Chemistry*, 61, 6516–6524.

Marangon, M., Van Sluyter, S. C., Waters, E. J. and Menz, R. I. 2014. Structure of haze forming proteins in white wines: *Vitis vinifera* thaumatin-like proteins. *PLoS ONE*, 9, e113757.

Marchal, R. and Jeandet, P. 2009. Use of enological additives for colloid and tartrate salt stabilization in white wines and for improvement of sparkling wine foaming properties. In: Moreno-Arribas, M. V. and Polo, M. C. (eds.). *Wine Chemistry and Biochemistry*. New York: Springer.

Martinez-Rodriguez, A. J. and Polo, M. C. 2003. Effect of the addition of bentonite to the tirage solution on the nitrogen composition and sensory quality of sparkling wines. *Food Chemistry*, 81, 383–388.

Mierczynska-Vasilev, A., Boyer, P., Vasilev, K. and Smith, P. A. 2017. A novel technology for the rapid, selective, magnetic removal of pathogenesis-related proteins from wines. *Food Chemistry*, 232, 508–514.

Moio, L., Ugliano, M., Gambuti, A., Genovese, A. and Piombino, P. 2004. Influence of clarification treatment on concentrations of selected free varietal aroma compounds and glycoconjugates in Falanghina (*Vitis vinifera* L.) must and wine. *American Journal of Enology and Viticulture*, 55, 7–12.

Moretti, R. H. and Berg, H. W. 1965. Variability among wines to protein clouding. *American Journal of Enology and Viticulture*, 16, 69–78.

Muhlack, R., Nordestgaard, S., Waters, E. J., O'Neill, B. K., Lim, A., and Colby, C. B. 2006. In-line dosing for bentonite fining of wine or juice: Contact time, clarification, product recovery and sensory effects. *Australian Journal of Grape and Wine Research*, 12, 221–234.

Muhlack, R. A. 2006. Development of alternative bentonite treatments for heat-unstable white wine. Doctor of philosophy thesis, University of Adelaide.

Murphey, J. M., Spayd, S. E. and Powers, J. R. 1989. Effect of grape maturation on soluble protein characteristics of Gewürztraminer and white Riesling juice and wine. *American Journal of Enology and Viticulture*, 40, 199–207.

Nordestgaard, S., Chuan, Y. P., O'neill, B., Waters, E., Deans, L., Policki, P. and Colby, C. 2007. In-line dosing of white wine for bentonite fining with centrifugal clarification. *American Journal of Enology and Viticulture*, 58, 283–285.

Okuda, T., Fukui, M., Takayanagi, T. and Yokotsuka, K. 2006. Characterization of major stable proteins in Chardonnay wine. *Food Science and Technology Research*, 12, 131–136.

Pérez-Magariño, S., Ortega-Heras, M., Martínez-Lapuente, L., Guadalupe, Z. and Ayestarán, B. 2013. Multivariate analysis for the differentiation of sparkling wines elaborated from autochthonous Spanish grape varieties: Volatile compounds, amino acids and biogenic amines. *European Food Research and Technology*, 236, 827–841.

Peynaud, E. 1984. *Knowing and Making Wine*, New York: John Wiley & Sons.

Pocock, K. F. and Waters, E. J. 2006. Protein haze in bottled white wines: How well do stability tests and bentonite fining trials predict haze formation during storage and transport? *Australian Journal of Grape and Wine Research*, 12, 212–220.

Pocock, K. F., Alexander, G. M., Hayasaka, Y., Jones, P. R. and Waters, E. J. 2007. Sulfatea candidate for the missing essential factor that is required for the formation of protein haze in white wine. *Journal of Agricultural and Food Chemistry*, 55, 1799–1807.

Pocock, K. F., Salazar, F. N. and Waters, E. J. 2011. The effect of bentonite fining at different stages of white winemaking on protein stability. *Australian Journal of Grape and Wine Research*, 17, 280–284.

Powers, J. R., Nagel, C. W. and Weller, K. 1988. Protein removal from a wine by immobilized grape proanthocyanidins. *American Journal of Enology and Viticulture*, 39, 117–120.

Ramey, D., Bertrand, A., Ough, C. S., Singleton, V. L. and Sanders, E. 1986. Effects of skin contact temperature on Chardonnay must and wine composition. *American Journal of Enology and Viticulture*, 37, 99–106.

Ribéreau-Gayon, P., Glories, Y., Maujean, A. and Dubourdieu, D. 2006. *Handbook of Enology*. Volume 2. *The Chemistry of Wine Stabilization and Treatments*, Chichester, West Sussex, England: John Wiley & Sons Ltd.

Rowe, J. D., Harbertson, J. F., Osborne, J. P., Freitag, M., Lim, J. and Bakalinsky, A. T. 2010. Systematic identification of yeast proteins extracted into model wine during aging on the yeast lees. *Journal of Agricultural and Food Chemistry*, 58, 2337–2346.

Salazar, F. N., Marangon, M., Labbé, M., Lira, E., Rodríguez-Bencomo, J. J. and López, F. 2017. Comparative study of sodium bentonite and sodium-activated bentonite fining during white wine fermentation: Its effect on protein content, protein stability, lees volume, and volatile compounds. *European Food Research and Technology*, 1–12.

Sanborn, M., Edwards, C. G. and Ross, C. F. 2010. Impact of fining on chemical and sensory properties of Washington State Chardonnay and Gewürztraminer wines. *American Journal of Enology and Viticulture*, 61, 31–41.

Sarmento, M. R., Oliveira, J. C., Slatner, M. and Boulton, R. B. 2000. Influence of intrinsic factors on conventional wine protein stability tests. *Food Control*, 11, 423–432.

Sauvage, F.-X., Bach, B., Moutounet, M. and Vernhet, A. 2010. Proteins in white wines: Thermo-sensitivity and differential adsorbtion by bentonite. *Food Chemistry*, 118, 26–34.

Somers, T. C. and Ziemelis, G. 1973. The use of gel column analysis in evaluation of bentonite fining procedures. *American Journal of Enology and Viticulture*, 24, 51–54.

Sommer, P., Wegmann-Herr, P. and Fischer, U. 2015. Correlating the need for bentonite fining in wine with anomalous weather patterns. *Journal of Wine Research*, 26, 29–39.

Tian, B., Harrison, R., Morton, J. and Deb-Choudhury, S. 2015. Proteomic analysis of Sauvignon blanc grape skin, pulp and seed and relative quantification of pathogenesis-related proteins. *PLoS ONE*, 10, e0130132.

Toland, T. M., Fugelsang, K. C. and Muller, C. J. 1996. Methods for estimating protein instability in white wines: A comparison. *American Journal of Enology and Viticulture*, 47, 111–112.

Vanrell, G., Canals, R., Esteruelas, M., Fort, F., Canals, J. M. and Zamora, F. 2007. Influence of the use of bentonite as a riddling agent on foam quality and protein fraction of sparkling wines (Cava). *Food Chemistry*, 104, 148–155.

Van Sluyter, S. C., Marangon, M., Stranks, S. D., Neilson, K. A., Hayasaka, Y., Haynes, P. A., Menz, R. I. and Waters, E. J. 2009. Two-step purification of pathogenesis-related proteins from grape juice and crystallization of thaumatin-like proteins. *Journal of Agricultural and Food Chemistry*, 57, 11376–11382.

Vela, E., Hernández-Orte, P., Castro, E., Ferreira, V. and Lopez, R. 2016. Effect of bentonite fining on polyfunctional mercaptans and other volatile compounds in Sauvignon blanc wines. *American Journal of Enology and Viticulture*, 68, 30–38.

Vernhet, A., Pellerin, P., Prieur, C., Osmianski, J. and Moutounet, M. 1996. Charge properties of some grape and wine polysaccharide and polyphenolic fractions. *American Journal of Enology and Viticulture*, 47, 25–30.

Vincenzi, S., Panighel, A., Gazzola, D., Flamini, R. and Curioni, A. 2015. Study of combined effect of proteins and bentonite fining on the wine aroma loss. *Journal of Agricultural and Food Chemistry*, 63, 2314–2320.

Vincenzi, S., Polesani, M. and Curioni, A. 2005. Removal of specific protein components by chitin enhances protein stability in a white wine. *American Journal of Enology and Viticulture*, 56, 246–254.

Voilley, A., Lamer, C., Dubois, P. and Feuillat, M. 1990. Influence of macromolecules and treatments on the behavior of aroma compounds in a model wine. *Journal of Agricultural and Food Chemistry*, 38, 248–251.

Waters, E. J., Wallace, W. and Williams, P. J. 1992. Identification of heat-unstable wine proteins and their resistance to peptidases. *Journal of Agricultural and Food Chemistry*, 40, 1514–1519.

Waters, E. J., Pellerin, P. and Brillouet, J.-M. 1994a. A *Saccharomyces* mannoprotein that protects wine from protein haze. *Carbohydrate Polymers*, 23, 185–191.

Waters, E. J., Pellerin, P. and Brillouet, J.-M. 1994b. A wine arabinogalactan-protein that reduces heat-induced wine protein haze. *Bioscience, Biotechnology, and Biochemistry*, 58, 43–48.

Waters, E. J., Peng, Z., Pocock, K. F. and Williams, P. J. 1995. Proteins in white wine, I: Procyanidin occurrence in soluble proteins and insoluble protein hazes and its relationship to protein instability. *Australian Journal of Grape and Wine Research*, 1, 86–93.

Waters, E. J., Shirley, N. J. and Williams, P. J. 1996. Nuisance proteins of wine are grape pathogenesis-related proteins. *Journal of Agricultural and Food Chemistry*, 44, 3–5.

Waters, E. J., Alexander, G., Muhlack, R., Pocock, K. F., Colby, C., O'neill, B. K., Høj, P. B. and Jones, P. 2005. Preventing protein haze in bottled white wine. *Australian Journal of Grape and Wine Research*, 11, 215–225.

Waters, E. J. and Colby, C. B. 2009. Proteins. In: Moreno-Arribas, M. V. and Polo, M. C. (eds.) *Wine Chemistry and Biochemistry*. New York: Springer.

Weetall, H. H., Zelko, J. T. and Bailey, L. F. 1984. A new method for the stabilization of white wine. *American Journal of Enology and Viticulture*, 35, 212–215.

Weiss, K. C. and Bisson, L. F. 2002. Effect of bentonite treatment of grape juice on yeast fermentation. *American Journal of Enology and Viticulture*, 53, 28–36.

Wigand, P., Tenzer, S., Schild, H. and Decker, H. 2009. Analysis of protein composition of red wine in comparison with rosé and white wines by electrophoresis and high-pressure liquid chromatography–Mass spectrometry (HPLC-MS). *Journal of Agricultural and Food Chemistry*, 57, 4328–4333.

Zoecklein, B. W., Fugelsang, K. C., Gump, B. H. and Nury, F. S. 1995. *Fining and Fining Agents. Wine Analysis and Production*. Boston, MA: Springer.

4 Wine Aging and Spoilage

Pilar Rubio-Bretón, Teresa Garde-Cerdán, Juana Martínez, Ana Gonzalo-Diago, Eva P. Pérez-Álvarez, and Matteo Bordiga

CONTENTS

4.1 TANK

4.1.1 Maturation

The transformation from simple grape juice to the complexity of wine represents the aim of modern winemaking technology. However, winemakers, in order to control the entire process, have a multitude of options available, depending on all sorts of factors including grape variety, tradition, wine style, and cost. Immediately after fermentation, wines present rough and unpleasant notes. A period of maturation is required (e.g., for tannins to soften). The choice of maturation vessel and the maturation period depend upon the style of wine to be produced, including quality and cost factors. There are several types of maturation vessels, including stainless steel tanks and wooden barrels. Stainless steel represents an ideal storage material because of its properties (e.g., impermeable to gases such as oxygen). The wine therefore is stored until required for the bottling process. Of course, stainless steel appears ideally suited to temperature control and in applications where long-term, oxygen-free storage is required. Conversely, most high-quality red wines undergo a period of barrel maturation, usually between 9 and 22 months. During this period, the wine will undergo a controlled oxygenation process and absorb some oak products, including wood tannins and vanillin. Barrel size affects the wine maturation; commonly, the smaller the barrel is, the quicker is the maturation. Temperature also plays an important part (the lower the temperature is, the slower the maturation will be). Usually, white wine is stored in stainless steel or concrete tanks until ready for bottling. It is important to exclude oxygen in this phase, maintaining the tanks completely full.

4.1.2 Stainless Steel Tanks

Fermenting and storing wine into stainless steel represents a common way winemakers have to preserve crispness, acidity, and the purity of fruit. While oak is somewhat porous, stainless prevents all oxidation without imparting any new flavors. In a way, what goes in is what comes out: wines showing fresh, clean, vibrant, and robust fruit notes.

4.1.2.1 Closed Cylindrical Tanks

Tanks are designed for fermentation of grape must and storage of wine under atmospheric pressure. Tanks can be realized with different finished surfaces, such as marbled, brushed, sandblasted, or mechanically polished. Perfectly finished welds together with smooth surfaces may prevent the formation of wine stone and accumulation of impurities inside the tank. Commonly, tanks can be equipped with temperature-regulating elements (e.g., laser-welded heat exchangers, plate heat exchangers, cooling coil, and electronic temperature monitoring). Moreover, according to a customer's specifications, other equipment may be included (e.g., different manways, valves, legs with adjustable base, agitators, etc.). Another type of tank can be arranged into stacks, in order to make best use of the wine cellar space. To be stacked, tanks must have the same diameters and the carrying capacity of the basic tank must be strictly observed. Other modified tanks may show two or multiple compartments. The use of different combinations of these characteristics represents a good compromise in optimization of wine cellar space.

4.1.2.2 Open-Top Cylindrical Tanks

Tanks designed with a floating lid represent another type of commonly used tank. The floating lid can be easily adapted to the level of wine in the tank. Inflating the floating lid gasket with the enclosed pump prevents the uncontrolled inflow of air into the tank. The reinforced upper rim of the storage tank (piped reinforcement) gives strength and stability, even preventing any deformations of the tank jacket. Several modifications and integrations may be included, such as temperature-regulating equipment (e.g., floating lid with cooling, pillow plate on the tank jacket, cooling coil). Other technical equipment, such as valves, adjustable legs and agitators, may be present. These tanks appear suitable for storing small/medium quantities of wine and may be designed as flat-bottomed tanks without legs or conical bottom tanks with legs. Available capacity ranges from 300 to 1,000 L.

4.1.2.3 Rectangular and Oval Tanks

The main advantage of these tanks is the ability to make good use of cellar space. They are very suitable for smaller and narrower cellars. All inner edges are rounded and without any angle welds, thus presenting a smooth and easy-to-clean surface. The bottom is inclined, allowing an easy liquid discharge. Rectangular tanks may be designed with or without edge reinforcements. Tanks with the same width and depth can be arranged into stacks without any connective fitting. These fittings also enable forklift transportation of empty storage tanks. Oval storage tanks represent a type that is very suitable for low-ceiling cellars. If specified by customers, these tanks can be equipped with temperature-regulating systems and other special technical equipment (previously reported).

4.1.2.4 Insulated and Pressure Tanks

Insulated storage tanks appear suitable for wine storage, fermentation, stabilization, and other oenological processes that require a container with thermoisolative characteristics. In the cylindrical part, laser-welded heat exchangers (pillow-plate) are included. Insulation material (polyurethane foam; thickness is 50 mm) covers the inner tank surface. This insulation is covered with the outer insulation jacket. Another type of tank is designed for sparkling wine production (e.g., the Charmant method). The related second fermentation is conducted in large, closed, pressurized tanks. For this reason, these tanks are equipped with a valve to achieve isobaric conditions and a decantation valve to remove sediment. Tanks also have a safety valve, a level indicator, a manometer, and an agitator.

4.1.3 Clay/Terra-Cotta Amphorae

For thousands of years, terra-cotta containers called amphorae were the only option available to transform grape juice into wine. They originated in the European region now known as Georgia about 6,000 years ago. Today, a small (but growing) number of producers have adopted amphorae of varying sizes and origins (Figure 4.1). Winemakers who have switched to amphorae say the vessels are able to valorize the expression of their grapes and vineyard areas. Due to the porous nature of the material, a gentle microoxygenation represents the main benefit claimed by clay, but without adding oak flavor to the wine. Microoxidation means exposing wine to microscopic amounts of oxygen, too low to cause actual oxidation, but able to positively fix color and soften aggressive tannins. Oxidation itself represents a damaging reaction for most wine (except a few deliberately oxidized ones like Madeira, for example), leading to losses of fruity notes but often gaining notes of varnish or prunes, and browning. The amphorae, excellent for red wines, can be stored in a cellar, or buried in cool, damp earth for months at a time. The effect on the wine is something related to earthy and mushroom-like notes. The amphorae are realized completely by hand, following the rules of an ancient technique known as "based work." The amphora stands still while the artisan goes round and round attaching new clay. The whole process takes several days (months) including a long, natural drying and a slow firing. The salts and the calcium carbonates represent two crucial (natural) components able to confer the right microporosity to terra-cotta, thus enabling the wine to "breathe" correctly. All these features, combined with the antioxidant and antibacterial action

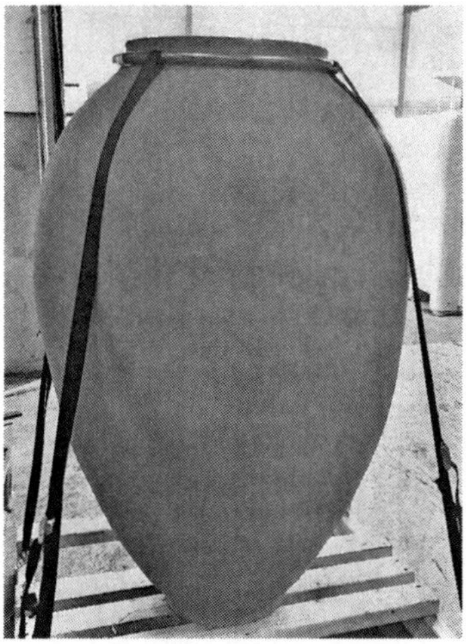

FIGURE 4.1 A brand new (handmade) terra-cotta amphora. (Courtesy of Beckham Estate Vineyard, Sherwood, OR.)

and the high thermal insulation capacity, make the terra-cotta amphora an ideal tool for wine production.

4.1.4 Concrete Tanks/Vats

The next closest material to clay in terms of its winemaking properties is concrete. Concrete has been used since the early twentieth century; traditionally it was the choice for huge vats and tanks in large-scale cooperative wineries. Concrete had the advantages of being cheap and easy to make into large vats to fill whatever space was available (Figure 4.2). Such vats appear quite neutral, especially if lined with glass tiles or epoxy resin. Concrete tanks/vats are characterized by a remarkable thermal inertia thanks to their thick walls. For this reason, temperature tends to stay quite steady, helping to avoid counterproductive spikes during fermentation. Disadvantages include being difficult to keep clean and lack of flexibility for producers who may need to work with smaller lots of wine (Figure 4.3). However, today's new-generation concrete tanks are often much smaller. Moreover, these are designed with hatches to allow access for cleaning and temperature control. Compared with

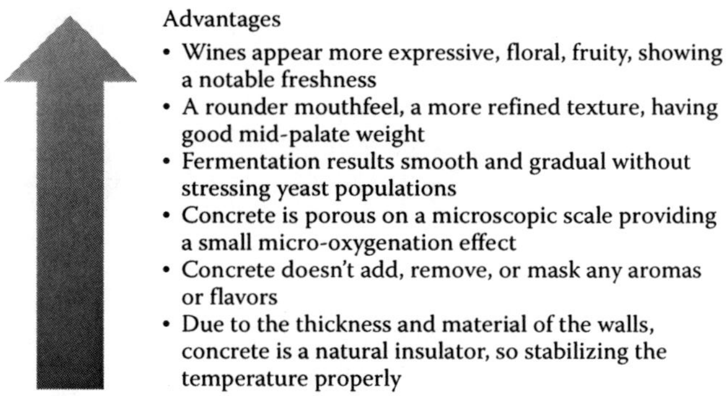

FIGURE 4.2 A summary of the advantages of making wines using concrete.

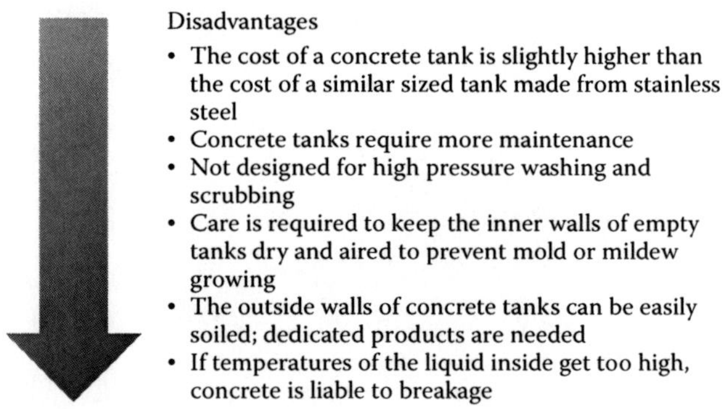

FIGURE 4.3 A summary of the disadvantages of making wines using concrete.

stainless steel, concrete tanks are more expensive. However, wine quality significantly improves, justifying the initial investment.

It must be highlighted that in the last decade, winemakers have rediscovered concrete tanks. It seems that Europe and the United States are on top in the uptake of larger format concrete fermenters and storage vessels. However, the last generation of concrete tanks needed scraping and rewaxing prior to vintage. This included allocating time and resources for intensive training sessions with cellar staff. There were also serious issues with tank walls cracking. Wine quality problems arose also if the wax was not applied evenly over the inside surface, causing dank or cement dust notes. In recent years, concrete tanks have changed into a fermenter and storage tank option, moving far beyond the tanks commonly used in the 1990s. Currently, any shape can be realized out of concrete (e.g., amphora, egg, teacup, pyramid, and elliptical). More conventional shapes also exist, such as square, cylinder, open-topped round, open-topped rectangular, and conical (Figures 4.4 and 4.5) Further modifications include optional ports for racking and cleaning, heating and cooling coils. In some cases a cooling system is hidden within the walls of the tank itself; others fit cooling plates to the inside of the tanks. New-generation tanks range from 500 up to 10,500 L.

Functionally speaking, concrete lies somewhere between stainless steel and oak barrels. Like stainless steel, these are neutral containers (though some claim they can impart a mineral note into a wine). As previously reported, one of concrete's biggest selling points is its thick walls, effective in stabilizing temperatures (especially useful during fermentation). New-generation tanks are moveable, offering further flexibility in winery layout changes as they are not fixed to the ground or built in-ground like many of the older generation ones. These are being used for fermentation and maturation of white and red wines. However, tanks used for red wines are not usually used for white wines, similarly to barrel use.

FIGURE 4.4 A typical example of some new-generation concrete tanks (NuBarrel™). (Courtesy of Sonoma Cast Stone, Petaluma, CA.)

FIGURE 4.5 Some brand new concrete tanks (NuBarrel™). (Courtesy of Sonoma Cast Stone, Petaluma, CA.)

4.1.4.1 Other Considerations

A typical suggestion from winemakers using concrete tanks to keep concrete tanks full at all times in order to keep the tanks "healthy." The combination of the porosity and slight unevenness of the inner surface may provide possible breeding sites for microbial contamination (e.g., *Brettanomyces*). By now, it is clear that there is a long, slow lag phase with the start-up of ferment in concrete compared with stainless steel. In terms of the performance of different shapes of the tanks, definitely the elliptical one (compared with egg one) is the best for white wine maturation due to a much larger surface area in contact with the lees. Wine quality was enhanced using this tank. Winemakers appreciate the easy cleaning protocols. The common concept of a concrete tank is that it is very rough and porous and thus very difficult to clean. The expectation is that extreme measures must be used to clean the surface. Conversely, for cleaning, usually 35°C water and a mild caustic cleaning solution, followed by a tartaric acid rinse, is used. Scrubbing the inner walls is not required. From an economic viewpoint, it seems evident there are benefits aside from wine quality that concrete has over stainless—mainly, less or no energy required to cool concrete down. Most importantly, the fermentation appears slower and more even in temperature, so cooling is not really needed. Moreover, the expected operational life for a concrete tank is 30 to 40 years or more.

4.1.5 Plastic Tanks for Wine Storage

Plastic tanks are made using food-grade polyethylene (Flextank®) with characteristics similar to those of a new wooden barrel in an oxygenation environment for wine maturation. These are designed for providing all barrel-like functions with more control by the winemaker, in a hygienic and long-lasting vessel. Of course, plastic tanks must not be considered as a substitute for stainless steel tanks. The best designs for wine storage tanks have important improvements over other production materials such as stainless steel, concrete, and wooden production materials. Figure 4.6 reports some of the features of tanks constructed with plastic. One of the most important attributes of plastic over stainless steel in the manufacture of tanks is its flexibility and relative strength compared to stainless steel. A plastic tank's flexibility enables it to hold liquids, changing volume based upon the thermal dynamics of the storage environment. Stainless steel tanks are rigid containers constructed with thin walls and the relative strength of such walls. Wine cannot be completely sealed in a stainless steel tank because the vast majority of winery tanks are not able to withstand the pressures due to thermal expansion.

In a winery, during the day, wine expands and contracts itself because of temperature variations. A water/alcohol mixture has a relatively steep percentage coefficient of thermal expansion per degree in the range of wine storage. Thus, the volume of wine (a water/alcohol mixture characterized by a rapid coefficient of thermal expansion per degree) increases as temperature goes up, and vice versa. Stainless steel tanks are stressed by this expansion and contraction process. Conversely, a plastic tank is more similar to a traditional oak barrel, better tolerating this phenomenon. Moreover, a plastic tank can be sanitized using 180°C water and/or strong chemicals. Thanks to a specific tank (22.7 hL), it is possible to make an efficient use of space in the cellar. These tanks can be stacked one on top of another and managed individually without moving them. Controlled permeation of oxygen through the walls appears as one of the most important aspects of plastic-designed tanks. Thanks to the polymers used in the construction of the tank, it is possible to have a controlled oxygen permeation similar to that of a 2-year-old wooden barrel.

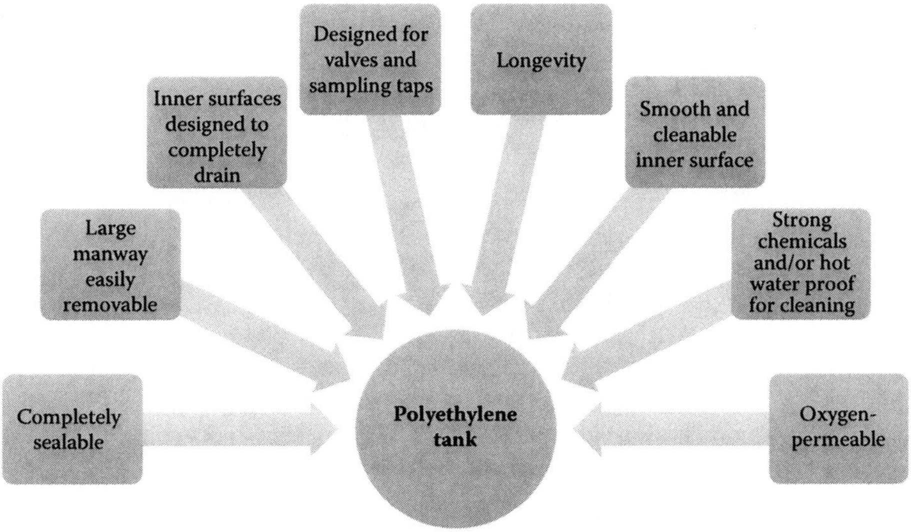

FIGURE 4.6 Features related to tanks constructed with plastic.

4.1.6 Square Barrel

Square barrels represent a recently developed alternative to traditional round barrels (Figure 4.7). These are characterized by a new design and manufacturing technology, due to the use of highest quality oak wood, able to affect the efficiency of production and quality of wine. These barrels achieve more efficient storeroom practice, owing to their angular shape, thus greatly increasing their storage capacity. Handling is simple due to the cubic shape, easily fitting on one Euro pallet. Stacked barrels do not need to be moved during use, and all kinds of activities can be done from the front. Space-saving transportation is another characteristic; in fact, barrels can be transported in a disassembled form where required. This appears very useful for large quantities and for shipping over long distances. Barrels are designed with a tapered angle of the floor and ceiling plate, thus allowing a quick filling and refilling at the same spot. Several levels of toasting are available for the construction of the barrel. The degrees of toasting can be adjusted to suit a customer's specifications thanks to new toasting technology. It is noteworthy that barrels may be recycled completely (e.g., reused by parquet flooring and furniture industries). The stainless steel bracket is also recycled.

4.1.7 Equipment for Oak Barrel Alternatives

As the use of oak barrel alternative products has become more popular, several companies are now designing specialized equipment to allow winemakers to incorporate oak staves and chips efficiently into their wine programs. The extraction process proceeds to a certain point before essentially stopping when chips or staves are suspended in wine (Figure 4.8). Oak alternatives are available in French or American oak in toasted or untoasted options as powder, chips, or tank staves. By running

FIGURE 4.7 A typical example of square barrels. (Courtesy of Azienda Agricola Erik Banti S.r.l., Scansano, Italy.)

FIGURE 4.8 Oak staves suspended within a concrete tank (NuBarrel™). (Courtesy of Sonoma Cast Stone, Petaluma, CA.)

wine through an extraction tank (a different format of barrel), the oak and wine interface is constantly changing. In this way, more flavor and aroma compounds are drawn from the oak. A typical 900 L (stainless steel) tank may be stacked and equipped with oxygen-permeable barriers slipped into two of the tank walls, thus providing diverse oxygen-transmission rates for wine maturation. The tank may be designed to work with staves, usually produced using staves from worn-out old barrels. Decommissioned barrels are broken down into staves. Successively these are stripped of any wine-saturated wood and retoasted with an infrared heat system. A fermentation tank may be developed with a stainless steel "basket" that can be filled with oak chips and staves. This appears to be a simple way to work and manage oak staves in tanks, especially because the "basket" can be removed with a forklift without draining the tank.

Winemakers are increasingly interested in continuing to discover the value of using oak alternatives including tannins, chips, and staves (maintaining quality while reducing costs). Moreover, growing interest is focusing on a program of microoxygenation during wine aging. Different microoxygenation systems may be programmed to use continual feedback and self-calibration to accurately dose oxygen by mass (not volume), thus ensuring the precise amount of oxygen needed. Using tannins in finished wines offers winemakers the opportunity to improve texture and structure and mask green characteristics, thus improving aromatic complexity and recapturing freshness. Due to their high antioxidant properties, tannins may also play a beneficial role in finished white wines. Low doses help maintain white wines' freshness, showing bright, vibrant flavors, especially when stored in tanks. Properly applied, tannins used in finished white wines may even help cut the overall levels of sulfur dioxide needed.

4.2 BARRELS

4.2.1 Origin

Oak has been used for centuries for the manufacture of barrels in which wine is preserved during the aging process. The predominance of oak in the manufacture of barrels is explained by its large presence in the forests of Europe, Asia, and North America and by its physical properties: resistance, flexibility, watertightness, natural durability (resistance to attacks by living organisms such as fungi or insects), and its ability to crack by splitting. In addition to its physical properties, the choice of this material was initially made because it was considered inert—that is, it does not add undesirable tastes to wine. Nowadays, it is known that oak barrels are not a mere container but contribute favorably to the organoleptic evolution of the great wines, especially red wines. The oak wood provides aromatic compounds to the wine that contribute shades of vanillin, carnation, toast, spices such as clove and pepper, smoked, licorice, etc. Also, the barrels allow the moderate oxidation of the wine that facilitates the joints between anthocyanins and tannins. These joints are necessary to the stabilization of the color and to the diminution of the astringency of the product. This oxygenation allows a favorable evolution of phenolic compounds, both visually and gustatorily. In spite of all these positive effects of oak on the quality of the wine, the aging period in barrels also implies loss of product by evaporation and by impregnation of the wood that forms the barrels, which influence, among other factors, the price of the wine.

4.2.2 Material

In oak wood, two distinct parts are distinguished: the components of the cellular wall and the extractable components.

Cellular wall components are

- Cellulose
- Hemicellulose
- Lignin

The extractable compounds are

- Phenolic compounds (lignans, coumarins, phenols, tannins)
- Aliphatic compounds (fatty acids, hydrocarbons)
- Terpene compounds
- Other compounds (lactones, carotenoids, norisoprenoids, etc.)

Cellulose is the most abundant component of vegetables. In oak it accounts for about 40% of the dry wood mass (Kolmann and Fengel, 1965). It is a homogeneous and linear polymer whose fundamental unit is a dimer, the diglucopyranose or cellobiose, formed by two units of glucose. It has rigid microfibrillary structure thanks to the hydrogen bonds established between the glucoses of the same chain or of

TABLE 4.1
Approximate Composition, in Percentage, of Oak Wood

Specie	Cellulose	Hemicellulose	Lignin
Quercus robur	41.1	22.2	29.6
Quercus sessilis	40.5	23.3	22.2
Quercus alba	50.0	22.0	32.0

Source: Masson, G., Puech, J.L., and Moutounet, M. *Bull. O.I.V.*, 1996, (785–786), 634–657.

neighboring chains (Barnoud, 1980). Cellulose is a very stable polymer, so it only reacts at high temperatures or using strong reagents. Hemicellulose accounts for about 25% of the dry wood mass (Kolmann and Fengel, 1965). The hemicellulose chains are attached to the cellulose fibers in the cellular wall. Hemicellulose is a complex polysaccharide consisting mainly of glucose, xylose, mannose, rhamnose, arabinose, and galactose. The hemicellulose can be partially hydrolyzed in aqueous media in the presence of catalysts (Zaitzeva and Nikitine, 1951). In the case of oak, acetic acid acts as a catalyst because oak wood contains about 3% of this acid that can be released (Nikitine, 1951). Lignin is the third largest constituent of wood and accounts for between 25% and 30% of the mass of dry wood (Kolmann and Fengel, 1965). It is a three-dimensional polymer formed by copolymerization of two phenylpropanoic alcohols: coniferyl alcohol and synapyl alcohol (Monties, 1987). The lignin of the oak wood is constituted by guaiacil and syringil units and is embedded in the cellulose fibers occupying the intercellular spaces, giving rigidity and impermeability to the cellular wall. Table 4.1 shows the proportion of these substances present in oak wood for the different species used in cooperage.

4.2.2.1 Extractable Compounds

Oak wood is rich in extractable organic compounds that account for up to 10% of the dry wood (Kolmann and Fengel, 1965). They are compounds of difficult classification since their nature is very varied; the phenolic compounds are the most abundant.

Phenolic compounds are classified as lignans, coumarins, simple phenols, and tannins. Lignans can be found in high concentrations in oak; linoresin is the most important compound in this group (Nabeta et al., 1987). Among the coumarins, scopoletin is the most abundant (Puech and Moutounet, 1988). Eugenol is the major composite among the volatile phenols of oak wood (Nishimura et al., 1983). Ellagic tannins are hexahydroxydiphenyl esters widely presented in dicotyledon angiosperms, especially in the genus *Quercus* (Seikel et al., 1971).

The aliphatic compounds are divided into hydrocarbons and fatty acids. The hydrocarbons present in the wood are alkanes and alkenes (Nishimura et al., 1983). The fatty acids, numerous in oak wood, are saturated and have an even number of carbon atoms (Nishimura et al., 1983). The terpene compounds found in this wood are two sesquiterpenes (Nabeta et al., 1986) and 23 terpenes (Nishimura et al., 1983). Other compounds also present in oak wood are lactones, the most important of which are the two isomers of β-methyl-γ-octalactone (also called oak lactone) (Masuda and Nishimura, 1971); among the carotenoids, lutein and β-carotene are the most important in oak wood (Masson, 1996). Regarding the norisoprenoids, Sefton et al. (1990) found up to 31 present in oak, among which β-ionone stands out.

4.2.3 Different Woods and Properties

Traditionally, the oak species most often used in cooperage are *Quercus alba* (American white oak), and two European species, *Quercus robur* L. (pedunculate oak) and *Quercus petraea* Liebl. (sessile oak). In general, American oak wood is richer than European in low molecular weight phenolic compounds, β-methyl-γ-octalactone, and some volatile compounds, but poorer in ellagitannins. Among European oak woods, *Q. robur* is richer in ellagitannins than *Q. petraea* (Prida and Puech, 2002). Recent studies have demonstrated that other botanical species, such as *Quercus pyrenaica* and *Quercus faginea* dispose phenolic and volatile composition comparable to chemical data reported for the most common oak wood species (*Q. robur* and *Q. petraea*) (Cadahía et al., 2001; Fernández de Simón et al., 2003). For instance, Gallego et al. (2012) showed that wine with chips and staves from *Q. pyrenaica* had higher aromatic intensity complexity—woody, balsamic, and cocoa—than American or French ones. Also, according to Chira and Teissedre (2015), wine aged in Slavonia oak wood showed characteristics similar and intermediate to the same wine aged in French and American oak woods, suggesting that this Slavonia oak wood is also suitable for barrel production of quality wine.

Oak species and their geographical origins play an important role in defining oak compositional differences and oxygen diffusion rates. For instance, the concentration of *cis*-lactone, which is one of the most odoriferous important compounds, is higher in American oak than in French oak (Garde-Cerdán, 2004; Sauvageot and Feuillat, 1999). Moreover, wines aged in American oak barrels have a higher *cis-*/*trans*-lactone ratio than those aged in French oak barrels (Garde-Cerdán and Ancín-Azpilicueta, 2006a; Garde-Cerdán et al., 2010; Gómez-Plaza et al., 2004). On the other hand, the natural rate of permeation of oxygen into oak barrels is lower for American oaks than for French oaks, probably due to the higher porosity of French woods (Nevares and del Álamo, 2008).

In a comparative study carried out with *Q. alba, Q. sessilis*, and *Q. robur* species, Chatonnet and Dubourdieu (1998), using synthetic wine and shavings of these oaks, found that the *Q. robur* species had a much higher content of extractable components than *Q. sessilis* and *Q. alba* species. Additionally, *Q. robur* species showed twice the ellagic tannin content as the *Q. sessilis* species. Likewise, *Q. sessilis* had twice the extractable polyphenol content as the *Q. alba* species. Regarding the concentration of eugenol, the authors found no significant differences between the three species mentioned. The *Q. alba* species presented greater aromatic potential than French oaks mainly due to its high content in oak lactones. Marco et al. (1994) conducted a study to establish relationships between the geographic origin of the oak and its chemical composition. For this purpose, they used 8 oak samples from the United States and 39 oak samples from six different regions of France. These authors found that French oak had higher concentrations of extractable phenols than American oak. They also observed that American oak presented greater wealth of oak lactones than most of the French oaks, with the exception of oak from the Nevers region, where the concentration of these compounds was comparable to that found in American oak. Therefore, wood composition is highly influenced by the oak species. French oak species have fewer differences among themselves than American oak species. Many authors (Chatonnet, 1995; Chatonnet and Dubourdieu, 1998; Feuillat, 1982; Feuillat et al., 1999; Miller et al., 1992; Monties, 1987; Pontallier et al., 1982; Sefton et al., 1993) agree that American oak is characterized by a smaller amount of tannin than French oak, as well as high concentrations of volatile compounds, mainly *cis*-oak lactone (Rubio-Bretón et al., 2013). The wood of *Q. robur* has a high content of extractable compounds and is low in volatile compounds, while *Q. sessilis* is richer in aromatic but poor in phenolic compounds (Chatonnet, 1992). In addition to the oak species, the geographic origin and silvicultural treatments together with the different processes, from the cutting of wood to the barrel is finished, have also significant influence in the wood composition.

4.2.4 Production Process

Barrel manufacture involves a first stage of selecting and cutting of the wood, followed by drying, shaping, toasting, and finishing (Figure 4.9).

4.2.4.1 Selecting, Cutting, and Drying

Only the lower trunk, free of branches and knots of straight trees, must be used to obtain the wood necessary to manufacture barrels. The trunks, which have been previously selected in the forest, must be processed to obtain the staves. First, the trunks are cut into pieces and the bark and sapwood removed, since in cooperage only the heartwood is used. Due to the different structures of its heartwood, American oak, more dense and with low porosity and permeability, can be sawed; however, the more porous European oak needs to be split in order to respect the natural grain. Sawing European oak would make it porous and affect its tightness (Navarro et al., 2016). This difference in the cut between European and American oaks will condition the evolution of the physical and chemical properties of the wood during drying and toasting (Cadahía et al., 2008; Fernández de Simón and Cadahía, 2007). Later, the staves must be subjected to a drying process, to ensure the mechanical resistance of the barrels and to eliminate some substances that contribute to bitterness and astringency of wines (Masson et al., 2000). Drying wood usually takes place naturally, during a variable time between 2 and 3 years in the open air. To make the drying period shorter, artificial drying processes have been introduced, in an oven with humidity and temperature controlled for 3–4 weeks, or mixed processes, starting with drying in the open air for a short period (12–15 months) and then finishing in an oven (Martínez et al., 2008).

4.2.4.2 Shaping

The shaping of the staves is the process by which the barrel adopts its characteristic form. For this process, heat and humidity application are required. The major

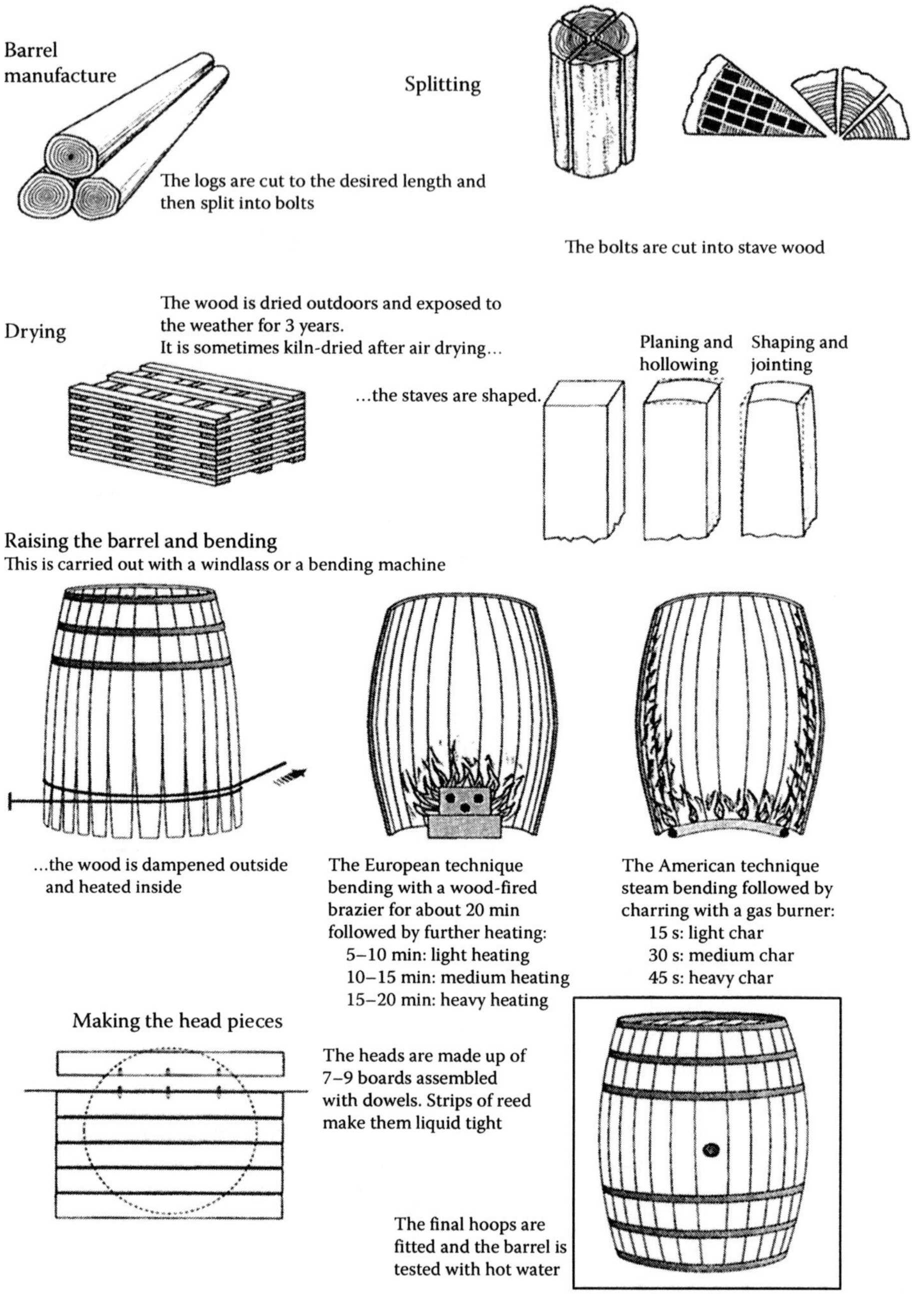

FIGURE 4.9 Diagram of barrel manufacture. (From Jackson, R.S. In Taylor, S.L., ed. *Wine Science. Principles and Applications.* New York: Academic Press, pp. 452–488, 2008.)

components of the wood can be deformed by the action of temperature or humidity: The heating acts on the plasticity of the lignin, which allows bending without breaking the staves by heat. Also, cellulose and hemicellulose can be deformed by humidity. Different heating systems, such as fire braziers, gas, steam or hot water, electric resistance, infrared waves, etc., exist. The nature, homogeneity, intensity, duration, and type of heating, as well as the periodicity of humidification, are parameters that decisively influence the quality of the barrel. On the other hand, there are companies that manufacture square oak barrels, in order to reduce the storage space and facilitate their cleaning. These square barrels contain the same quantity of oak as the traditional barrel and are toasted in order to emulate the results of the traditional cooperage process. But, because of their shape, the surface area of oak in contact with the wine is slightly smaller. This may not be undesirable but it is nonetheless different. A second difference concerns the lees—sediment that collects at the bottom of the barrel. The curved shape of the traditional barrel gives the lees a lesser contact surface area with the wine than in a flat-bottomed container. No scientific articles on the use of square barrels have been found, but experiences are known in some Spanish research centers (Instituto de Ciencias de la Vid y del Vino [ICVV] and Estación de Viticultura y Enología de Navarra [EVENA]) that work with these types of barrels. The results obtained with the square oak barrels were not appropriate, since the cleaning of the barrels was not easier, and wine leakage occurred in them.

4.2.4.3 Toasting

Once the shaping time is finished, in order to shape the barrel and to favor the formation of aroma substances, the barrel is subjected to a second toasting without humidity. The degree of toasting varies depending on the cooperage, but can be globally classified as light, medium, or heavy, depending on its duration and the temperature used; this obtains different sensory styles (Chatonnet, 1992; Rodríguez-Rodríguez and Gómez-Plaza, 2011). During this stage, the physical changes are accompanied by an evolution of the chemical composition of the wood, and this is the stage that provokes the most severe influence on these wood chemical composition modifications.

Therefore, the thermal treatment causes thermodegradation of some components of oak wood, which produces numerous volatile compounds. Furanic compounds are formed through thermal degradation of carbohydrates; volatile phenols come from the thermal degradation of lignin, and oak lactones are products of the dehydration of the acids present in wood. Medium toasting corresponds to the maximum synthesis of these volatile compounds (Koussissi et al., 2009). The chemical structures of the principal oak volatile compounds are shown in Figures 4.10 and 4.11. Additionally, with this thermal process there is a decrease in the soluble dry extract, which, linked to the degradation of the phenolic compounds, particularly of the ellagitannins (Chatonnet, 1992), minimizes the tannic contribution of the wood to the wines. On the other hand, toasting partially degrades the polymers of the wood, resulting in the formation of numerous aromatic compounds. In general, summarizing the aforementioned, the toasting of the wood produces a significant decrease of the astringency, accompanied by a remarkable increase of aromatic substances, some of them already present in the dry oak wood and others of new formation coming from the thermolysis of the heated wood (Martínez, 2004; Ribéreau-Gayon et al., 2003; Vivas, 2005).

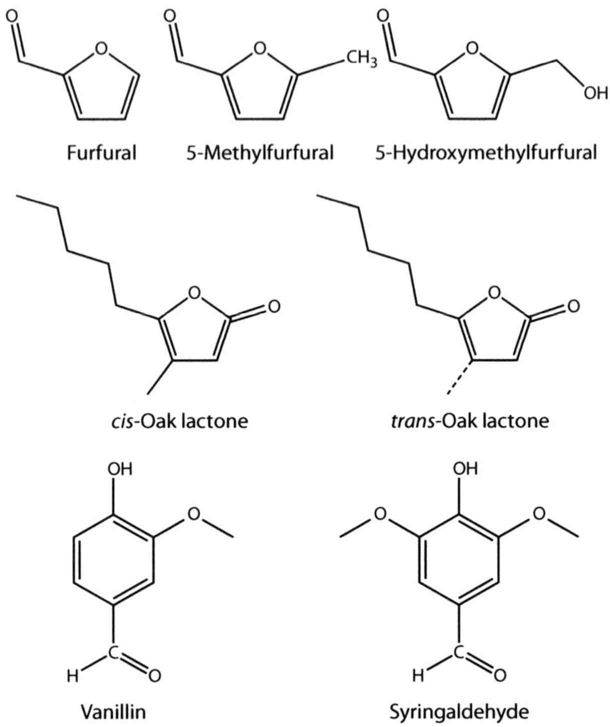

FIGURE 4.10 Chemical structures of the principal oak volatile compounds.

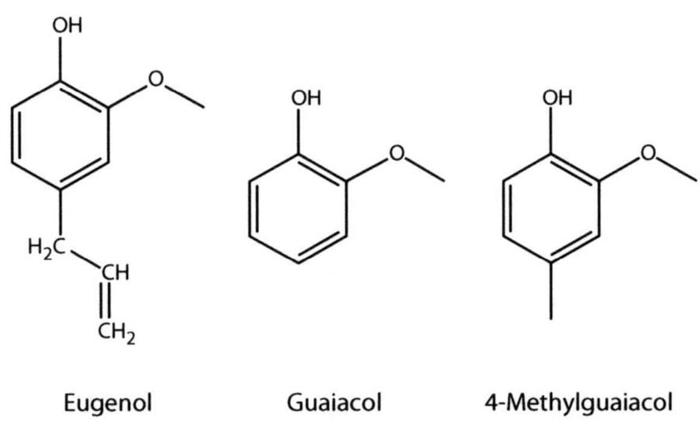

FIGURE 4.11 Chemical structures of the principal oak volatile compounds.

The β-methyl-γ-octalactone content in oak wood increases because of toasting; the ratio between *cis*- and *trans*-isomers is not affected by this process (Guichard et al., 1995). However, according to the results obtained by Bosso et al. (2008), the *cis*-/*trans*-oak lactones' ratio varied depending on the toasting conditions used in the production of the barrels, and their concentration tends to decrease as temperature increases. The same authors observed that the length of time and the temperature of the toasting process influenced the concentration of vanillin, furfuryl alcohol, *cis*-oak lactone, and 5-methylguaiacol. Herrero et al. (2016) evaluated the influence by different toasting levels in white wines aged in French oak barrels and observed that the maximum concentrations of furfurals, guaiacols, and vanillins were found in barrels at medium toasting levels. Chira and Teissedre (2013) observed different rates of extraction of volatile and nonvolatile compounds, depending mainly on the origin of the compounds in the wood (toasting or present in natural wood) as well as on the watering process during toasting. In the same way, wines in contact with wood subjected to watering during toasting were lower in ellagitannin concentrations. In addition, Chira and Teissedre (2015) observed that the watering process during toasting enhances furanic compounds, vanillin, and oak lactone extraction. Regarding sensory analysis, these authors observed that wines matured in medium toasting barrels were characterized with more vanilla flavor, and wines in light toasting barrels were perceived as less sweet and more astringent than the other samples.

Once the barrel is internally toasted, it is pressed, in order to acquire its definitive dimensions. Also, the caps or bottoms are placed. Then, the galvanized steel strips are placed and the filling or dug hole is drilled. Finally, watertightness tests with compressed air and external polishing are carried out. The barrel is put on the market after being cleaned, marked, and packed in a plastic film (Hidalgo, 2003).

4.2.4.4 Recycling

Since barrels contain finite extractable material, the amounts of volatile compounds available for extraction diminish with time (Garde-Cerdán et al., 2002; Garde-Cerdán and Ancín-Azpilicueta, 2006b; Peŕez-Prieto et al., 2002). As a consequence, most barrels are decommissioned after 5 or 6 years (Li et al., 2015). Used barrels can be rejuvenated by scraping a thin layer from the barrel's interior to expose fresh, unextracted wood (Mosedale et al., 1999). While this practice can prolong the life of a barrel, rejuvenated wood generally contains lower amounts of extractable material and contributes less aromatic compounds to the wines than new barrels. On the other hand, the high porosity of wood of the barrels encourages the proliferation of spoilage microorganisms and impedes their eradication (Renouf and Lounvaud-Funel, 2005). Washes performed between fills usually only remove lees and loose deposits; tartrate deposits remain, as would any microbe entrapped therein (Jiranek et al., 2008). Therefore, it is clear that, to obtain wine aged in wood with high-quality characteristics, proper sanitation of the barrel before filling is an essential step.

Nowadays, the most applied methods for barrel decontamination are based on the use of chemical additives and the application of physical processes (sulfur dioxide, ozone, aqueous steam, UV irradiation, microwaves, and ultrasound). However, these technologies are frequently borrowed from other contexts and thus lack specific validation, which brings to mind the peculiar features of barrels, such as wood composition, degree of cleanliness, and age (Guzzon et al., 2011). Among the chemical additives, sulfur remains the most widely used method, both through the burning of capsules of sulfur within the barrel and through washing with solutions of liquid sulfur dioxide (Costantini et al., 2016). In the study of Guzzon et al. (2011), four treatments were tested in order to know their efficacy in eradicating the yeasts inside the barrels, as well as to improve in this way the microbiological control of the yeasts: aqueous steam, UV irradiation, gaseous ozone, and aqueous ozone. The results indicated that aqueous steam and ozone were the most effective treatments, removing about 70% of the barrels' yeasts. UV appeared to be less effective, probably due to the porous nature of wood, which protects cells from direct irradiation. The results of Marko et al. (2005) and Malfeito-Ferreira et al. (2004) also support the use of ozone as a good alternative sanitizing agent for oak wine barrels. On the other hand, Falguera et al. (2013) proved that UV irradiation was effective in partially inactivating polyphenol oxidase and in reducing volatile acidity (an indirect spoilage measure) in white wines, but important aromatic compounds of wines were affected.

González-Arenzana et al. (2013) studied a microwave technology application investigating the ability of a new equipment prototype, based on high-frequency microwaves, to reduce microbial populations in oak wine barrels. In this study, it was demonstrated that microwave treatment was able to significantly decrease the main wine-associated microorganism populations from the first 8 mm of oak surfaces. Also, fouling of the interior surface of wine barrels by precipitated tartrate crystallization is a significant problem. Tartrate sediment forms a robust scale inhibiting oak compound extraction by wine, additionally harboring spoilage microorganisms and contaminants. Tartrate scale is typically removed by spray techniques, but often with not very good results. It was demonstrated by Porter et al. (2011) that high-power ultrasound is a novel cleaning application that can significantly remove tartrate deposits from the first 2 mm of oak surfaces; however, at depths of 2–8 mm, this cleaning technique was not reproducible under the parameters used in the study. Recently, Costantini et al. (2016) evaluated the effect on wine of dry ice blasting, used for the regeneration treatment of barrels, from microbiological, chemical, and sensory points of view. This technique allowed removal of deposits on wood, and therefore, it was able to regenerate the barrels. On the other hand, wine aged in a barrel treated with dry ice increased its aromatic profile and allowed a good sanitization of the wood, comparable to that of sulfur dioxide treatment.

4.2.5 Oak Alternatives

Wine aging in wooden barrels is one of the most common methods in the winemaking process that allows enhancing the properties of high-quality wines. During oak wood contact, wine undergoes different modifications, such as spontaneous clarification, CO_2 elimination, slow and continuous oxygen diffusion through the wood pores of the oak barrel, and extraction of wood compounds into the wine, which increases its stability and modulates its organoleptic quality and complexity such as aroma, structure, astringency, bitterness, and color persistence. However, due to the disadvantages of the traditional aging technology, such as the high price of oak barrels, the long aging time required, the large space taken up by barrels, and the labor that is required for the system maintenance, innovative aging technologies that artificially accelerate process and reduce costs have been developed.

One of these techniques consists of adding small pieces of oak wood, usually known as oak chips, to the wine. This practice was approved and legislated by the European Union (CE 2165/2005 and CE 1507/2006) and, consequently, the addition of oak chips has been generalized. Regulation of the European Union established that the size of the wood pieces should be such that at least 95% of them must be retained by a sieve of 2 mm. Some authors have studied the effect of chips on the phenolic composition of different wines, whereas others have focused on sensory analysis or on the volatile compounds, mainly on those extracted from the wood. Most studies conclude that the use of wood fragments can enhance the extraction rate of wood-related volatile compounds and accelerate red wine aging, stabilizing color in a shorter time than that necessary in the case of aging in barrels (Baiano et al., 2016; Chatonnet, 2007; Espitia-López et al., 2015; McCord, 2003; Rubio-Bretón, 2014). Nowadays, there is a great diversity of commercial products with different variables (botanical and geographical oak origin, size of fragments, toasting level, etc.) and different possibilities of use (moment of application, length of contact time, doses, joint application of microoxygenation, etc.), which will influence the changes, mainly due to the aromatic and polyphenolic compounds produced in the wine in contact with the wood.

As well as in the case of the cooperage, in the development of alternative products, wood from different geographical and botanical origins is used. Different shapes and sizes of oak wood pieces can be found on the market: chips, cubes or beans, powder, shavings or granules, dominoes, blocks, staves, tablets, or segments. The size of wood fragments is closely related to the extraction kinetics of wood-related compounds. Thus, with smaller oak pieces, the extraction of oak-derived compounds is much faster due to a larger contact surface (Campbell et al., 2006; Fernández de Simón et al., 2010). Although only the use of oak and chestnut woods is approved by the International Enological Codex of the International Organisation of Vine and Wine to wine aging (Basalekou et al., 2015), several studies have been carried out with other types of wood. Thus, Fernández de Simón et al. (2014) studied whether the nonanthocyanic phenolic profile could be a useful tool to identify wines aged in contact with cherry, chestnut, and acacia wood, either in barrels or with chips. There are differences in volatile composition of aged wines with woods of different botanical species or origins. In this way, a red wine macerated with Spanish chips (*Q. pyrenaica* and *Q. petraea*) was richer in furanic aldehydes and eugenol, whereas those macerated with American chips (*Q. alba*) contained higher concentrations of *cis*-oak lactone, vanillin, and methyl vanillate (Rodríguez-Bencomo et al., 2009). Other authors (del Álamo and Nevares, 2006) observed that with using French oak chips, the aging of wine was slower than with American or Hungarian oak chips.

Similarly to oak barrels, toasting is a conventional pretreatment of wood fragments of unquestionable importance to the generation of volatile substances that favorably enrich the aromatic profile of wines. The toasting level of wood fragments is related to the production of wood-related volatile compounds; thus, through controlling toasting level of wood fragments, desired sensory characters of wine can be obtained. According to Dumitriu et al. (2016), chips untoasted or with medium toast level showed a higher phenolic content, highlighting the phenolic families related to color properties and with astringency, compared to chips with medium-plus and heavy toasting degree. On the other hand, Guchu et al. (2006) and Rodríguez-Bencomo et al. (2009) detected that the concentration of aromatic compounds contributed by the wood fragments increased with the toasting degree. Furthermore, Chira and Teissedre (2013) showed that the oak chips with different types of toasting (untoast, light, light-plus, medium, medium-plus, and heavy) were very well discriminated according to their chemical composition and sensory assessment.

Regarding the time of application, oak alternative products can be used during the elaboration of the wines (white or red) and/or during the storage of the wines (in stainless steel tanks or used barrels). Gordillo et al. (2016) studied the application of oak chips during red must fermentation and observed that shorter maceration times enhanced phenolic extraction, color, and its stabilization in comparison to the traditional maceration. In the work of Baiano and de Gianni (2016), the oak chips were added before or after malolactic fermentation and no differences in the color parameters were found among control and treated wines. However, the wines treated before malolactic fermentation showed softer oak, spicy characters and more fruity notes than the wines treated after malolactic fermentation.

According to Gómez García-Carpintero et al. (2011), the use of wood chips gives rise to different sensory characteristics depending on at what point of the winemaking process the chips are added. The length of contact time of wine aging with oak chips usually ranges from a couple of weeks to several months. The contact time between wine and wood affects the sensory and chemical characteristics of wines, especially their wood-related volatile composition. The increased surface area of oak alternatives, compared to barrels, results in greater rates of wood-related volatile compound extraction, so both the quantity of oak and the length of contact time can be reduced in the presence of wood fragments compared to those of the barrels (Tao et al., 2014). Dumitriu et al. (2016) observed that time is the primary factor to differentiate among the wines obtained with chips of different degrees of toast and dose used. In this sense, no significant differences were obtained among wines aged for a short period of time (1.5 months), although at 3 months the wines can be clearly distinguished. The results showed by Kyraleou et al. (2015) allowed differentiating the samples according to the contact time based on their phenolic composition and color parameters, regardless of the types of wood chips added into the wines. On the other hand, the dose of alternative products mainly depends on the fragment's size. Usually, the dose used range between 1 and 5 g/L, but sometimes up to 20 g/L. Increasing the dose is usually accompanied by an increase in the wood notes provided to the wine (Gómez García-Carpintero et al., 2011). Moreover, Gordillo et al. (2016) found that increasing chip dose during red wine fermentation resulted in wines with lighter and less stable color. Otherwise, the oxidative process that occurs during wine aging in barrels does not take place when oak fragments are added to wines contained in stainless steel tanks.

For this reason, the use of wood alternatives is often combined with a microoxygenation process (addition of small and controlled amounts of oxygen into wine). Several studies claim that the effect of microoxygenation on wine color is similar to that which occurs during maturation in oak barrels (Atanasova et al., 2002; Cano-López et al., 2007). Oberholster et al. (2015) showed that microoxygenation in combination with wood alternatives such as oak chips and staves could be seen as relatively similar to wines aged in new oak barrels for a short time (6 months) in regard to sensory characteristics. The evolution of oxygen consumption in wine depends on the toasting levels, the origin, and the size of wood fragments. In this way, wines treated with French oak (*Q. petraea*) need more oxygen than those treated with American (*Q. alba*) and Spanish (*Q. pyrenaica*) oak. Moreover, the oxygen demands in wine is enhanced if wood fragments with bigger size are used during aging (Nevares and del Álamo, 2008; del Álamo et al., 2010).

On the other hand, Tao et al. (2016) studied the influence of high hydrostatic pressure processing in parallel with oak chip maceration on the physicochemical and sensory characteristics of a young red wine. The results demonstrated that high hydrostatic pressure processing was able to enhance the extraction of oak-related compounds from oak chips into wine and simultaneously to increase the wine antioxidant activity. There are several works that try to discriminate between wines aged in oak barrels and macerated with oak fragments. Basalekou et al. (2015) studied a method for the discrimination between wines aged in different barrels and in tanks with wood fragments for different time periods, using Fourier transform infrared spectroscopy (FT-IR) combined with discriminant analysis. Hernández-Orte et al. (2014) found that vanillin and acetovanillone are the compounds that explain the variance of wines treated with wood fragment, since they are present in higher concentrations than in wines aged in barrels, while the concentration of eugenol is significantly high in wines aged in barrels. So, the [(vanillin + acetovanillone)/eugenol] ratio is lower than 20 when wines are aged in barrels, while it is higher than 20 when the wines are macerated with wood fragments. A study evaluated the preferences of consumers for wines aged in oak barrels or for wines with oak chips (Pérez-Magariño et al., 2011). The authors observed considerable disparity in consumers' wine preferences, but since consumers did not significantly reject wines made with oak chips, they concluded that markets exist for wines using both oak maturation regimes. The application of wood fragments can be regarded as a good alternative to barrels for producing short-term aged wines with satisfactory quality and also may be an appropriate option for use during fermentation processes (Rubio-Bretón, 2014). However, in some cases the sensory quality of wines aged with wood fragments cannot be as good as those long-term aged wines in new barrels (Tao et al., 2014).

4.3 BOTTLES

4.3.1 Preparing Wine for Bottling

Different treatments may be undertaken to ensure that clarity remains stable once wine has been bottled. Even if treatments should be kept to a minimum, it has been previously reported that racking is one of the operations necessary to achieve clear wine, reducing the risk of off-flavors. It is important to protect wine from oxidation during the racking process (e.g., spreading vats or barrels with inert gases). Different operations may be undertaken to ensure this result.

4.3.1.1 Fining

Colloids represent the remaining lighter matter suspended in the wine after the coarse sediment has been removed by racking or centrifuge. Colloids are capable of passing through any filter and, if not removed, may cause the wine to look "hazy," forming a deposit. These are electrostatically charged compounds and can be removed by adding another colloid with the opposite charge. Egg whites, gelatin, isinglass (obtained from swim bladders of fish), and bentonite represent examples of fining agents. Fining may also be used to remove excess tannin, thus enhancing wine taste.

Polyvinylpolypyrrolidone (PVPP) is able to absorb phenolic compounds. This may be used at the fining stage to remove color from white wines, thus preventing a browning process.

4.3.1.2 Filtration

Filtration is the process used to remove solid particles. This may take place at different stages in winemaking (e.g., must or lees filtration). However, one of its main applications is in preparation for bottling. The processes of fining and filtration appear not to be interchangeable. The filtration process requires care and expertise. However, some winemakers choose to use minimum filtration to avoid depleting the "body" of the wine. There are three principal categories of filtration, which may be applied at different stages in the winemaking process.

4.3.1.3 Sheet Filtration

A series of specially designed, perforated steel plates are held in a frame. Sheets of filter medium (cloth or paper) are suspended between the plates. Successively, plates are tightly squeezed together by screws or hydraulic methods. Different ranges of porosity exist for the filter sheets. Then, wine is pumped between pairs of plates to pass through the filter sheets into a cavity. The fibers of the filter media are able to keep yeast cells and other matter contained in the wine.

4.3.1.4 Earth Filtration

Usually this approach is used for initial rough filtration and can remove large quantities of gummy solids, which consist of dead yeast cells and other matter from the grapes. The filtration takes place in two different stages. Firstly, a coarse-grade earth called diatomaceous (or kieselguhr), which is commonly used as the filter medium, is deposited on a supporting screen within a filter tank. A mixture of water and diatomaceous earth may be used to develop the filter bed. This is known as precoating. Secondly, further earth is mixed with wine, forming a slurry used to replenish the filtration surface through which the wine passes (continuously). During the passage of wine through the filter, it is quite frequent that the latter becomes clogged. In this case, diatomaceous earth must be completely replaced with fresh material. The rotary vacuum filter consists of a large horizontal cylinder characterized by a perforated screen covering the curved surface. A sophisticated filter may be applied to the enclosed rotary vacuum filter, reducing the risk of oxidation damaging the wine, but with this typology, the fouled earth must be removed manually.

4.3.1.5 Membrane Filtration

If applied, this is the final filtration process and is used just before bottling. The wine needs to have been properly clarified by other filtration approaches before using this process. Membrane filters are made of porous materials (plastic or ceramic), allowing water and compounds of small molecular size (e.g., alcohol, color, and flavor compounds) to pass through. Wine is passed under pressure across the surface to filter through, leaving undesirable materials trapped on the membrane (large compounds such as proteins). The process is not usually applied to full-bodied red wines as this may reduce body and flavors.

4.3.1.6 Stabilization

Stabilization may be carried out to prevent tartrate crystals forming after the wine has been bottled. The tartrates (either potassium or calcium salts of tartaric acid) are harmless. Sometimes these are found on the cork or as sediment in the bottle, causing unwarranted concern to consumers. Usually, wine is chilled to –4°C to inhibit the precipitation of tartrate crystals in bottles. In about a week the crystals are formed, thus allowing the cleared wine to be bottled. Another approach consists of reducing the temperature of the wine to approximately 0°C and then adding finely ground tartrates, followed by vigorous stirring. The small crystals are able to attract further ones, completing the entire process in about 24 hours.

4.3.1.7 Adjustment of Sulfur Dioxide Levels

Sulfur dioxide represents the winemaker's generally used antioxidant and disinfectant. However, only a certain percentage of this compound, the "free" SO_2, acts in this way. The other percentage remains "bound" with the wine and is not active. Before bottling, the free SO_2 levels should be properly adjusted to between 25 and 35 mg/L. Higher levels are required for sweet wines to inhibit further fermentation of the sugars.

4.3.1.8 Bottling

Several treatments may be carried out immediately prior to bottling to ensure the wine's final stability. These include pasteurization or sterilization. Four available systems of treatment are commonly used (Figure 4.12).

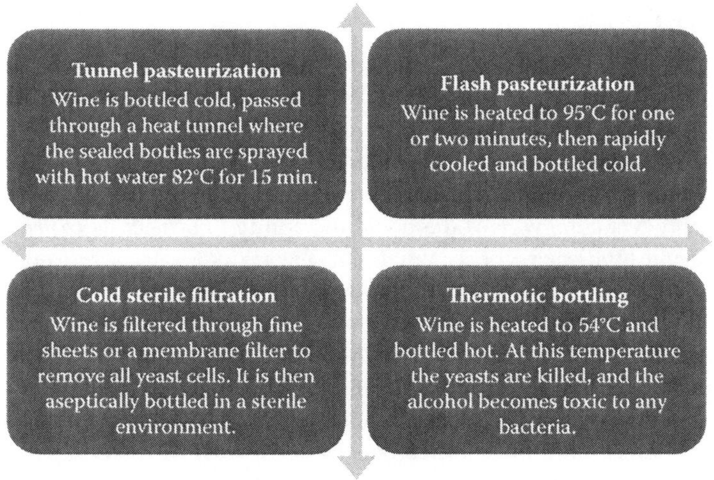

FIGURE 4.12 Available systems of treatment commonly used.

4.3.2 Glass

Glass bottles represent the most popular way to package wine today. Glass is a noncrystalline amorphous substance based on the chemical compound silica (silicon dioxide, or quartz), the primary constituent of sand. Once the bottles are formed, the inner surfaces are chemically treated to make the bottles nonporous (Athens, 2009; Colman and Paster, 2007). Apart from the function of containing, the bottle is what affects the aging process of wine. Glass protects the quality of the wine by reducing oxygen permutation. Conversely, other containers (e.g., more economical ones) often do not preserve the quality of wine like glass bottles. For example, plastic containers made from PET (polyethylene terephthalate) result in lighter and "greener" than traditional glass bottles. Even if in the first period after bottling wine in a PET container the wine appears comparable to a wine bottled in a glass bottle, after about 6 months to a year, the taste drastically changes. PET allows a significantly greater amount of oxygen to enter the wine, thus causing severe oxidation and alteration of the aroma profile. Although most wine today is consumed within a year after bottling, some wines are stored for a longer period. For wines that potentially need an aging period longer than a year, it seems reasonable to bottle them in glass.

4.3.3 Lightweight Glass Bottles

In the last decade, glass manufacturers have begun to slim down their bottles, producing a greener and more economical bottle. They have come up with a novel technique, denominated "light-weighting," which allows reduction of the overall amount of glass required for the bottle. Light-weighting is realized by reducing the wall thickness and removing the punt or the indention normally placed on the bottom of the bottle. By implementing this technique, manufacturers were able to decrease about 15% of glass usage. The overall cost savings was about 10%. Moreover, manufacturers are also using a greater proportion of recycled glass in the final product.

4.3.4 PET Bottles

In 2007, Constar International Inc. (leading producers of PET plastic containers) developed a new line of lightweight PET bottles, using MonOxbarTM technology. This comprises a monolayer design, offering several advantages over the traditional multilayer PET plastic containers. The monolayer PET bottles appear tough and do not delaminate during filling and distribution compared to multilayer counterparts. The manufacturing technology gives the bottle the ability to protect its oxygen-sensitive contents during prolonged storage durations. These bottles are able to protect their contents by chemically reacting with the oxygen present in the bottle. This reaction occurs while the permutation of oxygen slows down through the walls of the bottle. PET bottles show some advantages over traditional glass wine bottles (Figure 4.13). They are unbreakable, offer a greater flexibility, and are lighter than traditional glass bottles. The production of PET plastic bottles produces 50% fewer greenhouse gases compared to that of a traditional glass one (same volume) (Constar, 2004). However, PET bottles do not provide the same amount of protection as that of traditional glass bottles. Air is still able to pass through the PET lining, thus causing oxidation.

FIGURE 4.13 A typical example of PET bottles. (Courtesy of EnVino®, Burlingame, CA.)

4.4 CLOSURES

Cork is the traditional way of closing wine bottles (Figure 4.14). Cork is a natural product, coming from the bark of a type of evergreen oak tree, *Quercus suber.* The main quality of cork is its elasticity (if compressed, it is able to regain its original size). Hence, when used as a closure, it will provide a tight seal so long as it is kept humid. Wine bottles should therefore be kept lying flat to prevent the cork from drying out. However, because corks are a natural product, the quality is variable, thus affecting the wine sometimes. Synthetic materials have been developed as an alternative to natural cork (e.g., lookalike products), manufactured from a closed-cell polymer mixture, showing similar abilities to natural cork, including porosity to gases and water vapor and allowing the wine to "breathe." Another widespread

FIGURE 4.14 Traditional cork closures.

typology is represented by a metal screw cap closure (Stelvin®), considered as the best means for excluding oxygen. Wine closed with cork suffers from cork taint, an undesirable effect of the compound TCA (2,4,6,-trichloranisole), which develops during cork processing. The screw cap closure represents a valuable option for prevention of cork taint. But, although this may preserve the aroma and freshness, it is known that a screw cap does not allow the wine to peacefully and quietly mature.

4.4.1 Cork

Cork is harvested primarily from the cork oak (*Quercus suber*) that is endemic to the western Mediterranean countries. Once the trees are about 25 years old, the cork may be stripped from the trunks every 9 years. Cork is composed of suberin, a hydrophobic substance. Thanks to its impermeability, flexibility, and elasticity, cork is most commonly used for wine stoppers. The cork industry (mainly located in Spain, Portugal, and Italy) is generally regarded as environmentally friendly. The sustainability and the easy recycling represent two of its most distinctive aspects.

There are two types of corks: the whole cork stopper and the technical cork stopper. Hand-punching results in a more consistent quality (e.g., a skilled worker can choose the best sections in the cork strip). Much of the corkwood left over from punching is sent for granulation and utilized for technical corks. Technical corks are designed corks, commonly combining a body (agglomerated natural cork granules) and a fine natural cork disk (in contact with the wine) (Figure 4.15). Although it is one of the oldest stoppers for wine bottles, due to the fact that natural cork might taint the wine, use of new alternatives (synthetic stoppers or aluminum screw caps) is spreading. Cork taint can affect wines indifferently by price and quality. Its production mainly results from naturally occurring airborne fungi presented with chlorophenol compounds (industrial pollutants found in many pesticides and wood preservatives) and then converted into chloroanisole. Even if random oxidation

FIGURE 4.15 Different cork closures.

remains a problem, it must be noted that new manufacturing approaches have been developed by industry.

Currently, cork planks now undergo a technically refined boiling process to deliver cleaned ones with the proper humidity for punching in the shortest possible time (about 3 days rather than weeks). However, recent experiments have proved that vaporization by steam, for a second time, may significantly reduce the levels of TCA precursors, also reducing the amount of water used in the process. New research has developed a coating film applicable to the cork stopper (e.g., Bacchus barrier). This represents a high-performance film (a highly moisture-resistant polymer) with low permeability to oxygen and other gases that is permanently bonded to the ends of wine corks. This handling reduces cork taint, strengthens the cork ends, and improves the seal with the bottle. It is also claimed to be highly effective in extending shelf life of the wine.

4.4.2 Synthetic or Plastic Stopper for Wine Bottles

The first generation of plastic corks was largely rejected as these were made of solid plastic, which caused difficulties in insertion and extraction, leaving gaps around the edges (oxygen leaks). Conversely, new plastic corks (extruded or foamed) are made using materials with substantially no taste and odor.

4.4.3 Neocork

In 1996, Supreme Corq Inc. (Kent, WA) started up the commercialization of its coextruded corks, known as Neocork™. A foam core of extremely neutral low-density polyethylene (LDPE) represents the inner part. A substance (Thermoplastic Elastomer) showing elastic properties of natural rubber represents the outer layer. LDPE was chosen not only for its well known elastic properties, but also primarily for its purity, neutrality, and cleanliness—features also appreciated in food and medical packaging. Similarly, the thermoplastic elastomer, which is well known for its purity and cleanliness and commonly used in many food and medical applications, was chosen.

4.4.4 Nomacorc

At the same time, in 1996, Nomacorc LLC (Zebulon, NC) started up the commercialization of its coextruded, foam-core polyethylene corks. The innovative process was to realize corks with two types of extruded plastic. A firm inner core would hold the shape of the cork and a spongy exterior would make it fit better. The new corks have the feel of natural cork and are easier to remove using a corkscrew. The core is designed for optimum oxygen transfer rates through the closure, while the outer skin provides an accurate and consistent seal. The Nomacorc process coextrudes in two stages. First, raw materials are mixed, melted, and extruded to create a long, foamed cylinder (closure's core). Then a second extrusion process applies a flexible outer skin (thermally bonded to the inner cylinder). The shape is fixed by cooling (water) before the cutting operation of the closure.

4.4.5 Swan–Kork

In this process each cork is formed individually in a separate form-mold, using the injection-pressing technique. Due to this technique, smaller dimension tolerances and a closed surface are guaranteed. Its core is homogeneous (fine cell structure) by using a specially developed material composition. The injection-pressing technique ensures a higher density of the material.

4.4.6 Synthetic Stopper for Controlled Permeability

KORKED™ represents synthetic stoppers with controlled permeability, realized by an injection procedure in a monocavity mold using high-density polymers. The result is a stopper of about 8 g weight (the highest density in the market). This assures perfect adherence to the bottle neck and a higher level of elastic return. A controlled permeability membrane represents the novelty. This solution positively affects wine structure, ensuring an extremely low oxygen addition.

4.4.7 Screw Caps for Wine Bottles

These closures, made using aluminum, are also known as Stelvin®. They consist of a cap with threads fitting the bottleneck and a liner made of plastic, cork, rubber, or other soft material in order to make a seal with the mouth of the bottle. These may form a very tight seal (insufficient oxygen ingress), thus sometimes resulting in sulfur-like aromas and flavor suppression. Reduction resulting from extremely low oxygen being present during winemaking and aging increases the presence of unfavorable sulfur-based components (e.g., rotten eggs). Some varietals and winemaking styles are more susceptible to reduction and closures that allow minimal oxygen ingress.

4.4.8 Vino-Seal

The introduction of Vino-Seal closures into the market dates back to 2003. These closures are realized using glass with an inert O-ring (Elvax®). This sealing ring makes it possible to create a hermetic seal that prevents oxidation and TCA contamination (Figure 4.16). Due to its nature (glass), it is recyclable and lends the weight of being

FIGURE 4.16 Vino-Seal glass closures characterized by the typical inert O-ring.

environmentally friendly. Fitted securely in a circular groove on the glass stopper, the sealing ring (elastic, durable, halogen- and plasticizer-free polymer) is insoluble in alcohol and resistant to weak acids and other wine constituents. Vino-Seal closures appear aesthetically pleasant and can be easily opened without the use of other implements. The stopper can be pulled out by hand and sits tightly in the bottle when reinserted.

4.5 PACKAGING

4.5.1 Tetra Pak

Tetra Pak® carton packages (multilayers) are designed by using different packaging materials (paperboard, polyethylene, and aluminum foil). The main material in packages is paperboard, which is a renewable material made from wood that is able to provide stability, strength, and smoothness to the printing surface. It is applied in just enough quantity to make the package stable, without adding unnecessary weight. Polyethylene protects against outside moisture and enables the paperboard to stick to the aluminum foil, which protects against oxygen and light, maintaining the nutritional value and flavors at ambient temperatures. These containers offer several advantages over traditional glass bottles. While both types of containers are recyclable and may be produced by recycled material, usually glass bottles used in the wine industry are virgin bottles. Furthermore, 70% of the material used in Tetra Pak® containers comes from recycled material. Containers also have a significant weight advantage over traditional glass bottles. As previously noted, they primarily use packaging materials made from paper and weigh about 40 g compared to glass bottles (about 750 g). Compared to glass bottles, they are responsible for about 90% less packaging material, 50% less energy over the life cycle of the container, 80% fewer greenhouse gases, and 60% less solid waste volume. Moreover, these containers are easy to stack during transport and storage. However, considering that microoxidation is absent using this typology, containers are not suitable for long-term storage or aging. In conclusion, Tetra Pak® mainly offers a cheaper and greener alternative for wineries that tend to produce table wines.

4.5.2 Wine in Portable, Single-Serve Packaging

To meet the requirements of the new generation of wine drinkers, the packaging needs continually to evolve. Several occasions—from the dinner table down to the backyard party with friends, including festivals and sports events—represent a good starting point for new typologies. Outdoor events are the primary reason why people demand on-the-go packaging.

4.5.2.1 Wine in a Can

The Vinsafe™ packaging technology was developed by Barokes Wines in Australia; it was first introduced to the world as wine in a can without the typical tinny or aluminum taste. This result was possible due to the unique can lining, which ensures no contact between the wine and the aluminum material. Its consumption is rising worldwide; Japan, the world imperium of vending machines, is the main country in which this innovative technology has gathered highest appreciation.

4.5.2.2 Single-Serve Wine Package

Several new packaging solutions for the single-serve wine market are present today. "Couple" glasses are made entirely of PET. These represent one of the latest variations on a packaging concept for single-serve wine. Couple interlocks two wine glasses using an integrated stem for efficient stacking and shrink wrapping. The foot of the glass serves as a lid to protect the foil seal and also serves as a coaster. In 2012 "staked wines" were launched as a portable, single-serve wine package, presented in stacks of four interlocking, stackable, stemless PET wine glasses bundled in a shrink-sleeve label. Each glass contains 187 mL of wine and a four-pack has the same volume as a traditional bottle of wine (750 mL). Central to the stacked wine concept is a technology developed called Vinoware, which combines packaging constructed of oxygen-barrier PET with a filling and sealing process able to eliminate oxygen from the package headspace. This technology enables about a 12- to 14-month shelf life. The structure of the single-serve PET container (stemless wine glass) offers consumers a more sophisticated drinking experience than that offered by some other options. The concept of four stacked wine cups has been rather popular in the market. In 2014, Stormhoek Wine (South Africa) introduced it into the market. The "Stormhoek Stack" comprises four individual 187 mL PET stemless glasses that are 50% lighter with a shelf life of 96 weeks. Each stemless glass is heat sealed with peal-back foil lids. The packaging is 100% recyclable, lightweight, and shatterproof, making it ideal for outdoor consumption occasions.

4.5.3 The Evolution of the Bag-in-the-Box (BIB)

Bag-in-the-Box® (BIB) first made its way onto the wine market in the 1960s, when it was traditionally used to package generic bulk wines. This type of packaging comprises one or more layers of high-barrier flexible films known as the bladder, which is nestled in a paperboard container. The evolution of the BIB over the years reflects the changing of social trends of the modern wine drinker. The new generation of BIB can offer several advantages over a glass bottle, including improved distribution, cost-efficient packaging, and a container that is easy to open and pour. Several sizes exist, such as 1.5, 3, and 5 L containers—equivalent to two, four, and ten 750 mL bottles of wine. Although its design does provide manufacturers and consumers with a lighter weight product, the internal bladder that holds the wine may cause the wine to age (even if unopened) because the plastic packaging is not hermitically sealed. It was determined that the overall quality of the white wine packaged in a BIB container was significantly affected by the oxygen transmission rate (OTR) of the package (Fu et al., 2009). Color, free and total sulfur dioxide, total aldehyde, and total phenol content were all correlated with an increase of OTR. Compared to a traditional glass bottle, wine stored in a BIB container for over a year tends to degrade substantially. The new BIB systems, developed ready to use with the tap (e.g., Vitop) on the outside of the box, are more practical, as the end user does not have to remove the tap from the box; this is also more intuitive for the consumer.

4.6 SPOILAGE (DEFECTS, PREVENTION, TREATMENTS)

The association of microorganisms with the fermentation of alcoholic beverages dates back to past times. The winemaking process takes place in a complex ecological environment where the biochemistry and interaction of yeasts, bacteria, and fungi play a crucial role in the final product. To understand the conditions under which a specific microorganism can cause spoilage appears fundamental, as does understanding the off-flavors, odors, and color changes associated with the specific spoilage condition.

4.6.1 Origins of Wine Spoilage Microorganisms

There are three stages at which microorganisms can enter the winemaking process and affect the quality of the end product. The first involves the raw material (grapes) that is in direct contact with the winery equipment. For this reason when crushers, presses, tanks, pipes, pumps, or filtration units are not properly sanitized, the equipment will serve as an inoculant of the grape juice. Usually, the grapes delivered to a winery are not all in a healthy state. This situation affects the natural biodiversity of the microorganisms present in the juice. External conditions (such as grape variety, the state of grapes at harvest, the health of the grapes) indirectly affect the natural microflora. Similar considerations concern temperature, rainfall, soil, and the use of insecticides and fungicides (Pretorius et al., 1999). The second stage may occur during fermentation. At this point the grape juice contains both the natural flora and that hosted by the wine cellar and its equipment. The addition of sulfur dioxide (SO_2) to the juice and its composition (low pH and high sugar and acid content) selectively act on the development of yeasts and bacteria during alcoholic fermentation. *Saccharomyces cerevisiae* represents the dominant yeast during fermentation and, due to the increase of ethanol concentrations, the development of certain fungi and bacteria results in further suppression. Postfermentation represents the third stage at which the product may be susceptible to spoilage. In this situation, spoilage may occur in the bottle or during storage (e.g., oak barrels). Critical factors include good cellar sanitation, exclusion of oxygen, and the correct dosage of antimicrobial agents. At this stage, fungi and species of *Actinomyces* and *Streptomyces* present in the corks or oak barrels can also affect wine quality.

4.6.2 Spoilage by Yeasts (Table 4.2)

4.6.2.1 Refermentation

Saccharomyces is regarded as a spoilage organism only if it is present in the wrong place at the wrong time, thus causing refermentation. *Schizosaccharomyces pombe* has been associated with wine spoilage by forming sediment at the bottom of the bottle (Boulton et al., 1996). *Saccharomycodes ludwigii*, highly tolerant to ethanol and resistant to SO_2 and sorbate, produces high levels of acetaldehyde. This has been isolated as a slimy flocculent mass in bottles.

TABLE 4.2
Spoilage of Wines by Yeasts

Yeasts	Spoilage
Brettanomyces intermedius Anamorph: *Dekkera intermedia*	Produces volatile phenols causing medicinal, phenolic, horsey notes; mousy off-flavor results from isomers of tetrahydropyridines; produces high level of acetic acid
Candida spp. *C. vini* *C. stellata* *C. pulcherrima* *C. krusei*	Wine exposed to air will develop film layers; high concentrations of acetaldehyde, volatile acids, and esters
Hanseniaspora uvarum Anamorph: *Kloeckera apiculata*	High levels of acetic acid and its esters
Hansenula anomala (Pichia anomala)	High levels of acetic acid; large amounts of ethyl acetate, isoamyl acetate, and methyl butyl acetate, and development of film layer
Metschnikowia pulcherrima	Film layer and high levels of ethyl acetate and acetaldehyde
Pichia spp. *P. farinosa* *P. membranaefaciens* *P. vini*	Chalky film layer and high levels of acetaldehyde
Saccharomyces cerevisiae	Refermentation of wine with residual sugars
Saccharomycodes ludwigii	High concentrations of acetaldehyde; flocculent masses settle as chunks
Schizosaccharomyces pombe	Refermentation of bottled wine; deacidification
Zygosaccharomyces bailii	Secondary fermentation of wine with large amounts of CO_2; turbidity and sediment; high levels of acetic acid and esters

4.6.2.2 Ester Formation

Wines showing concentrations >200 mg/L of ethyl acetate and 0.6 mg/L of acetate are considered spoiled. *Hansenula anomala, Kloeckera apiculata,* and *Hanseniaspora uvarum* are largely associated with the ester taint of faulty wines. These species are associated with grape juice and result in spoilage at the early stages of alcoholic fermentation (Boulton et al., 1996).

4.6.2.3 Hydrogen Sulfide and Volatile Sulfur Compounds

Sulfur-containing compounds play a crucial role in wine flavor, mainly due to their high volatility (low threshold values) and reactivity. These compounds are responsible for off-flavors, showing notes comparable to rotten eggs, rubber, onion, garlic, and cabbage (Boulton et al., 1996). Environmental factors, fermentation temperature, and wine pH may influence the ability of yeasts to produce hydrogen sulfide (H_2S) (Henschke and Jiranek, 1993). H_2S may react with other wine components, producing mercaptans, thiols, and disulfides. Commonly, these compounds are perceived as onion or cabbage off-flavors. It is thus crucial to select *S. cerevisiae* yeast strains that produce limited amounts of H_2S in order to reduce the related risks.

4.6.2.4 Volatile Acidity

Yeasts involved in the acetification of wine include *Brettanomyces* (anamorph Dekkera), *Pichia anomala, Kloeckera apiculata,* and *Candida krusei.* Acetic acid is the major volatile acid present in wine (>90%). A wine becomes objectionable when acetic acid reaches a value higher than 1.1 g/L (Zoecklein et al., 1995). Acetic acid is formed as a by-product during the early stages of alcoholic fermentation.

4.6.2.5 Formation of Volatile Phenols

Wines contaminated by *Brettanomyces* usually show notes comparable to mousy, horsey, tar, tobacco, leathery, and pharmaceutical. The compounds responsible for these features are volatile phenols. They are produced by decarboxylation and reduction of hydroxycinnamic acids such as *p*-coumaric and ferulic acid. The more representative molecules are 4-vinylguaiacol and 4-vinylphenol, among vinylphenols, and 4-ethylguaiacol and 4-ethyl phenol, among ethylphenols, respectively (Boulton et al., 1996).

4.6.2.6 Film Formation

Some yeasts may form a film layer on top of stored wine; *Candida, Metschnikowia,* and *Pichia* species have been associated with this trait. This situation appears to be detrimental (not only visual) to wine quality, imparting an oxidized flavor due to the production of acetaldehyde.

4.6.2.7 Deacidification

Malic and tartaric acid accounts for about 90% of the titratable acidity of grapes. The acidity is essential, having a direct impact on the wine flavor and indirectly affecting the pH, color, and stability. Deacidification of wine can be obtained by the biological conversion of malic acid to lactic acid and carbon dioxide. Lactic acid bacteria (especially *Oenococcus oeni*) mediate this process, known as malolactic fermentation (Henick-Kling, 1993). A secondary effect of deacidification is represented by the loss in red color due to an increase in pH. If over-deacidification occurs and the pH has increased, it must be noted that color may change from full red to a bluish hue. Similarly, this situation negatively affects wine quality due to a loss of acidity, indirectly encouraging spoilage organisms.

4.6.2.8 Formation of Ethyl Carbamate

The typical reaction between ethanol and urea produces ethyl carbamate, which is considered a carcinogen (Monteiro and Bisson, 1991). The secretion of urea by yeasts is enhanced at elevated fermentation temperatures. Young wines contain L-arginine (the precursor), which is degraded to ornithine, ammonia, and carbon dioxide. Urea is produced as an intermediate. High levels of ethyl carbamate may occur in wine during aging or storage at elevated temperatures.

4.6.2.9 Acid Formation

In wine, lactic acid bacteria (LAB) can increase the acid content by producing lactic acid and acetic acid. The D-lactic acid is associated with spoilage, and the L-lactic acid is produced during malolactic fermentation. The formation of D-lactic acid arises from the reduction of pyruvic acid. Homofermentative species of Lactobacilli and Pediococci are responsible for its pathway. Conversely, heterofermentative species such as *Lactobacillus* and *Oenococcus* produce D-lactic acid and acetic acid through the 6-phosphogluconate pathway. In addition to its sensorial effect, acetic acid produced by LAB has been implicated in stuck fermentations. Acetic acid is able to inhibit the growth of *Saccharomyces*, thus affecting ethanol production.

4.6.2.10 Mannitol

Mannitol is produced by heterofermentative Lactobacilli, with the reduction of fructose or fructose-6-phospate. The spoilage problems are not directly related to this compound but instead to the production of acetic acid, D-lactic acid, propanol, butanol, and diacetyl. In this case, wine typically shows features perceived as viscous and sweetish.

4.6.2.11 Ropiness

Viscosity and a slimy appearance called "ropy" are related to the production of extracellular polysaccharide, composed of D-glucan. Usually *Pediococcus* have been implicated in ropiness. The production of extracellular polysaccharides was induced by ethanol. Lowering the pH under 3.5 may effectively control ropiness in wine.

4.6.2.12 Mousiness

Heterofermentative lactobacilli together with yeast *Brettanomyces* are responsible for notes that are reminiscent of mouse urine or acetamide. The compounds able to affect wine are the ethyl amino acid (lysine) derivatives, 2-acetyl-1,4,5,6-tetrahydropyridine and its isomer, 2-acetyl-3,4,5,6-tetrahydropyridine (Boulton et al., 1996). Their production is related to ethanol or propanol, and these are therefore associated with wine rather than grape juice.

4.6.2.13 Organic Acid Utilization

Organic acids (e.g., citric, tartaric, and sorbic), may be metabolized by certain LAB to affect wine quality. Diacetyl represents the most important metabolite, giving buttery, nutty, and/or toasty notes to the wine. Certain LAB may metabolize sorbic acid, and the resulting defect is known as "geranium tone," an off-odor typical of crushed geranium leaves. Tartaric acid, even if considered microbiologically stable, is the target of few lactobacilli (e.g., *plantarum* and *brevis*) that show the ability to degrade this compound under wine conditions. Acetic acid represents the product of tartaric acid metabolization. This product might increase the volatile acidity of wine, rendering it unacceptable.

4.6.2.14 Acrolein

Acrolein formation has been associated with species of the genera *Lactobacillus*, *Oenococcus*, and *Pediococcus*. However, it is definitely strain dependent. Acrolein is produced during bacterial degradation of glycerol. As a single component it is not an issue. However, when it reacts with the phenolic groups of anthocyanins it produces an unpleasant, bitter flavor.

4.6.2.15 Biogenic Amines

Biogenic amines together with ethyl carbamate do not spoil wine with off-flavors, but do pose health issues for the consumer. The production of biogenic amines in wines through LAB should be considered a crucial criterion in the selection of starter cultures. Biogenic amines are produced by specific amino acid decarboxylation from their respective amino acids (precursors). If taken in high concentration, histamine can cause headaches, hypotension, and digestive problems, whereas tyramine and phenylethylamine are associated with migraines and hypertension. Histamine, tyramine, putrescine, cadaverine, and phenylethylamine represent the most important biogenic amines produced in winemaking conditions. The formation of these amines is highest during malolactic fermentation (Soufleros et al. 1998). Commonly, the histamine content may increase during aging and storage.

4.6.3 Spoilage by Acetic Acid Bacteria (Table 4.3)

TABLE 4.3
Spoilage by Lactic Acid Bacteria

Bacteria	Spoilage
	Lactic Acid Bacteria
Lactobacillus brevis	Produces ethyl carbamate precursors; tartaric acid utilization; acidification of wine through the production of acetic and lactic acids; mousy taints
Lactobacillus cellobiosus *Lactobacillus hilgardii*	Mousy taints from tetrahydropyridine; bitterness arising from glycerol metabolism
Lactobacillus kunkeei	Production of high levels of acetic acid; stuck fermentations
Lactobacillus plantarum	Tartrate degradation; high level of diacetyl
Oenococcus oeni	Degrades arginine to produce ethyl carbamate precursors; produces histamine; stuck fermentation; buttery flavor due to increased level of diacetyl
Pediococcus damnosus	Produces histamine; polysaccharide synthesis
Pediococcus parvulus	Acrolein formation from glycerol; bitterness
Pediococcus pentosaceus	Produces polysaccharides; increased viscosity
	Acetic Acid Bacteria
Acetobacter aceti *Acetobacter pasteurianus* *Gluconobacter oxydans*	Oxidation of ethanol to acetaldehyde and acetic acid; production of ethyl acetate; production of acetoin from lactic acid
	Endo-spore-forming Bacteria
Bacillus spp. *Clostridium*	Increased acidity (butyric acid); sediment

4.6.3.1 Volatile Acidity

Acetic acid bacteria (AAB) belong to the family of Acetobacteriaceae, commonly known as the vinegar bacteria. AAB are Gram-negative, aerobic, catalase-positive microorganisms able to utilize glucose with acetic acid as a product. Acetic acid is the major component in wine associated with volatile acidity. The legal limit for acetic acid in wine is about 1.2 g/L, after which the wine becomes unpleasant. However, acetic acid is able to modify the perception of other important wine constituents. For example, tannins may be intensified.

4.6.3.2 Acetaldehyde

Wines containing high amounts of AAB may contain significant amounts of acetaldehyde. This molecule represents an intermediate metabolite in the production of acetic acid from ethanol under low oxygen concentrations. The descriptors of this defect range from nutty and sherry-like to being reminiscent of overripe apples.

4.6.3.3 Acetoin

Acetoin has a characteristic aroma described as "butter-like." Its levels in wine range from 3 to about 30 mg/L. In addition to affecting sensorial quality, acetoin may bind the free SO_2, thus reducing its antimicrobial activity. Strains of *Acetobacter* and *Gluconobacter* are able to oxidize lactic acid to acetoin under low-oxygen conditions.

4.6.4 Spoilage by Molds

The infection of grapes by molds (filamentous fungi) before harvest can be detrimental to wine quality (Figure 4.17). Molds detected on grapes include species of the genera *Alternaria, Aspergillus, Botrytis, Cladosporium, Mucor, Oidium, Penicillium, Plasmopara, Rhizopus,* and *Uncinula* (Fleet, 1998). Molds are sensitive to ethanol concentrations (above 3%), low pH, SO_2 and anaerobiosis. In these conditions, molds

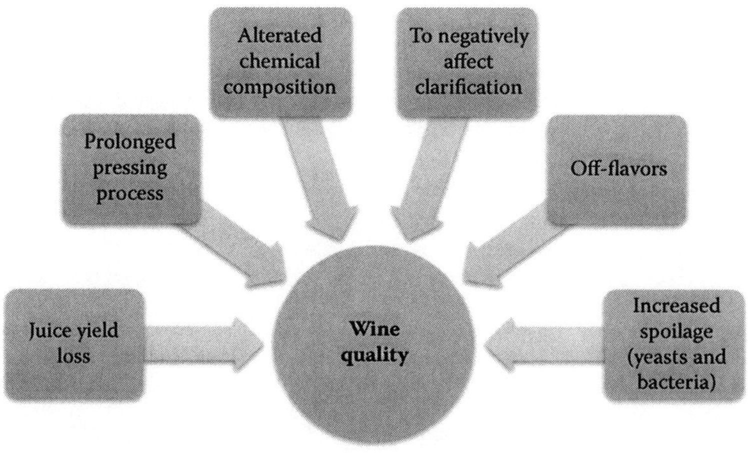

FIGURE 4.17 Molds can affect the wine quality in one of the following manners. (Pearson and Goheen, 1994.)

are unable to survive in wine. Molds produce antiyeast metabolites that might affect alcoholic and malolactic fermentations, altering the chemical composition of the grape juice through their enzymatic secretions. Molds can also produce mycotoxins (regarded as carcinogens)—a matter of great concern. *Aspergillus* and *Penicillium* represent two genera of molds associated with infected grapes able to produce these mycotoxins (e.g., aflatoxins, patulin, and ochratoxin A).

4.6.4.1 Cork Taint

Microbiological contamination of corks can affect the wine quality by producing typical off-flavors. *Aspergillus, Cladosporium, Monilia, Paecilomyces, Penicillium,* and *Trichoderma* represent the common genera associated with cork taint. The major compound responsible for the cork taint is 2,4,6-trichloroanisole, perceived as earthy or musty off-flavor. Other compounds are also implicated in cork taint, such as 1-octen-3-ol, 1-octen-3-one, 2-methylisoborneol, geosmin, and guaiacol.

4.6.5 Preservation

A rigorous quality control is required to assure the microbiological stability of the winemaking process. Chemical preservatives are used in general to inhibit specific populations of microorganisms. The techniques applied for food preservation have a long history. These include chilling, fermentation or acidification, addition of chemical preservatives, heat pasteurization, and sterilization. The techniques applied in the winemaking process include (1) procedures that prevent access of microorganisms, (2) inactivation of unwanted microorganisms when the first step failed, and (3) procedures that slow or inhibit their growth.

4.6.5.1 Chemical Preservation

Preservatives used in the wine industry are most effective in their undissociated form, prevalent at a low pH (Zoecklein et al., 1995). Sulfur dioxide is one of the oldest compounds used in the food and beverage industries due to its antioxidative and antimicrobial properties. Sulfite is present in three forms in an aqueous solution and their equilibria are pH dependent. At low pH values sulfite exists mainly as molecular SO_2, at intermediate pH as bisulfite ions, and at high pH as sulfite ions. At wine pH values (3.0–4.0), the major proportion are bisulfite ions (95%), with only 5% in the active molecular form (Romano and Suzzi, 1993). SO_2 antimicrobial activity may be diminished by binding to wine compounds, thus rendering it inactive. The use of SO_2 appears to be a tricky process, as the concentration added to the different stages of vinification should be sufficient to inhibit only the unwanted species, preserving the good ones responsible for alcoholic fermentation or malolactic fermentation. Sorbic acid (antifungal agent) is a short chain unsaturated fatty acid generally used in the food and beverage industries. It is used in sweetened wines to inhibit refermentation by *S. cerevisiae*. However, sorbic acid is not an effective inhibitor of LAB, AAB, or yeasts such as *Brettanomyces* and *Saccharomycodes*. Certain LAB produce geranium-like off-flavors that may utilize sorbic acid. The common name of dimethyldicarbonate (DMDC) is Velcorin®. DMDC represents a sterilant being lethal against yeasts and bacteria, ensuring a controllable fermentation with the inoculated starter cultures. The way of action is the denaturation of the fermentative pathway enzymes (e.g., glyceraldehyde-3 phosphate dehydrogenase and alcohol dehydrogenase).

4.6.5.2 Biopreservatives

Although the traditional preservatives previously reported are still widely used, there exists a growing demand from consumers for alternative methods, such as natural antimicrobial compounds from plants, animals, and microorganisms, as biological preservatives. Bacteriocins (e.g., nisin, pediocin, and plantaricin) produced by some LAB are small polypeptides, showing inhibitory activity to other bacterial species. These polypeptides act on the cell wall of bacteria to induce cell lysis (Bruno et al., 1992). Species of *Lactobacillus* and *Pediococcus* are more resistant to nisin than *Oenococcus oeni* strains. Pediocin and plantaricin have been shown to successfully kill *Oenococcus oeni* cells (Nel et al., 2002). More recently, a bacteriocin-like inhibitory substance has been shown to be effective against wine *Lactobacillus* species (Yurdugül and Bozoglu, 2008). Lysozyme is an enzyme with bactericidal properties, commonly used as a preservative in the food industry. Lysozyme can be added at various steps throughout grape vinification to inhibit LAB. The aroma of wine is not affected by the addition of lysozyme (Bartowsky et al., 2004). Similarly, with all wine treatments, the addition of lysozyme must be considered carefully. This compound is able to bind with tannins and polyphenols in red wines, typically resulting in a slight decrease in wine color or formation of a wine haze (Bartowsky et al., 2004).

4.6.5.3 Alternative Technologies to Eliminate Bacteria from Wine

A range of emerging technologies have been used successfully in several food and beverage industries that are focused on eliminating microorganisms, wine included. These include ultrahigh-pressure processing, ultrasound, ultraviolet irradiation, and pulsed electric fields. Ultrahigh-pressure treatment was recognized as a potential preservation approach. The applied pressure causes inactivation of microorganisms and enzymes but does not affect flavor molecules and vitamins. Its effect is primarily a result of cytoplasmic membrane damage. High-power ultrasound uses frequencies in the range of 20 to 100 kHz, showing the ability to cause the formation and collapse (cavitation) of high-energy microbubbles. This approach may be used in food processing to inactivate microbes (cell membrane thinning) (Piyasena et al., 2003). Ultraviolet irradiation has been shown to significantly reduce LAB populations, including *Lactobacillus* species. A further reported activity showed the ability to kill fungi on harvested grapes (Valero et al., 2007). Pulsed electric field technology has been utilized as a means of sterilizing the product in beverage industries. It has been studied as an alternative to pasteurization in fruit drink production, which can lead to losses in nutritional and organoleptic qualities. This approach involves application of short pulses (1 to 10 ms) of higher or low-intensity electric fields to foods. The latter can be placed between two electrodes in batches, or in continuous flow systems, at low-processing temperatures (<50°C). Recently, this approach, used in combination with natural antimicrobials, was able to enhance the microbiocidal protection of fruit juices (Mosqueda-Melgar et al., 2008).

REFERENCES

Atanasova, V.; Fulcrand, H.; Cheynier, W.; Moutounet, M. Effect of oxygenation on polyphenol changes occurring in the course of wine-making. *Anal. Chim. Acta*, 458, 15–27, 2002.

Athens, K. Bottler caters to regional wineries. Sustainable Industries, January, 2009.

Baiano, A.; de Gianni, A. Timing of the treatment with oak chips: The case of Nero di Troia wine. *Eur. Food Res. Technol.*, 242, 1343–1353, 2016.

Baiano, A.; de Gianni, A.; Mentana, A.; Quinto, M.; Centonze, D.; del Nobile, M.A. Effects of the treatment with oak chips on color related phenolics, volatile composition, and sensory profile of red wines: The case of Aglianico and Montepulciano. *Eur. Food Res. Technol.*, 242, 745–767, 2016.

Barnoud, F. La cellulose. In *Les polymères végétaux*. Ed. B. Monties; Paris, pp. 66–86, 1980.

Bartowsky, E.J.; Costello, P.J.; Villa, A.; Henschke, P.A. The chemical and sensorial effects of lysozyme addition to red and white wines over six months' cellar storage. *Aust. J. Grape Wine Res.*, 10, 143–150, 2004.

Basalekou, M.; Pappas, C.; Kotseridis, Y.; Strataridaki, A.; Geniatakis, E.; Tarantilis, P.; Kallithraka, S. Monitoring wine with Fourier transform infrared spectroscopy (FT-IR). BIO Web of Conferences, 2015.

Bosso, A.; Petrozziello, M.; Santini, D.; Motta, S.; Guaita, M.; Marulli, C. Effect of grain type and toasting conditions of barrels on the concentration of the volatile substances released by the wood and on the sensory characteristics of Montepulciano d'Abruzzo. *J. Food Sci.*, 73, S373–S382, 2008.

Boulton, R.B.; Singleton, Y.L.; Bisson, L.E.; Kunkee, R.E. *Principles and Practices of Winemaking.* New York: Chapman & Hall, 1996.

Bruno, M.E.; Kaiser, A.; Montville, T.J. Depletion of the proton motive force by nisin in *Listeria monocytogenes* cells. *Appl. Environ. Microbiol.*, 58, 2255–2259, 1992.

Cadahía, E.; Fernández de Simón, B.; Poveda, P.; Sanz, M. Utilización de *Quercus pyrenaica* Willd. de Castilla y León en el envejecimiento de vinos: Comparación con roble francés y americano. Ed. Instituto Nacional de Investigación y Tecnología Agraria y Alimentaria. Ministerio de Ciencia e Innovación. Madrid, 2008.

Cadahía, E.; Muñoz, L.; Fernández de Simón, B.; García-Vallejo, M.C. Changes in low molecular weight phenolic compounds in Spanish, French, and American oak woods during natural seasoning and toasting. *J. Agric. Food Chem.*, 49, 1790–1798, 2001.

Campbell, J.I.; Pollnitz, A.P.; Sefton, M.A.; Herderich, M.J.; Pretorius, I.S. Factors affecting the influence of oak chips on wine flavour. *Aust. New Zeal. Wine Ind. J.*, 21, 38–42, 2006.

Cano-López, M.; Pardo-Minguez, F.; López-Roca, J.M.; Gómez-Plaza, E. Chromatic characteristics and anthocyanin profile of a micro-oxygenated red wine after oak or bottle maturation. *Eur. Food Res. Technol.*, 225, 125–132, 2007.

Chatonnet, P. *Le Bois et la Qualité des Vins et des Eaux-de-Vie.* Martillac: Vigne et Vin Publications Internationales, 1992.

Chatonnet, P. Principales origines et caractéristiques des chênes destines à l'élevage des vins. *Rev. Oenol.*, 75, 15–18, 1995.

Chatonnet, P. Productos alternativos a la crianza en barrica de los vinos. Influencia de los parámetros de fabricación y de uso. *Revista Enología*, 4, 2–24, 2007.

Chatonnet, P.; Dubourdieu, D. Comparative study of the characteristics of American white oak (*Quercus alba*) and European oak (*Quercus petraea* and *Quercus robur*) for production of barrels used in barrel aging of wines. *Am. J. Enol. Vitic.*, 49, 79–85, 1998.

Chira, K.; Teissedre, P.L. Chemical and sensory evaluation of wine matured in oak barrel: Effect of oak species involved and toasting process. *Eur. Food Res. Technol.*, 240, 533–547, 2015.

Chira, K.; Teissedre, P.L. Relation between volatile composition, ellagitannin content and sensory perception of oak wood chips representing different toasting processes. *Eur. Food Res. Technol.*, 236, 735–746, 2013.

Colman, T.; Paster, P. Red, white, and "green": The cost of carbon in the global wine trade. *AAWE Working Papers*, 9, 1–17, 2007.

Costantini, A.; Valdano, E.; Cravero, M.C.; Petrozziello, M.; Piano, F.; Bernasconi, A.; Garcia-Moruno, E. Dry ice blasting, a new tool for barrel regeneration treatment. *Eur. Food Res. Technol.*, 242, 1673–1683, 2016.

Constar International Inc. PET oxygen barrier effectiveness: A comparison of alternatives. Presented at PACK EXPO, 2004.

del Álamo, M.; Nevares, I.; Gallego, L.; Fernández de Simón, B.; Cadahía, E. Micro-oxygenation strategy depends on origin and size of oak chips or staves during accelerated red wine aging. *Anal. Chim. Acta*, 660, 92–101, 2010.

del Álamo-Sanza, M.; Nevares-Domínguez, I. Wine aging in bottle from artificial systems (staves and chips) and oak woods: Anthocyanin composition. *Anal. Chim. Acta*, 563, 255–263, 2006.

Dumitriu, G.D.; López de Lerma, N.; Cotea, V.V.; Zamfir, C.I.; Peinado, R.A. Effect of aging time, dosage and toasting level of oak chips on the color parameters, phenolic compounds and antioxidant activity of red wines (var. *Feteasca neagra*). *Eur. Food Res. Technol.*, 242, 2171–2180, 2016.

Espitia-López, J.; Escalona-Buendía, H.B.; Luna, H.; Verde-Calvo, J.R. Multivariate study of the evolution of phenolic composition and sensory profile on mouth of Mexican red Merlot wine aged in barrels vs. wood chips. *CyTA—J. Food*, 13, 26–31, 2015.

Falguera, V.; Forns, M.; Ibarz, A. UV–Vis irradiation: An alternative to reduce SO2 in white wines? *LWT-Food Sci. Technol.*, 51, 59–64, 2013.

Fernández de Simón, B.; Cadahía, E. Tratamiento de la madera de roble para tonelería, I.1: 9–23. In: *Crianza en Barricas y Otras Alternativas*. Ed. Fundación para la Cultura del Vino. Madrid. 2007.

Fernández de Simón, B.; Cadahía, E.; del Álamo, M.; Nevares, I. Effect of size, seasoning and toasting in the volatile compounds in toasted oak wood and in a red wine treated with them. *Anal. Chim. Acta*, 660, 211–220, 2010.

Fernández de Simón, B.; Cadahía, E.; Jalocha, J. Volatile compounds in a Spanish red wine aged in barrels made of Spanish, French, and American oak wood. *J. Agric. Food Chem.*, 51, 7671–7678, 2003.

Fernández de Simón, B.; Sanz, M.; Cadahía, E.; Martínez, J.; Esteruelas, E.; Muñoz, A.M. Polyphenolic compounds as chemical markers of wine ageing in contact with cherry, chestnut, false acacia, ash and oak wood. *Food Chem.*, 143, 66–76, 2014.

Feuillat, F.; Keller, R.; Sauvageot, F.; Puech, J.L. Characterization of French oak cooperage (*Quercus robur* L., *Quercus petraea* L.). Research of the study group on barrel-aging Burgundy wines. *Am. J. Enol. Vitic.*, 50, 513–518, 1999.

Feuillat, M. L'élevage des vins de Bourgogne en fûts de chêne. *Rev. Fr. Oenol.*, 88bis, 17–29, 1982.

Fleet, G.H. The microbiology of alcoholic beverages. In: Wood, B.J.B. (ed). *Microbiology of Fermented Foods*, vol. 1. , London: Blackie Academic & Professional. pp. 217–262, 1998.

Fu, Y.; Lim, T.; McNicholas, P. Changes on enological parameters of white wine packaged in bag-in-box during secondary shelf life. *J. Food Sci.*, 74(8), c608–c618, 2009.

Gallego, L.; del Álamo, M.; Nevares, I.; Fernández, J.A.; Fernández de Simón, B.; Cadahía, E. Phenolic compounds and sensorial characterization of wines aged with alternative to barrel products made of Spanish oak wood (*Quercus pyrenaica* Willd.). *Food Sci. Technol. Int.*, 18, 151–165, 2012.

Garde-Cerdán, T. Estudio de la composición volátil de vinos tintos envejecidos en diferentes tipos de barricas de roble. Thesis. Universidad Pública de Navarra, Spain. 2004.
Garde-Cerdán, T.; Ancín-Azpilicueta, C. Effect of oak barrel type on the volatile composition of wine. Storage time optimization. *LWT-Food Sci. Technol.*, 39, 199–205, 2006b.
Garde-Cerdán, T.; Ancín-Azpilicueta, C. Review of quality factors on wine ageing in oak barrels. *Trends Food Sci. Technol.*, 17, 438–447, 2006a.
Garde-Cerdán, T.; Lorenzo, C.; Carot, J.M.; Esteve, M.D.; Climent, M.D.; Salinas, M.R. Effects of composition, storage time, geographic origin and oak type on the accumulation of some volatile oak compounds and ethylphenols in wines. *Food Chem.*, 122, 1076–1082, 2010.
Garde-Cerdán, T.; Rodríguez-Mozaz, S.; Ancín-Azpilicueta, C. Volatile composition of aged wine in used barrels of French oak and of American oak. *Food Res. Int.*, 35, 603–610, 2002.
Gómez García-Carpintero, E.; Gómez-Gallego, M.A.; Sánchez-Palomo, E.; González-Viñas, M.A. Sensory descriptive analysis of Bobal red wines treated with oak chips at different stages of winemaking. *Aust. J. Grape Wine Res.*, 17, 368–377, 2011.
Gómez-Plaza, E.; Pérez-Prieto, L.J.; Fernández-Fernández, J.I.; López-Roca, J.M. The effect of successive uses of oak barrels on the extraction of oak-related volatile compounds from wine. *Int. J. Food Sci. Technol.*, 39, 1069–1078, 2004.
González-Arenzana, L.; Santamaría, P.; López, R.; Garijo, P.; Gutiérrez, A.R.; Garde-Cerdán, T.; López-Alfaro, I. Microwave technology as a new tool to improve microbiological control of oak barrels: A preliminary study. *Food Control*, 30, 536–539, 2013.
Gordillo, B.; Baca-Bocanegra, B.; Rodriguez-Pulído, F.J; González-Miret, M.L.; García-Estévez, I.; Quijada-Morín, N.; Heredia, F.J.; Escribano-Bailón, M.T. Optimisation of an oak chips-grape mix maceration process. Influence of chip dose and maceration time. *Food Chem.*, 206, 249–259, 2016.
Guchu, E.; Díaz-Maroto, M.C.; Pérez-Coello, M.S.; González-Viñas, M.A.; Cabezudo, M.D. Volatile composition and sensory characteristics of Chardonnay wines treated with American and Hungarian oak chips. *Food Chem.*, 99, 350–359, 2006.
Guichard, G.; Fournier, N.; Masson, G.; Puech, J.L. Steroisomers of β-methyl-γ-octalactone. I. Quantification in Brandies as a function of wood origin and treatment of the barrels. *Am. J. Enol. Vitic.*, 46, 419–423, 1995.
Guzzon, R.; Widmann, G.; Malacarne, M.; Nardin, T.; Nicolini, G.; Larcher, R. Survey of the yeast population inside wine barrels and the effects of certain techniques in preventing microbiological spoilage. *Eur. Food Res. Technol.*, 233, 285–291, 2011.
Henick-Kling, T. Malolactic fermentation. In: Fleet, G.H. (ed). *Wine Microbiology and Biotechnology.* Chur, Switzerland: Harwood Academic Publishers. pp. 289–326, 1993.
Henschke, P.A.; Jiranek, V. Yeasts—Metabolism of nitrogen compounds. In: Fleet, G.H. (ed). *Wine Microbiology and Biotechnology.* Chur, Switzerland: Harwood Academic Publishers. pp. 77–164, 1993.
Hernández-Orte, P.; Franco, E.; González-Huerta, C.; Martínez-García, J.; Cabellos, M.; Suberviola, J.; Orriols, I.; Cacho, J. Criteria to discriminate between wines aged in oak barrels and macerated with oak fragments. *Food Res. Int.*, 57, 234–241, 2014.
Herrero, P.; Sáenz-Navajas, M.P.; Avizcuri, J.M.; Culleré, L.; Balda, P.; Antón, E.C.; Ferreira, V.; Escudero, A. Study of Chardonnay and Sauvignon blanc wines from D.O.Ca Rioja (Spain) aged in different French oak wood barrels: Chemical and aroma quality aspects. *Food Res. Int.*, 89, 227–236, 2016.
Hidalgo, J. *Tratado de enología.* Ed. Mundi-Prensa. Madrid. 2003.
Jackson, R.S. Oak and cooperage. In ed. Taylor, S.L., *Wine Science. Principles and Applications.* New York: Academic Press, pp. 452–488, 2008.

Jiranek, V.; Grbin, P.; Yap, A.; Barnes, M.; Bates, D. High power ultrasonics as a novel tool offering new opportunities for managing wine microbiology. *Biotechnol. Lett.*, 30, 1–6, 2008.

Kolmann, F.; Fengel, D. Ändrungen der Chemischen Zusammen Setung von Holz durch Thermische Behandlun. *Holz Roh-Werkst.*, 23, 461–468, 1965.

Koussissi, E.; Dourtoglou, V.G.; Ageloussis, G.; Paraskevopoulos, Y.; Dourtoglou, T.; Paterson, A.; Chatzilazarou, A. Influence of toasting of oak chips on red wine maturation from sensory and gas chromatographic headspace analysis. *Food Chem.*, 114, 1503–1509, 2009.

Kyraleou, M.; Kallithraka, S.; Chira, K.; Tzanakouli, E.; Ligas, I.; Kotseridis, Y. Differentiation of wines treated with wood chips based on their phenolic content, volatile composition, and sensory parameters. *J. Food Sci.*, 80, 2701–2710, 2015.

Li, S.; Crump, A.M.; Grbin, P.R.; Cozzolino, D.; Warren, P.; Hayasaka, Y.; Wilkinson, K.L. Aroma potential of oak battens prepared from decommissioned oak barrels. *J. Agric. Food Chem.*, 63, 3419–3425, 2015.

Malfeito-Ferreira, M.; Laureano, P.; Barata, A.; D'Antuono, I.; Stender, H.; Loureiro, V. Effect of different barrique sanitation procedures on yeasts isolated from the inner layers of wood. *Am. J. Enol. Vitic.*, 55, 3, 2004.

Marco, J.; Artajona, J.; Larrechi, M.S.; Rius, F.X. Relationship between geographical origin and chemical composition of wood for oak barrels. *Am. J. Enol. Vitic.*, 45, 192–200, 1994.

Marko, S.D.; Dormedy, E.S.; Fugelsang, K.C.; Gump, B.; Wample, R.L. Analysis of oak volatiles by gas chromatography-mass spectrometry after ozone sanitization. *Am. J. Enol. Vitic.*, 56, 46–51, 2005.

Martínez, J. Incidencia del origen de la madera de roble en la calidad de los vinos de Tempranillo de la D.O.Ca. Rioja durante la crianza en barrica. Thesis. Universidad de La Rioja, Spain. 2004.

Martínez, J.; Cadahía, E.; Fernández de Simón, B.; Ojeda, S.; Rubio, P. Effect of the seasoning method on the chemical composition of oak heartwood to cooperage. *J. Agric. Food Chem.*, 56, 3089–3096, 2008.

Masson, E.; Baumes, R.; Moutounet, M., Puech, J.L. The effect of kiln-drying on the levels of ellagitannins and volatile compounds of European oak (*Quercus petraea* Liebl.) stave wood. *Am. J. Enol. Vitic.*, 51, 201–214, 2000.

Masson, G. Composition chimique du bois de chêne de tonnellerie. Essai de discrimination des espèces et des forets. Thesis. Université de Montpellier II, France, 1996.

Masson, G.; Puech, J.L.; Moutounet, M. The chemical composition of barrel oak wood. *Bull. O.I.V.*, 785–786, 634–657, 1996.

Masuda, M.; Nishimura, K. Branched nonalactones from some *Quercus species*. *Phytochem.*, 10, 1401–1402, 1971.

McCord, J. Application of toasted oak and micro-oxygenation to ageing of Cabernet Sauvignon wines. *Aust. New Zeal. Grapegrower and Winemaker*, 43–53, 2003.

Miller, D.P.; Howel, G.S.; Michaellis, C.S.; Dickmann, D.I. The content of phenolic and aldehyde flavor components of white oak as affected by site and species. *Am. J. Enol. Vitic.*, 43, 333–338, 1992.

Monteiro, F.F.; Bisson, L.F. Amino acid utilization and urea formation during vinification fermentations. *Am. J. Enol. Vitic.*, 40, 199–208, 1991.

Monties, B. Composition chimique des bois de chêne: Composés phenoliques, relations avec quelques propietés physiques et chimiques susceptibles d´influencer la qualité des vins et des eaux-de-vie. *Conn. Vigne Vin*, 21, 169–190, 1987.

Mosedale, J.R.; Puech, J.L.; Feuillat, F. The influence on wine flavor of the oak species and natural variation of heartwood components. *Am. J. Enol. Vitic.*, 50, 503–512, 1999.

Mosqueda-Melgar, J.; Raybaudi-Massilia, R.M.; Martin-Belloso, O. Combination of high-intensity pulsed electric fields with natural antimicrobials to inactivate pathogenic microorganisms and extend the shelf-life of melon and watermelon juices. *Food Microbiol.*, 25, 479–491, 2008.

Nabeta, K.; Yonekubo, J.; Miyake, M. Analysis of volatile constituents of European and Japaneses oaks. *Mokuzai Gakkaishi*, 32, 921–927, 1986.

Nabeta, K.; Yonekubo, J.; Miyake, M. Phenolic compounds from the heartwood of European oak (Quercus robur L.) and brandy. *Mokuzai Gakkaishi*, 33, 408–415, 1987.

Navarro, M.; Kontoudakis, N.; Gómez-Alonso, S.; García-Romero, E.; Canals, J.M.; Hermosín-Gutíerrez, I.; Zamora, F. Influence of the botanical origin and toasting level on the ellagitannin content of wines aged in new and used oak barrels. *Food Res. Int.*, 87, 197–203, 2016.

Nel, H.A.; Bauer, R.; Wolfaardt, G.M.; Dicks, L.M.T. Effect of bacteriocins pediocin PD-1, plantaricin 423, and nisin on biofilms of *Oenococcus oeni* on a stainless steel surface. *Am. J. Enol. Vitic.*, 53, 191–196, 2002.

Nevares, I.; del Álamo, M. Measurement of dissolved oxygen during red wines tank aging with chips and micro-oxygenation. *Anal. Chim. Acta*, 621, 68–78, 2008.

Nikitine, N.I. In *Wood Chemistry*; the Academy of Science U.S.S.R., Moscow, 1951.

Nishimura, K.; Ohnishi, M.; Masuda, M.; Koga, K.; Matsuyama, R. Reactions of wood components during maturation. In *Flavour of distilled beverages: Origin and development.* Ed. J.R. Piggott; Ellis Horwood, Chichester, pp. 241–255, 1983.

Oberholster, A.; Elmendorf, B.L.; Lerno, L.A.; King, E.S.; Heymann, H.; Brenneman, C.E.; Boulton, R.B. Barrel maturation, oak alternatives and micro-oxygenation: Influence on red wine aging and quality. *Food Chem.*, 173, 1250–1258, 2015.

Pearson, R.E.; Goheen, A.C. *Compendium of Grape Diseases.* St. Paul, MN: APS Press, 1994.

Pérez-Magariño, S.; Ortega-Heras, M.; González-Sanjosé, M.L. Wine consumption habits and consumer preferences between wines aged in barrels or with chips. *J. Sci. Food Agric.*, 91, 943–949, 2011.

Pérez-Prieto, L.J.; López-Roca, J.M.; Martínez-Cutillas, A.; Pardo-Minguez, F.; Gómez-Plaza, E. Maturing wine in oak barrels. Effects of origin, volume, and age of the barrel on the wine volatile composition. *J. Agric. Food Chem.*, 50, 3272–3276, 2002.

Piyasena, P.; Mohareb, E.; McKellar, R.C. Inactivation of microbes using ultrasound: A review. *Int. J. Food Microbiol.*, 87, 207–216, 2003.

Pontallier, P.; Salagoïty, M.H.; Ribéreau-Gayon, P. Intervention du bois de chêne dans l'evolution des vins rouges élevés en barriques. *Conn. Vigne Vin*, 16, 45–61, 1982.

Porter, G.W.; Lewis, A.; Barnes, M.; Williams, R. Evaluation of high power ultrasound porous cleaning efficacy in American oak wine barrels using x-ray tomography. *Innov. Food Sci. Emerg. Technol.*, 12, 509–514, 2011.

Pretorius, I.S.; Van Der Westhuizen, T.J.; Augustyn, O.P.H. Yeast biodiversity in vineyards and wineries and its importance to the South African wine industry. *S. Afri. J. Enol. Vitic.*, 20, 61–74, 1999.

Prida, A.; Puech, J.L. Influence of geographical origin and botanical species on the content of extractives in American, French, and East European oak woods. *J. Agric. Food Chem.*, 54, 8115–8126, 2006.

Puech, J.L.; Moutounet, M. Liquid chromatographic determination of scopoletin in hydroalcoholic extract of oak wood and distilled alcoholic beverages. *J. Assoc. Anal. Chem.*, 71, 512–514, 1988.

Renouf, V.; Lounvaud-Funel, A. Incidence microbiologique de l'usage de barriques neuves et/ou des barriques usagées. *Rev. Fr. Oenol.*, 211, 10–14, 2005.

Ribéreau-Gayon, P.; Glories, Y.; Maujean, A.; Dubourdieu, D. Tratado de enología. Tomo II: *Química del vino. Estabilización y tratamientos.* Ed. Mundi-Prensa. Madrid. 2003.

Rodríguez-Bencomo, J.J.; Ortega-Heras, M.; Pérez-Magariño, S.; González-Huerta, C. Volatile compounds of red wines macerated with Spanish, American, and French oak chips. *J. Agric. Food Chem.*, 57, 6383–6391, 2009.

Rodríguez-Rodríguez, P.; Gómez-Plaza, E. Effect of volume and toast level of French oak barrels (*Quercus petraea* L.) on cabernet sauvignon wine characteristics. *Am. J. Enol. Vitic.*, 62, 359–365, 2011.

Romano, P.; Suzzi, G. Sulfur dioxide and wine microorganisms. In: Fleet, G.H. (ed). *Wine Microbiology and Biotechnology*. Chur, Switzerland: Harwood Academic Publishers. pp. 373–393, 1993.

Rubio-Bretón, P. Utilización de fragmentos de roble en los procesos fermentativos y durante el envejecimiento de vinos tintos. Thesis. Universidad de La Rioja, Spain. 2014.

Rubio-Bretón, P.; Lorenzo, C.; Salinas, M.R.; Martínez, J.; Garde-Cerdán, T. Influence of oak barrel aging on the quality of red wines. In C. A. Chuteira, and A.B. Grão (Eds.), *Oak: Ecology, Types and Management*, pp. 59–86. New York: Nova Science Publishers, Inc., 2013.

Sauvageot, F.; Feuillat, F. The influence of oak wood (*Quercus robur* L., *Q. petraea* Liebl.) on the flavor of Burgundy Pinot noir. An examination of variation among individual trees. *Am. J. Enol. Vitic.*, 50, 447–455, 1999.

Sefton, M.; Francis, Y.; Williams, P. Volatile norisoprenoid compounds as constituents of oak woods used in wine and spirit maturation. *J. Agric. Food Chem.*, 38, 2045–2049, 1990.

Sefton, M.A.; Francis, I.L.; Pocock, K.J.; Williams, P.J. The influence of natural seasoning on the concentrations of eugenol, vanillin and *cis*- and *trans*-ß-methyl-γ-octalactone extracted from French and American oakwood. *Sci. Aliments*, 13, 629–643, 1993.

Seikel, M.K.; Hostettler, F.D.; Niemann, G.J. Phenolics of *Quercus rubra* wood. *Phytochem.*, 10, 2249–2251, 1971.

Soufleros, E.; Barrios, M-L.; Bertand, A. Correlation between the content of biogenic amines and other wine compounds. *Am. J. Enol. Vitic.*, 49, 266–278, 1998.

Tao, Y.; García, J.F.; Sun, D.W. Advances in wine aging technologies for enhancing wine quality and accelerating wine aging process. *Crit. Rev. Food Sci. Nutr.*, 54, 817–835, 2014.

Tao, Y.; Sun, D.W.; Górecki, A.; Błaszczak, W.; Lamparski, G.; Amarowicz, R.; Fornal, J.; Jelinski, T. A preliminary study about the influence of high hydrostatic pressure processing in parallel with oak chip maceration on the physicochemical and sensory properties of a young red wine. *Food Chem.*, 194, 545–554, 2016.

Valero, A.; Begum, M.; Leong, S.L.; Hocking, A.D.; Ramos, A.J.; Sanchis, V.; Marin, S. Effect of germicidal UVC light on fungi isolated from grapes and raisins. *Lett. Appl. Microbiol.*, 45, 238–243, 2007.

Vivas, N. *Manual de tonelería*. Ed. Mundi-Prensa. Madrid. 2005.

Yurdugül, S.; Bozoglu, F. Effects of a bacteriocinlike substance produced by *Leuconostoc mesenteroides* subsp. *cremoris* on spoilage strain *Lactobacillus fructivorans* and various pathogens. *Int. J. Food Sci. Technol.*, 43, 76–81, 2008.

Zaitzeva, A.F.; Nikitine, N.I. Hydrolyse aqueuse du bois de chêne. *J. Chem. App.*, 24, 392, 1951.

Zoecklein, B.W.; Fugelsang, K.E.; Gump, B.H.; Nury, E.S. *Wine Analysis and Production*. New York: Chapman & Hall, 1995.

5 The Stability and Stabilization of Beer

Paul S. Hughes

CONTENTS

5.1 SUMMARY

The improved management of oxygen in the brewing process has substantially affected the incidences and types of microbiological infections in the brewery, reduced the rate of beer flavor deterioration in-pack, and contributed to the retardation of haze formation during storage. Innovations in terms of further microbiological stabilization have largely focused on rapid and sensitive detection, identification, and enumeration of microorganisms, with modern molecular biology techniques generally providing more information than classical techniques. The maintenance of hygiene in dispense lines, though, is challenging, particularly if biofilms form, as they can be difficult to remove. The further stabilization of beer flavor today looks to better logistics for the product from brewery to consumer, both in terms of agitation and minimizing temperature rises and temperature cycling. The challenges are greatest for those products that have an extended distribution chain, such as exported products. However, as we have seen, the complexity of beer chemistry makes it unlikely that there will ever be an infinitely stable beer.

The brewer arguably has all the tools required today to manage the formation of haze in beer. Tried-and-tested palliatives, such as protein and/or polyphenol binders, are effective and the use of proline endoproteases to cleave proline-rich domains in haze-active proteins has clearly demonstrable benefits. Because of the challenges in maintaining a stable beer, many brewers take the view that fresh = good, an attitude that is reflected, for instance, with dates of production included on labels in addition to "best before" dates. Fortunately for the brewing industry, old beer is not dangerous, merely less palatable than its fresh counterpart.

5.2 INTRODUCTION

The creation of drinks that can be classified as beer has venerable origins. Beer gained popularity not only because of its alcohol content but also because beer has historically been a reliable source of potable water. Indeed, John Snow is credited with the first study in epidemiology when he was able to identify a well in central London as being the source of cholera, noting during his investigation that local brewery employees were able to escape cholera, because they were drinking beer as part of their diet (Snow, 1855). We now know that beer generally does not support pathogen growth and, as production invariably requires a boiling operation, is effectively sterilized. Nevertheless as we will see in this chapter, beer is generally not infinitely stable in the face of bacterial onslaught, and this is arguably still the most pressing problem in terms of beer stability.

As beer production became more industrialized and centralized, particularly from the eighteenth century onwards, there was increasing demand for a longer shelf-life, with biological instability being the limiting parameter. However, with the development of techniques such as pasteurization, biological stability was brought under control so that by the mid-1930s, oxygen-related damage became more apparent, expressed initially as haze formation, and then flavor, foam, and light instability (Figure 5.1). Here, then, we outline the four potential stability issues that can

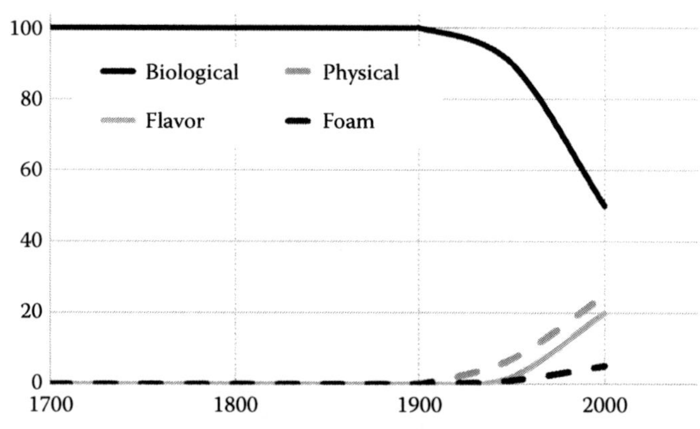

FIGURE 5.1 An estimate of relative efforts to manage beer stability as a function of time.

compromise quality in the market—microbiological stability, flavor stability, physical stability, color/light stability—and discuss the solutions employed to minimize or eliminate specific stability risks. Finally, we consider potentially new approaches to control product quality more closely.

5.3 MICROBIOLOGICAL STABILITY

Of all the stability issues that face brewers, none have been more problematic to the industry in general than biological instability, a subject that has been amply reviewed by Priest and Campbell (2003). It is well known that conventional beer does not provide ideal conditions for the proliferation of beer spoilage bacteria or "wild yeasts," given the presence of ethanol, hop-derived bitter acids (active against many Gram-positive bacteria), carbon dioxide, acid pH values (3.7–4.6), and low levels of nutrients. Some authors point to low oxygen levels as a bacterial and mold inhibitor (e.g., Hill, 2009), but it is worth noting that improved oxygen control throughout the beer production process has coincided with the rise in microaerophilic and anaerobic bacterial infections, with *Megasphaera* and *Pectinatus* species being the most prominent examples (Paradh et al., 2011). Additionally, the extension of the beer category into low-hopped and low- or nonalcoholic variants has contributed to increased risk of microbiological spoilage of beer in-pack.

Despite the broadly inhibitory conditions that exist in conventional beers, they are susceptible to infection from a broad range of spoilers. Indeed, the adoption of hops, at least in England, was driven by the recognized preservative qualities that hops had on beer (Scot, 1574). More recently, a survey of 500 samples from all parts of the brewery by Thelen et al. (2004, 2006) showed that around 10% of the samples were contaminated by various lactic acid bacteria (Figure 5.2), with a preponderance of *Lactobacillus brevis*, which is generally considered to be the most common

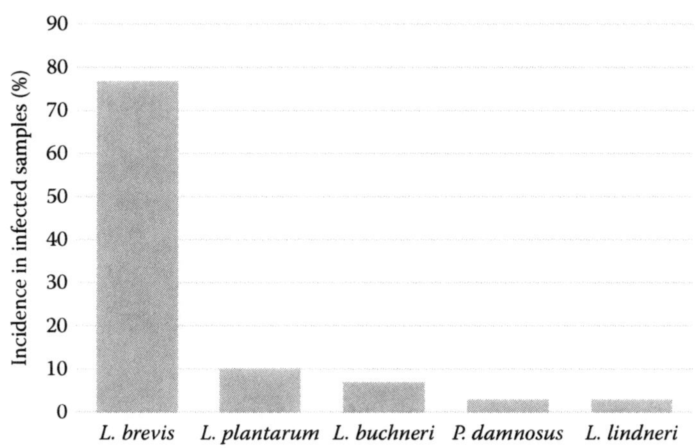

FIGURE 5.2 Relative levels of various infecting lactic acid bacteria. (After Thelen, K. et al., Master Brewers' Association of the Americas, *Technical Quarterly*, 2004, 41, 115–119; Master Brewers' Association of the Americas, *Technical Quarterly*, 2006, 43, 31–35.)

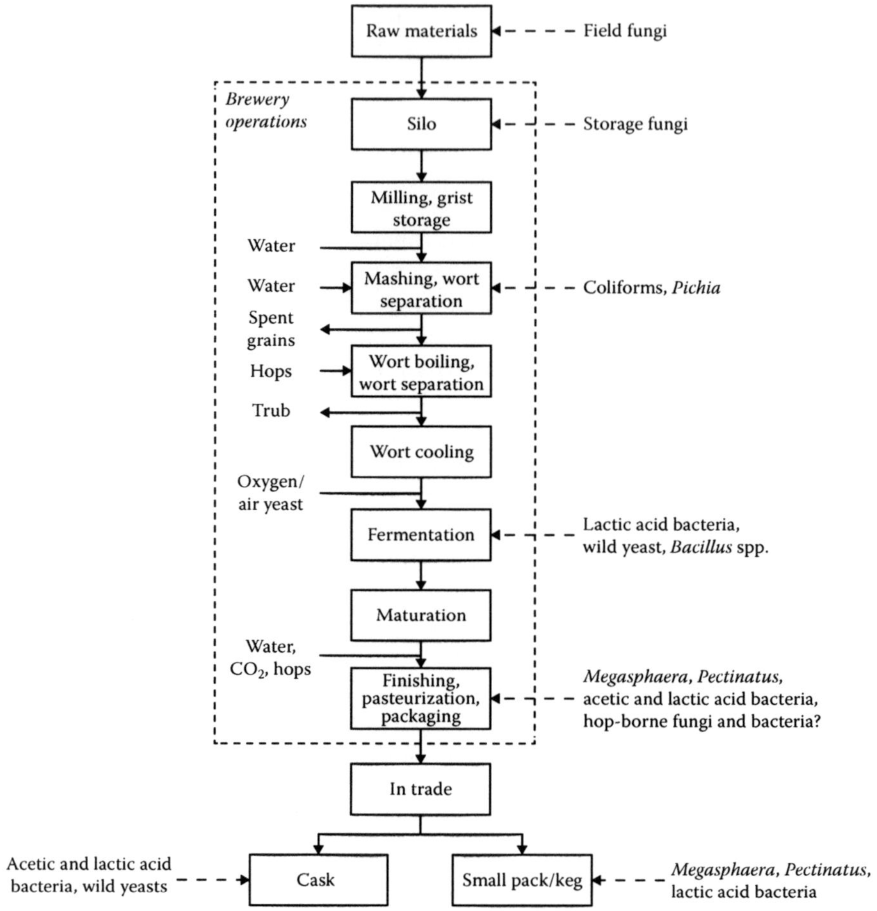

FIGURE 5.3 Microbial risk throughout the brewing process.

beer spoilage bacterium. Wild yeasts (i.e., yeasts that are not deliberately introduced into the process) occur as contaminants in the pitching yeast culture, from air or introduced with the raw materials. The increasing use of dried yeasts during beer production can arguably elevate the risk of bacterial and wild yeast infections during fermentation. To reduce the risk of microbial infection during beer production, control and management is required through all steps of the process, from raw materials intake and storage up to the point of consumption, whether that is from kegs, casks, or small pack (Figure 5.3).

5.3.1 Raw Materials

Insufficient hygiene and raw material contaminants are the biggest single routes to microbial infections in the brewery. Indeed, many brewers emphasize the hygiene aspect by such statements as "brewing is 90% cleaning" and "a brewer is almost perpetually wet!" So, the focus of the process, over and above producing consistently good-quality beer, is to kill off these contaminants before they damage the final product. Raw materials, such as barley and malt, are commonly infected by both field and storage fungi, while the presence of damaging bacteria on barley is less common. There are two main issues associated with fungal contamination of raw materials: the development of potentially harmful mycotoxins and the stimulation of uncontrolled release of carbonation, resulting in excessive foaming known as gushing. Additionally, recent work suggests that malt and barley microflora impact the production of malted barley (Juste et al., 2014).

While soaking barley (known as "steeping") can reduce levels of mycotoxins on the grain, the resulting waste water will be enriched in these mycotoxins. Mycotoxins are fungal metabolites that are harmful to animals and/or humans. Over 500 mycotoxins have been identified, and aflatoxin B1, ochratoxin A, zearalenone, deoxynivalenol (DON), the fumonisins, H-2 and HT-2 toxin have all been found in beer. Indeed, despite raw materials having limits set on specific mycotoxin levels by, for instance, the European Union, recent reports suggest that DON and HT-2 toxin in particular pose a potential health threat to heavy consumers of beer (Rodriguez-Carrasco et al., 2015). The transfer of mycotoxins from raw materials to final beer has been amply demonstrated (e.g., Schwarz et al., 1995; Scott, 1996), but there is also a risk to livestock if contaminated spent grains are passed on to the feed chain (Gonzalez Pereyra et al., 2011).

In production terms, some reports suggest that mycotoxins influence malting (e.g., Wolf-Hall, 2007) and also affect fermentation (e.g., Hill, 2009). The gushing phenomenon is typically sporadic and is associated with *Fusarium* contamination in the field. Mechanistically, gushing is now generally attributed to the fungal production of intensely hydrophobic polypeptides known as hydrophobins, which substantially enhance the foamability of beer (Sarlin et al., 2005). Arguably, the most important raw material for a brewery is its water, not only for the product itself but also for cooling and cleaning duties. As the water introduced into the brewing process is heated (during mashing) and then boiled (during hopped wort production), the main contamination risk is postfermentation, when water is often added to adjust beer to the appropriate level for sale, and if rinse water for cleaning is microbially contaminated. There are a range of methods available for treating brewing water. Membrane filtration and treatment with chlorine dioxide are both effective means of reducing or eliminating microbial load, and it is worth noting that while carbon filtration efficiently removes a wide range of chemical species, such as chlorine and salts, it is generally ineffective as a process for removing bacteria. Public water supplies are rigorously screened for fecal coliforms, introduced from sources including farm livestock and sewage effluent. Pathogens such as *Salmonella* and *Vibrio cholerae* do occur occasionally, but *Aerobacter aerogenes* is considered to be the most detrimental to brewed wort quality, not least because it is able to cleave α-1,6-glycosidic bonds in amylopectin and limit dextrins.

Algae and fungi also cause problems within a brewery, both in terms of producing flavor taints and blinding filters. Pitching yeast is the most significant source of bacterial contamination in the brewery. In particular cropping and repitching of contaminated yeast helps to propagate any infection into successive fermentations. Wild yeasts have a range of influences on final beer flavor (Back, 1987), including the development of so-called phenolic off-flavor (POF), caused by the ability of yeasts that have the POF gene to effectively decarboxylate phenolic acids such as ferulic acid to form, in that case, the highly flavor-active 4-vinylguaiacol. It is worth noting, though, that what one brewer considers to be a wild yeast is not necessarily the same as the view of another brewer. Thus, *Brettanomyces* "wild yeast" is in fact a desirable organism for some beers, including the Belgian Trappist ale, Orval. Bacterial contamination at pitching can also affect the rest of the beer production process. Lactic acid bacteria, such as *Pediococcus damnosus*, slow down fermentations and are also effective diacetyl producers. Gram-negative bacteria, such as *Hafnia proteus*, have similar effects and also increase the levels of dimethyl sulfide in beer. A common remedial step for removing bacterial contamination from yeast is to include an acid washing step. Typically around pH 2, this treatment is effective, but does not affect any wild yeasts that may be present.

The introduction of hops seems to play little part in terms of contribution of spoilage organisms into the brewing process. Conventionally, hops are boiled with the wort, effecting a sterilization. Nonetheless, the practice of substantial hop additions for dry hopping in some beer styles postfermentation may risk the longer term microbial stability of beers because, here, typically unprocessed or minimally processed hops are used and there is no subsequent stabilization or boiling step. The extension of brewing capacity by adding sugars or syrups to increase the concentration of wort sugars (euphemistically known as process intensification) does not generally add additional microbiological risk to the process. The osmotic strength of the added syrups precludes most spoilage, but they may contain sporulated bacteria, which can withstand boiling. Additionally, areas in storage tanks where there is relatively little osmotic stress (e.g., condensate at the top of storage tanks) can be a haven for bacteria and fungi and a potential source of flavor contamination. Sugars and syrups added to final beer for priming or sweetening should, of course, be free of contamination as their dilution into final beer will reduce osmotic pressure and allow any potential spoilers present to grow.

5.3.2 Wort and Fermentation

Sweet wort is produced with the express intention of being fermented by microorganisms (very often *Saccharomyces cerevisiae*), so it is both sugar- and nutrient rich. *H. proteus*, an enterobacterium, can cause unwanted off-flavors in the final product and, interestingly, is only found in breweries. Other microbial contaminants can yield off-flavors, such as hydrogen sulfide, acetaldehyde, acetic acid, diacetyl, phenolics, and estery (fruity/solvent) attributes, as well as cause haze. Once boiled, any microbes present will be largely destroyed, although some, such as *Bacillus* spp., can sporulate and resist heat treatment, persisting into final beer.

During fermentation, the major microbial culprits are lactic acid bacteria. Also, *H. proteus* and *Rahnella aqualitis* can survive at significant ethanol levels, so they can be harvested with cropped yeast and repitched. Aerobic acetic acid bacteria, such as *Acetobacter* and *Zymomonas* spp., as well as the anaerobic *Megasphaera cerevisiae* and *Pectinatus* spp., can result in off-flavors and substantial haze formation.

High-gravity fermentation can provide protection to some extent, with *H. proteus* and *R. aqualitis* eliminated by the higher alcohol levels achieved here, although other spoilers, such as lactic acid bacteria and *Acetobacter*, can survive elevated alcohol levels (Magnus et al., 1986). In recent years, low(er) calorie beers have become increasingly popular. The combination of lower levels of residual sugars and robust alcohol contents makes such beers less susceptible to spoilage. In contrast, low- or nonalcoholic beers are, in principle, more susceptible to spoilage, as the inhibitory "hurdle" of the presence of alcohol is reduced or missing. Nevertheless, one method of low-alcohol beer formation, that of producing a wort of low fermentability and fermenting the modest levels of simple sugars present, does restrict opportunities for microbial activities to establish themselves.

5.3.3 Beer Finishing

As has already been mentioned, beer generally has antimicrobial properties, but these are far from perfect, and various bacteria can cause off-flavors if they gain access to storage and/or finishing tanks. Most commonly, these are the acetic acid and lactic acid bacteria. The aerobic bacteria can, at least, be held in check by avoiding air or oxygen ingress. Nitrogen or carbon dioxide atmospheres in tanks, deaeration of all water postfermentation, and keeping pipelines filled with deaerated water when not in use can all assist in reducing the risk of microbial contamination of the final product. A common area for microbial contamination is in the packaging hall. The complexity of packaging equipment provides many opportunities for microbial proliferation and biofilm formation. In common with other surfaces, including dispense lines, biofilms develop to protect the microbial community and can require substantial deep cleaning to remove once they are well established. A mere few cells per bottle can result in a spoiled beer reaching the consumer. Beer spoiling lactics result in acidity and haze formation, while *Pectinatus* spp. and *Megasphaera* sp. spoil unpasteurized beer by producing mercaptans including hydrogen sulfide (rotten egg odors), dimethyl sulfide (sweetcorn aromas), and substantial levels of turbidity. Additionally, there can be issues with overattenuation, where bacteria continue to produce alcohol in-pack, diacetyl (butterscotch) formation, and the formation of bacterial "rope." Dispense systems are prone to spoilage, especially in terms of biofilm formation. Here, lines that are not cleaned sufficiently well or regularly become increasingly restricted due to the initial adhesion of bacteria and wild yeasts and then the formation of biofilms. In cask beers, where air is present in the headspace of the beer, aerobic bacteria can thrive, as can be detected in "old" casks by the clear presence of acetic acid (vinegar off-flavor).

5.3.4 Methods of Detection

It is not the intention here to review in detail the methods that can be used to detect and enumerate microorganisms in the brewery or the final product. Briefly, traditional methods of detection and identification require sampling and microscopy. Samples can be used to inoculate selective growth media and are allowed to grow for up to 3 days. Such methods are not completely satisfactory, especially if a higher degree of characterization is required. The methodology is also time consuming. Scientific and technological developments and time constraints, alongside a trend to reduce or eliminate pasteurization, have stimulated efforts to find better and faster diagnostic techniques. These can be classified as physical methods, biochemical methods, and molecular methods (Table 5.1).

TABLE 5.1
Summary of Rapid Microbial Detection and Identification Techniques Available

Class	Method	Approach
Physical methods	Impedance/ conductance	Measurement of impedance or conductance directly related to cell growth curves
	Microcalorimetry	Relies on bacterial growth releasing heat that can be detected using this technique; uncommon in brewing
	Turbidimetry	Based on increased visible light absorbance of samples containing cells; requires 2–4 days' growth
	Flow cytometry	Laser light scattering from individual cells, in the presence or absence of fluorochromes used to enumerate mixtures of bacterial cells
	Microcolony	Uses microscopy and selective staining to detect colonies before they become visible
Biochemical methods	Direct epifluorescence filter technique (DEFT)	Enhancement of the microcolony method; combines filtration with epifluorescent microscopy to yield results within 30 min
	ATP bioluminescence	Used mainly for surface swabbing; uses firefly bioluminescence chemistry to detect bacterial ATP
	Protein fingerprinting	Essentially a laboratory technique to separate and quantify protein portfolio to identify bacterial strains
	Immunoassay (ELISA)	Use of mono- or polyclonal antibodies and ELISA to enumerate specific bacterial species
	Gas chromatography	Measurement of secondary metabolites, such as organic acids, to detect bacteria, especially lactics and *Pectinatus*
Molecular methods	Molecular probes (DNA/RNA)	Molecular probes are small DNA or RNA segments that recognize complementary DNA or RNA sequences to identify specific sequences from an organism
	Ribotyping/ riboprinting	Similar to DNA molecular probes, but used to replicate ribosomal RNA (rRNA) to identify organisms
	Karyotyping	Use of electrophoresis to separate individual chromosomes; often used for yeast characterization
	End-point PCR	Use of specific DNA primers to replicate DNA sequences specific to selected bacteria
	qPCR	Also known as real-time PCR, where DNA amplification is monitored throughout cycling rather than at the end, to enable (semi)-quantitation of DNA molecules

After: Hill, A.E., In *Beer—A Quality Perspective*, eds. Russell, I., Bamforth, C.W. and Stewart, G., Elsevier, London, 2009, pp. 163–183.

5.3.5 Reducing the Risk of Microbial Contamination

While it is essential to understand where microbial contamination might occur in the brewery and final product, prevention is, in principle, to be favored over a cure. Rigorous and regular cleaning is *de rigueur*, despite the impact it has on water, cleaning agent, and staff use. Restricting the presence of air in vessels and pipework—for instance, by filling with an inert gas such as nitrogen (that has previously been sterile filtered)—reduces the incidence of all but the anaerobic bacteria. We have already stated that pitching yeast is a significant source of contamination, so keeping yeast anaerobic and acidic helps it to compete more successfully with contaminating bacteria.

Traditionally the stabilization of finished beer has been affected by the application of pasteurization. There are two broad types of pasteurization: tunnel pasteurization, where the packaged beer is subject to heating and cooling cycles over a period of minutes, and flash pasteurization, where the beer is heated to elevated temperatures for a period of seconds before filling into the package. Tunnel pasteurization has remained a popular technique, despite high capital and running costs, as well as the perceived loss in product sensory quality, because it is a reliable microbial deactivation method. The use of nonpasteurizable packaging, such as certain plastics, means that either flash pasteurization or alternative approaches are also required. These include sterile filtration in clean rooms through membranes with nominal pore sizes of 1.2 or even 0.22 μm. Other, more experimental techniques such as pulsed electric fields and ultrahigh-pressure treatment have been evaluated. Some manufacturers also resort to the addition of preserving agents. Sorbate, benzoate, chitosan, and even bacteriocins have all been used in trials.

There is a need for diligence throughout the brewing process if the risks of microbial infection are to be minimized or eliminated. A combination of preventative strategies throughout production, and assiduous monitoring of the plant and the raw material inputs is required, as infections can result in problems ranging from product nonconformance to complete brewery shutdown for deep cleaning and sanitization. On-premise dispensing offers additional challenges as maintaining sterility of dispense lines is often outside the control of the brewer, relying on the diligence of the retail outlet management.

5.4 FLAVOR STABILITY

The development of approaches to the reduction of microbial spoilage in beer has arguably helped to make other areas of beer instability more apparent. Early discussions focused more on the damage that oxygen did to beer in storage, with the development of haze being considered more troublesome than flavor changes (de Clerck, 1934). Today, though, there are a range of effective palliatives for restricting the formation of haze in beer (see later discussion), but the situation is not so straightforward for flavor instability. There are various reasons for this. Firstly, the range of chemical reactions that can occur in beer, whether stimulated by oxygen/oxidation or not, is complex and diverse (Figure 5.4; for a comprehensive review of the chemistry of beer flavor instability see Vanderhaegen et al., 2006). While not

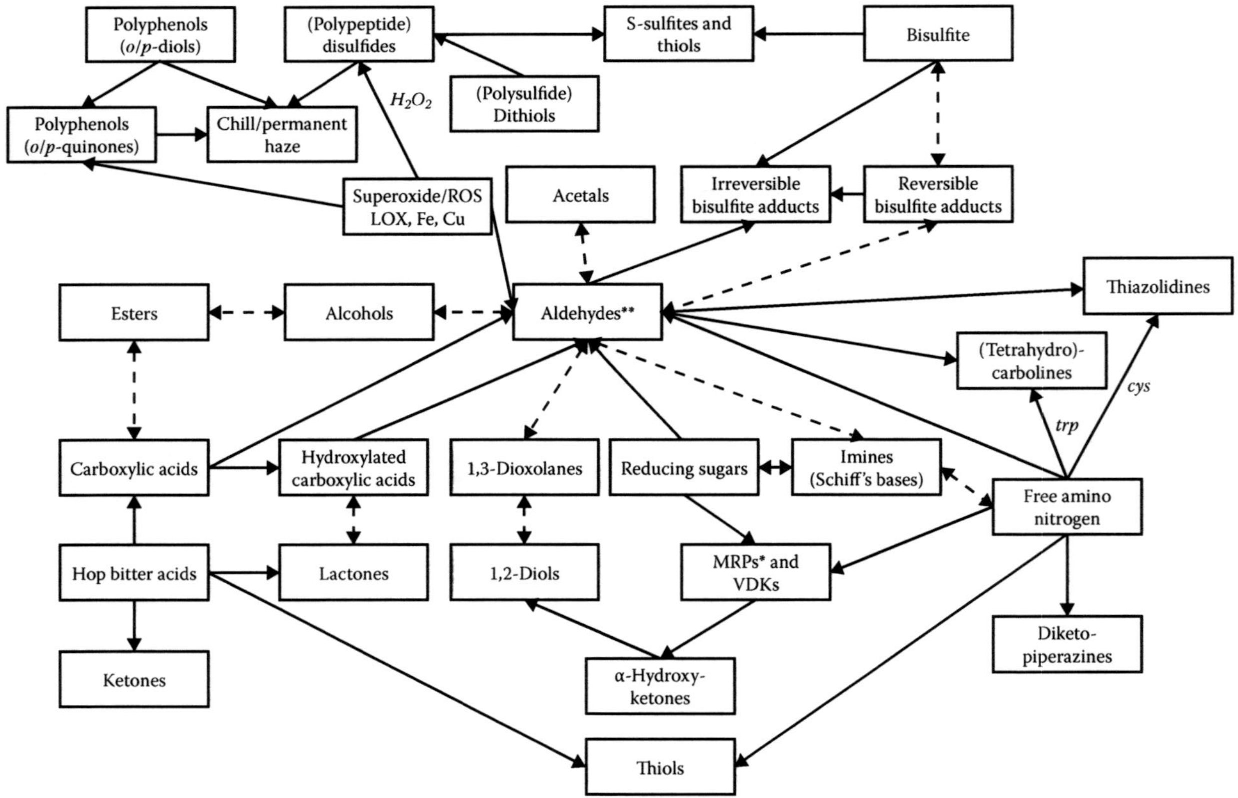

FIGURE 5.4 Simplified scheme illustrating the chemical complexity and diversity of beer chemistry. Many if not all of these pathways influence the flavor of beer in some way. (Dashed lines indicate equilibrium processes; cys = cysteine; trp = tryptophan; * MRPs = Maillard reaction products; ** certain aldehydes can react with themselves to form aldol condensation products.)

all of the pathways highlighted here necessarily result in flavor changes in beer, many can. Additional complexity arises because the way in which a beer's flavor will change is dependent on storage temperature (Kaneda et al., 1995), and the balance of the degradation reactions in one beer will vary from that of another (Bamforth and Lentini, 2009). So, in essence, beer flavor changes in-pack are dependent on both beer composition and storage conditions. Historically, the compositional aspect that has received the most attention, due to its clearly demonstrable impact on both beer flavor and beer haze formation, is oxygen.

5.4.1 Oxygen and Reactive Oxygen Species (ROS)

Oxygen itself is sparingly soluble in beer, but has marginally more solubility in beers containing more alcohol and less dissolved solids. Cooler temperatures also increase oxygen solubility in beer. However, as we will see later, the presence of oxygen during mashing can begin a cascade of reactions resulting in the oxidation of lipids to highly flavor-active aldehydes such as E-2-nonenal, with molecular oxygen presumably partitioning into lipid phases in the mashing system preferentially due to its lipophilic nature. Oxygen itself is relatively unreactive. It exists in a triplet ground state, but can be converted into "active" oxygen by various agents, including iron and copper. From molecular oxygen, a range of ROS can be formed (Figure 5.5). For instance, iron(II) can donate an electron to singlet oxygen to generate the superoxide or hydroperoxyl radical (depending upon pH; at beer pH hydroperoxyl will predominate). Copper(II) can regenerate oxygen from superoxide but the copper(I) formed can reduce superoxide to peroxide. Another set of reactions mediated by redox metals are those where iron(II) reacts with hydrogen peroxide through a series of reactions to form the hydroxyl radical, a process known as the Fenton reaction. The hydroxyl radical is the most reactive ROS and is rather unspecific in its

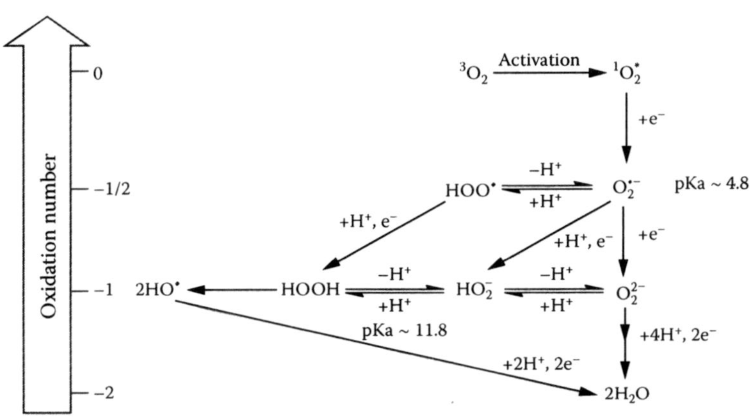

FIGURE 5.5 Summary of the relevant ROS that can be formed in beer during production and aging. The hydroxyl and peroxyl radicals, together with hydrogen peroxide, are the principal culprits in beer, while, at higher pH values, such as those during mashing, the superoxide radical anion plays a more significant role (e.g., in lipid oxidation).

reactivity, being able to react with a wide variety of molecular entities. This is in contrast to, say, the hydroperoxyl/superoxide radical, which exhibits more selective reactivity and is a key element in the oxidation of unsaturated fatty acids, which in turn can lead to the formation of flavor-active aldehydes.

5.4.2 Sulfur

It is fair to say that the role of sulfur in beer flavor instability is not yet fully understood. Muller (1997) has shown that oxidation of sulfhydryl groups during mashing leads to the formation of disulfide bridges and the reduction of oxygen to hydrogen peroxide. Additionally, reactive sulfur thiolyl radicals can be produced from free thiols in the presence of copper(II), which can then proceed on to other nonselective reactions, not least of which is the activation of molecular oxygen.

5.4.3 Sources of Flavor Changes in Beer

As we have seen previously, there is a wide range of chemical reactions that can occur to cause beer flavor changes over time. This is not only due to the formation of undesirable flavor notes, but can also be due to the loss of desirable flavors. With both occurring simultaneously, there is also a threat to the overall balance of the beer. The main (bio)chemical pathways to the flavor instability of beer are summarized in Table 5.2. Areas for controlling beer flavor stability are indicated in Figure 5.6.

Much has been made in the past of the importance of E-2-nonenal to the expression of beer flavor changes in-pack. This is understandable given its flavor potency (flavor threshold of around 0.1 μg/L in beer; Meilgaard, 1975; or <1 pM), so that the chemistry for its formation does not need to be high yielding and can be a mere leak in a reaction pathway. Indeed, the only reason to consider compounds such as E-2-nonenal seriously is precisely because it is so readily detectable by the human olfactory system. However, as we have seen, the chemistry of beer is complex and it is overly simplistic to focus only on E-2-nonenal as a sole source of beer flavor changes in-pack. Other aldehydes, particularly the Strecker aldehydes, also make a significant contribution to flavor changes in-pack and cannot be ignored in any detailed discussion on beer flavor stability.

From Figure 5.4 it is apparent that many other beer components affect the levels of aldehydes directly, either through reversible adducts, such as Schiff's bases, bisulfite adducts, and acetals/dioxolanes, and irreversibly through the formation of compounds with certain amino acids such as tryptophan and cysteine. In turn these components that directly affect the activity of aldehydes are themselves subject to competing reactions, with bisulfite demonstrating pseudo-first-order depletion kinetics in a wide range of beers (Ilett and Simpson, 1995). In addition to aldehydes, other potentially flavor-relevant species include lactones, acetals, and changes in ester and sulfur compound profiles. It is difficult to see how many of these chemical reactions can be completely prevented in the quest for a perfectly flavor-stable beer. A pragmatic approach is to consider infinite flavor stability as a notional goal, in order to guide a path towards a product that is sufficiently flavor stable. Appraisal of flavor

TABLE 5.2
Main Routes of the Chemical Changes in Beer during Aging

Chemistry	Summary
Fatty acid oxidation: enzymic	Lipoxygenases catalyze the formation of fatty acid hydroperoxides, primarily from linoleic (18:2) and linolenic (18:3; Doderer et al., 1992) acids. This starts a cascade of enzymic and chemical reactions that culminates in the formation of flavor-active aldehydes, including the infamous E-2-nonenal. Barleys bred to have low- or no LOX are considered by some to yield more flavor-stable beers (Hirota et al., 2005).
Fatty acid oxidation: nonenzymic	Arguably, the major contributing factor to the oxidative degradation of barley-borne fatty acids. Oxygen, activated by iron to produce superoxide and hydroperoxyl radicals, induces the reaction sequence resulting in staling aldehydes (Bamforth, 1999).
Hop acid oxidation	Oxidation of iso-α-acids in model systems yielded a broad range of unsaturated aldehydes (Hashimoto et al., 1979). β-Acid oxidation generates a series of furanones (Moir, 1994) and hulupones. Additionally, hop oxidation can yield significant levels of short-chain fatty acids. During beer aging, hop bitter acids are depleted (Hughes et al., 1997).
Higher alcohol oxidation	Earlier work suggested that melanoidins in beer acted as hydrogen acceptors to oxidize alcohols to aldehydes (Hashimoto, 1972). Later work (Irwin et al., 1991) suggests that this mechanism is likely to be an insignificant contributor to beer flavor instability.
Strecker degradation of amino acids	The initial reaction of amino acids with reducing sugars to form Amadori products (α-aminoketones) leads directly to the formation of 1,2-dicarbonyls, which in turn react with amino acids to form so-called Strecker aldehydes. The main amino acids here are methionine, valine, leucine, isoleucine, and phenylalanine. The resulting aldehydes are considered to be significant contributors to beer flavor instability.
Aldol condensation	This classical reaction occurs between an enolizable carbonyl compound and a ketone or aldehyde. The resulting β-hydroxyketone ("aldol") undergoes dehydration (if there is an α-hydrogen present) to form an α,β-unsaturated carbonyl compound. In principle, E-2-nonenal could form from heptanal and acetaldehyde (Hashimoto and Kuroiwa, 1975).
1,3-Dioxolane formation	These cyclic acetals are formed between 1,2-diols (e.g., butane-2,3-diol) and aldehydes (Peppard and Halsey, 1981). The reaction product between butane-2,3-diol and acetaldehyde has a pronounced sherry/dried fruit character.
Carbonyl binding: bisulfite	The reversible addition of bisulfite to carbonyls is a well-known chemical reaction. In beer this reaction is likely to reduce the flavor activity of aldehydes, especially the simpler aldehydes, which have a greater tendency to form adducts. Nyborg et al. (1999) and Dufour et al. (1999) showed that bisulfite can also add irreversibly to the carbon–carbon double bond of α,β-unsaturated aldehydes.
Carbonyl binding: amino acids and proteins	Similarly, free amino nitrogen can form reversible imines with aldehydes and indeed this is the same chemistry that ultimately gives rise to the initial Amadori products during the formation of Maillard reaction products.

(*Continued*)

TABLE 5.2 (CONTINUED)
Main Routes of the Chemical Changes in Beer during Aging

Chemistry	Summary
Carbonyl binding: thiazolidines	Recent work (Baert et al., 2015) has shown that aldehydes can be bound into cyclic thiazolidines by reaction with cysteine.
Carbonyl binding: carbolines	A range of carbonyl compounds have been shown to react with tryptophan and tryptamine to form tricyclic carbolines and tetrahydrocarbolines. The significance of this reaction in beer has yet to be established.
Carbonyl reduction: yeast	The presence of yeast can result in the reduction of carbonyl compounds. For instance, the maturation of beer to reduce diacetyl to acetoin, then butane-2,3-diol, and the reduction of acetaldehyde to ethanol, can be considered positive.
Enzymic flavor release	Yeast can also enzymically hydrolyze glucosidases, such as those with hop compounds (Goldstein et al., 1999). If beer remains unpasteurized, then such enzymes may persist into final beer and can cause age-related flavor changes.
ROS scavenging	The antioxidant activity of species such as polyphenols and phenolic acids allows for the scavenging of a range of ROS, including superoxide/hydroperoxyl and hydroxyl radicals (Walters et al., 1997a,b). Whether Maillard reaction products (MRPs) are antioxidant or pro-oxidant is a matter for debate, and is likely to be due to the detailed chemistry of the broad range of reactivities that go to make up the broad range of MRPs.
Removal of hydrogen peroxide	Hydrogen peroxide can be generated by the ROS-mediated cross linking of protein thiols to disulfide bridges during mashing (Muller, 1997), but this in turn can be removed by the presence of peroxidases and polyphenols (Clarkson et al., 1992). Thus the damage that hydrogen peroxide can cause is dependent on the relative contributions of the two competing (bio)chemical reactivities.
VDK release	If maturation of beer does not proceed to completion, residual acetolactate and acetohydroxybutyrate can slowly generate VDKs in-pack.
Changes in sulfur compounds	The existence of reducing conditions in beer can result in the reduction of DMSO to dimethyl sulfide (Bamforth and Parsons, 1985) and the reduction of bisulfite to hydrogen sulfide (Walker and Simpson, 1994). Other sulfur-derived flavor defects, such as ribes (blackcurrant leaf) can also develop on storage.
Changes in esters	Esters are in dynamic equilibrium with their component acids and alcohols. Thus if beer has any of these components out of equilibrium, during storage there will be a tendency to restore that equilibrium. A good example of this is the presence of malt-derived nicotinic acid in fresh beer, which develops ethyl nicotinate during storage (Narziss et al., 1999).
Flavor scalping	Certain packaging materials (e.g., some crown seals) are able to adsorb flavor-active compounds, especially hydrophobic species such as hop aroma impact characters, which can affect beer flavor during storage.

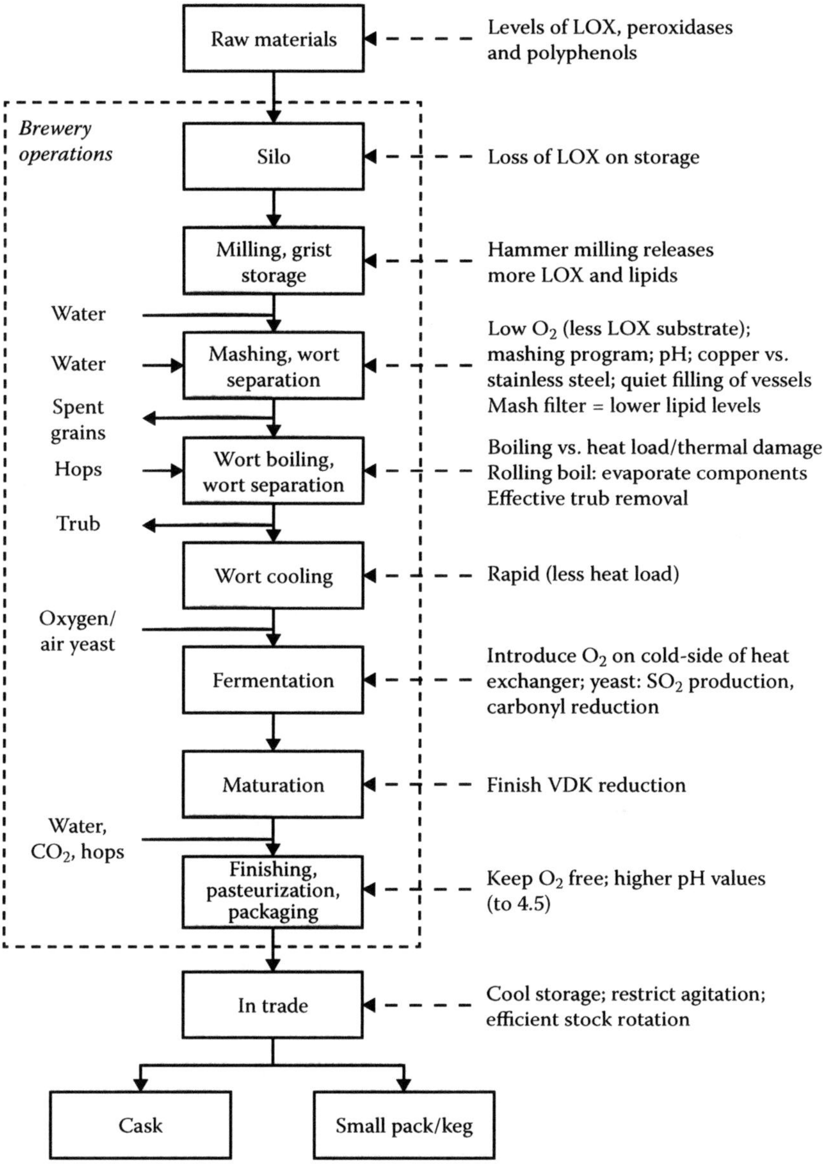

FIGURE 5.6 Simplified summary of the flavor stability issues throughout the brewing process. (After Bamforth, C.W. and Lentini, A., in *Beer—A Quality Perspective*, eds. Russell, I., Bamforth, C.W. and Stewart, G., Elsevier, London, 2009, pp. 85–109.)

instability at a specific time point does not necessarily convey the trajectory of beer flavor changes. Thus, in principle, product evaluation after a fixed aging period—say, 3 months—does not necessarily reflect the whole product timeline (Figure 5.7) and, indeed, if assessments are made on the basis of a sensory panel, data noise can make such determinations problematic.

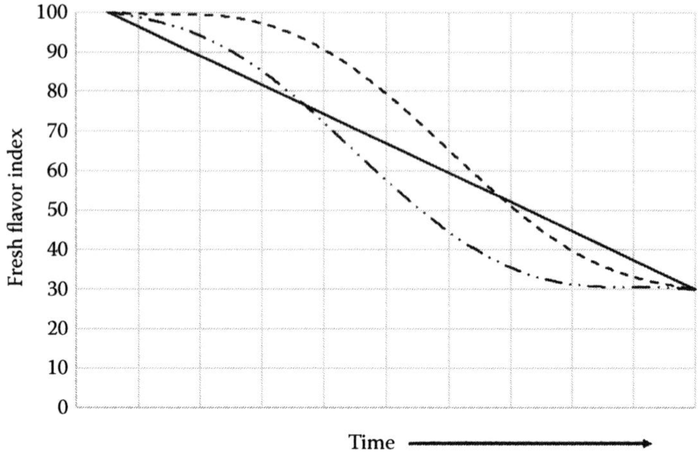

FIGURE 5.7 Hypothetical flavor trajectories for aging beers. Deciding which beer is the most stable depends on when that assessment is made. The beer showing a linear flavor deterioration generally performs less well than that which slowly deteriorates initially, but after a longer aging time performs better.

5.4.4 Determination of Beer Flavor Instability

The methods used for the assessment of beer flavor stability are generally analytical- or sensory based. With the analytical methodology, the focus is either on specific analytes, such as aldehydes or other staling indicators (e.g., Narziss et al., 1999) or on a measure of some property of the beer itself, such as evaluation of the antioxidant status of a beer over time. The choice of methodology is dependent on the required outcome, although from a consumer acceptance perspective, it is always prudent for a brewer to know how his or her products perform in trade from a taste perspective. Analytical methodology has focused on flavor-active compounds, particularly aldehydes (e.g., Vesely et al., 2003), and the determination of traces of redox metals such as iron and copper, as these two can cause rapid and demonstrable flavor changes in beer.

More recently, some brewers have adopted electron spin resonance (ESR) or, more accurately, electron paramagnetic resonance (EPR) methods to determine the concentration of free radicals in beer and during the brewing process (Uchida et al., 1996). The best-known published application is the measurement of the "lag time." Here, a spin-trap, typically PBN (*N*-*t*-butyl-α-phenylnitrone), is added to a beer sample and the whole sample is heated, typically at 60°C. Initially, there is no change in the total free radical concentration in the sample as measured by ESR, as those free radicals formed are considered to be bound by radical scavengers present in the beer. Eventually, the free radical signal begins to rise linearly with time, as scavenging activity is depleted. The time taken for the signal to become time dependent is termed the lag time and has been shown to correlate with a range of parameters, such as soluble iron and bisulfite activities. Bench-top ESR instruments are now installed in some breweries to maintain a check on potentially damaging shortening of lag times. Which radicals the spin-trap binds to is unknown, although hydroxyethyl radicals derived from ethanol seem to be likely candidates.

The quest for a perpetually flavor-stable beer is likely to be in vain. Nevertheless, brewers have made good progress in prolonging flavor shelf-life, something that is particularly pertinent to brewers that distribute their product widely or even internationally.

5.5 PHYSICAL STABILITY

As mentioned earlier, the early studies on the interaction of beer with oxygen focused primarily on the formation of haze. Arguably, haze is more of a concern than flavor deterioration in that it is an easier shared experience: Holding up a glass of beer, it is apparent to most whether a beer is hazy or not, whereas flavor perception is a more individual assessment and endeavor. Although haze has a minor impact on taste perception, beer consumers will make initial judgments based on what they see, and unexpected haze can negatively affect consumer expectations.

Beer contains significant levels of polyphenols and polypeptides, so it is perhaps unsurprising that beer is prone to the formation of particles with time. With the exception of certain niche beer types, such as wheat beers or agitated bottle fermented beers, this formation of haze is not only unacceptable to the consumer but also a further point of concern for the brewer. The common forms of haze are described next.

5.5.1 Polyphenol–Polypeptide Hazes

Reactions between polyphenols and polypeptides are widespread in society. The traditional process of leather tanning is the irreversible reaction of hide-bound proteins with added polyphenols. Polyphenolic compounds are often termed astringent. This is due to the reaction of polyphenols with salivary proteins, which are predominantly the so-called proline-rich proteins. Such hazes are the most widespread sources of haze in the brewing industry. The interaction is initially an intermolecular, presumably hydrogen-bonded, complex between the polyphenol and the protein. Further oxidative processes can then result in covalent linkages between the protein and polyphenol moieties. The formation of the initial complex is reversible, while the covalently linked aggregates are likely to persist.

There is scope for secondary hydrophobic interactions between polypeptides and polyphenols. Many commercial beers can develop so-called chill haze when kept cool for even relatively short periods of time. This disappears (redissolves) when the beer is allowed to warm up. If the beer is stored for longer, the formation of haze becomes increasingly irreversible, giving rise to permanent haze. The presence of chill haze potentiates the formation of permanent haze, so while the formation of chill haze is reversible, its propensity to form should be minimized to ensure good physical stability. The structure of haze formed in beer has received some attention. The best-known hypothesis is that put forward by Siebert and Lynn (2008; Figure 5.8). The model is based on the idea that only proline-rich proteins can interact with polyphenols to form chill haze and that there is a fixed number of binding

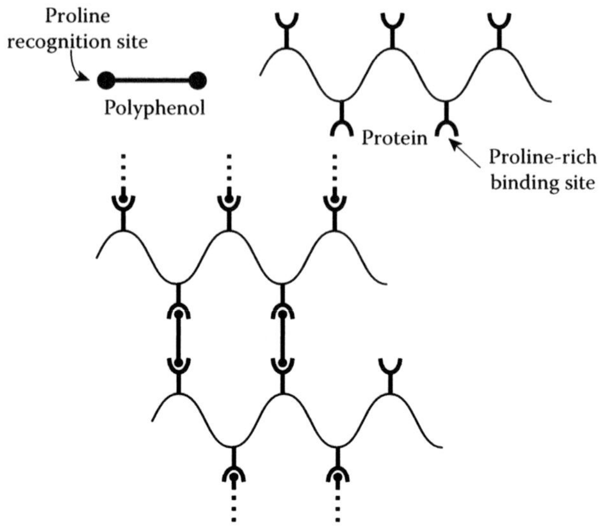

FIGURE 5.8 Model of polypeptide–polyphenol haze formation, taking account of the observed stoichiometry of such hazes. (After Siebert, K.J. and Lynn, P.Y., *Journal of the American Society of Brewing Chemists*, 2008, 66, 48–54.)

sites, implying a more or less definite stoichiometry for haze formation. For the formation of large aggregates, the model also needs to account for cross linking, so the polyphenols must have at least two protein-binding sites.

More recently, molecular models of polyphenols with a segment of polyproline indicate that it is precisely the open coil structure of polyproline domains that are a good steric and electronic fit to more complex polyphenols (Figure 5.9). Various downstream operations have been devised to minimize or prevent the formation of protein–polyphenol hazes. Removal of one or the other of the reactive species

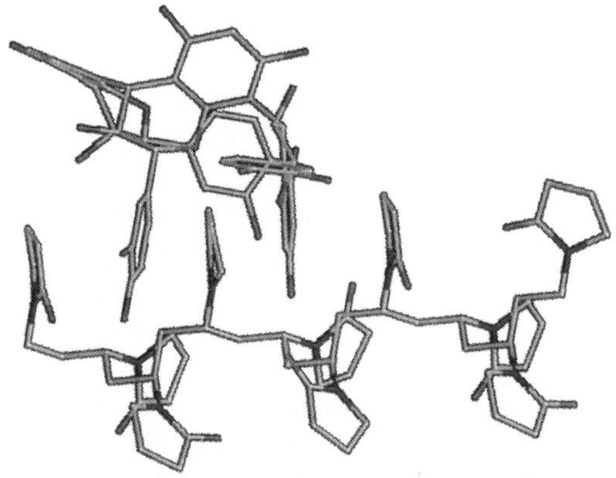

FIGURE 5.9 Molecular model illustrating the steric fit of a polyphenol, to a decaproline residue (Hughes, unpublished results). Note the intercalation of the phenol rings with the imide rings of the proline.

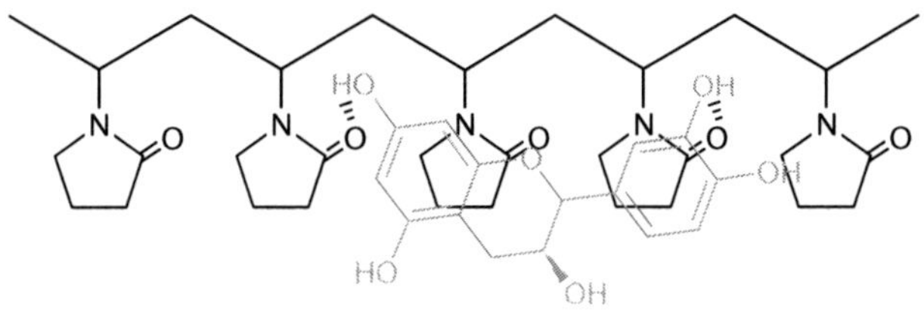

FIGURE 5.10 Two-dimensional representation of a PVP segment overlaid with epicatechin. (Rehmanji, M. et al., *Technical Quarterly*, 2005, 42, 332–338.)

is one option. Indeed, there are commercially available silicas that can adsorb beer polypeptides in an effort to reduce their final levels in beer. Many of these products are designed specifically to remove haze-forming polypeptides, as there is a concern that treatment with excess quantities of silica could compromise beer foam stability. More recently, various enzymes have been developed that specifically cleave proline–proline peptide bonds. Enzymes such as Brewer's Clarex™ have been shown to be effective in retarding the formation of hazes in beer.

Another approach is to remove the polyphenols present. Here again initial research showed that nylon polymers could do this effectively. Today, polymers of vinylpyrrolidone (PVP and its cross-linked variant, PVPP; Figure 5.10) are used that can effectively reduce the polyphenol content of beer typically by a factor of two to three. The key feature of both nylon and PVP/PVPP is the presence of amide linkages so that they retain the hydrogen bonding sites that polyphenols favor on polypeptides. As with polypeptide removal, the removal of polyphenols should be done judiciously, as these compounds may provide protection against the deterioration of beer flavor. Often a combination of treatments is employed in the expectation of finding a compromise solution. The model of polypeptide–polyphenol haze is, though, a little simplistic as metal ions have been shown to accumulate in haze particles and recent reports show that sugars are also present.

Finally, cold filtration is often effective for enhancing the physical shelf-life of beer. Beers freeze at subzero temperatures, the exact temperature being dependent on the rate of cooling and the specific beer composition. Ideally, filtration should be carried out close to the freezing point of a particular beer (typically around –2°C), especially if the beer has been stored cold so that any chill haze has had the opportunity to form. Removal of as much chill haze as possible prior to packaging has been observed to retard the formation of polypeptide–polyphenol hazes. However, in practical terms, it is important to minimize the risk of beer freezing on the filter, which limits how cold the beer can be filtered.

5.5.2 Calcium Oxalate

Oxalic acid is released from malt during mashing. It forms sparingly soluble salts with calcium, the major divalent cation in beer. Its reported solubility is 0.67 mg/L in water, although Molzon (1976) reports that in 10% (v/v) ethanol in water, calcium oxalate solubility is more than halved. If there is insufficient calcium upstream, the oxalate is not effectively precipitated and is a potential "time bomb" in-pack. Such crystals not only cause haze but also can provide a multitude of nucleation sites for the nucleation of dissolved gases in beer. On opening the package, the rapid gas nucleation results in uncontrolled foaming or "gushing" and, in such situations, can result in substantial beer loss. This effect is amply demonstrated if small particles of glass are added to a bottle of beer, although it is best if this is performed over a sink!

5.5.3 Carbohydrates

Starch (α-glucan) itself should be degraded upstream by the amylolytic activity during mashing. Any starch remaining, though gelatinized, undergoes a process known as retrogradation when it cools. Retrograded starch is more resistant to enzymic breakdown than conventional starch and, when formed, can result in haze. This haze may be clearly visible or, alternatively might be deemed "invisible." The latter, while not immediately apparent to the eye, nonetheless scatters light in conventional haze measurements. Similarly, β-glucans, which should also be substantially absent from the final beer, can also give rise to visible and invisible hazes. Wheat is used as an adjunct for the production of certain beers and is relatively rich in pentosan sugars. Again, such biopolymers can give haze problems in the final product.

5.5.4 Other Sources of Haze in Beer

Some other phenomena can result in haze in the final beer, although they are generally rare. The exogenous foam stabilizer propylene glycol alginate (PGA) can give rise to haze problems under two circumstances. Firstly, it should not be added downstream simultaneously with the chill-proofing enzyme papain as they tend to react together, forming haze. Secondly, PGA can react with beer proteins if the beer experiences elevated temperatures (>45°C) for a few days. The latter might seem inconceivable, but exported beers can be exposed to such temperatures during distribution. Dead bacteria from malt, carbohydrates and proteins from damaged yeast, and lubricants used in canning or on packaging conveyors can also give haze in the final beer.

5.6 COLOR AND LIGHT STABILITY

The sensitivity of beer to light is well known and has been studied for well over 100 years. While many chemical reactions in beer are initiated by UV and visible light, the two most important are those that lead to the formation of methanethiol (MT) and 3-methyl-2-butene-1-thiol (MBT). These compounds are highly flavor active in a typical beer, with the threshold of MT reported as 2 µg/L and for MBT as low as 2 ng/L (Meilgaard, 1975). With flavor descriptors of rotting vegetables and drains for MT and skunky for MBT, it is clear that these compounds can cause substantial damage to beer flavor when formed. The mechanism of MBT formation invokes initial photolysis of the α-hydroxyketone bond, a Norrish type I reaction. The radical subsequently forms decarbonylates and releases a stabilized dimethylallyl radical that abstracts a sulfydryl radical to form MBT (Figure 5.11). MT forms in small

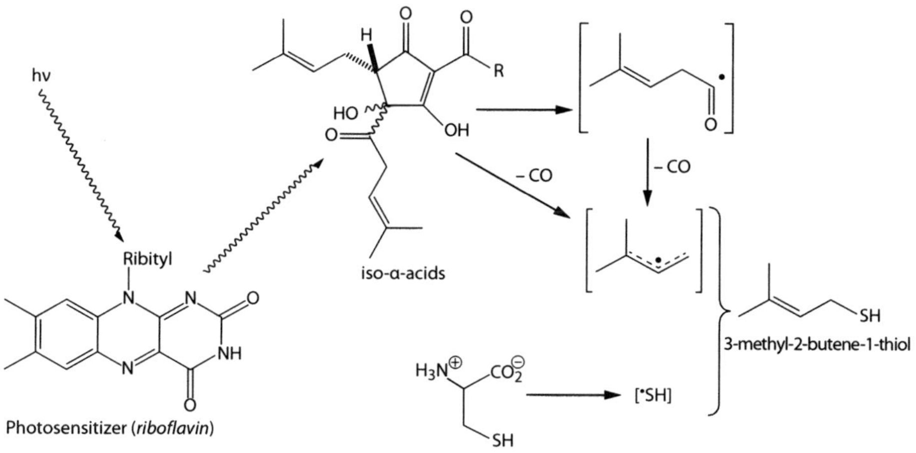

FIGURE 5.11 Proposed scheme for the light-activated formation of 3-methyl-2-butene-1-thiol in beer. The source of the sulfur "radical" is unknown.

quantities during fermentation via the S-adenosylmethionine cycle, but in terms of light stability, photosensitized degradation of methionine yields MT, a reaction that also occurs in milk.

REFERENCES

Back, W., Detection and identification of foreign yeast in brewing, *Brauwelt*, 1987, 127, 735–737.

Baert, J., De Clippeleer, J., De Cooman, L. and Aerts, G., Exploring the binding behavior of beer staling aldehydes in model systems, *Journal of the American Society of Brewing Chemists*, 2015, 73, 100–108.

Bamforth, C.W., Enzymic and non-enzymic oxidation on the brewhouse: A theoretical consideration, *Journal of the Institute of Brewing*, 1999, 105, 237–242.

Bamforth, C.W. and Lentini, A., The flavor instability of beer, in *Beer—A Quality Perspective*, eds. Russell, I., Bamforth, C.W. and Stewart, G., Elsevier, London, 2009, pp. 85–109.

Bamforth, C.W. and Parsons, R., New procedures to improve the flavor stability of beer, *Journal of the American Society of Brewing Chemists*, 1985, 43, 197–202.

Clarkson, S.P., Large, P.J. and Bamforth, C.W., A two-substrate kinetic study of peroxidase cationic isoenzymes in barley malt, *Phytochemistry*, 1992, 31, 743–749.

de Clerck, J., rH and its relevance in brewing, *Journal of the Institute of Brewing*, 1934, 40, 407–419.

Doderer, A., Kokkelink, I., Van Der Veen, S., Valk, B.E. and Douma, A.C., Purification and characterization of two lipoxygenase isoenzymes from germinating barley, *Biochimica et Biophysica Acta*, 1992, 1120, 97–104.

Dufour, J.-P., Leus, M., Baxter, A.J. and Hayman, A.R., Characterization of the reaction of bisulfite with unsaturated aldehydes in a beer model system using nuclear magnetic resonance spectroscopy, *Journal of the American Society of Brewing Chemists*, 1999, 57, 138–144.

Goldstein, H., Ting, P., Navarro, A. and Ryder, D., Water-soluble hop flavor precursors and their role in beer flavor, *Proceedings of the 27th Congress of the European Brewery Convention*, Cannes, 1999, 53–62.

Gonzalez Pereyra, M.L., Rosa, C.A., Dalcero, A.M. and Cavaglieri, L.R., Mycobiota and mycotoxins in malted barley and brewer's spent grain from Argentinean breweries, *Letters in Applied Microbiology*, 2011, 53, 649–655.

Hashimoto, N., Oxidation of higher alcohols by melanoidins in beer, *Journal of the Institute of Brewing*, 1972, 78, 43–51.

Hashimoto, N. and Kuroiwa, Y., Proposed pathways for the formation of volatile aldehydes during storage of bottled beers, *Proceedings of the American Society of Brewing Chemists*, 1975, 104–111.

Hashimoto, N., Shimazu, T. and Eshima, T., Oxidative degradation of isohumulones in relation to beer flavor, *Report of the Research Laboratory of the Kirin Brewery Company*, 1979, 22, 1–10.

Hill, A.E., Microbiological stability of beer, in *Beer—A Quality Perspective*, eds. Russell, I., Bamforth, C.W. and Stewart, G., Elsevier, London, 2009, pp. 163–183.

Hirota, N., Kuroda, H., Takoi, K., Kaneko, T., Kaneda, H., Yoshida, I., Takashio, M., Ito, K. and Takeda, K., Development of novel barley with improved beer foam and flavor stability—The impact of lipoxygenase-less barley in brewing industry, *Proceedings of the 30th European Brewery Convention Congress*, Prague, 2005, 46–52.

Hughes, P.S., Menneer, I.D., Walters, M.T. and Marinova, G., Differential behavior of *cis*- and *trans*-iso-α-acids, *Proceedings of the 26th European Brewery Convention Congress*, Maastricht, 1997, 231–238.

Ilett, D.R. and Simpson, W.J., Loss of sulfur dioxide during storage of bottled and canned beers, *Food Research International*, 1995, 28, 393–396.

Irwin, A.J., Barker, A.L. and Pipasts, P., The role of copper, oxygen and polyphenols in beer flavor instability, *Journal of the American Society of Brewing Chemists*, 1991, 49, 140–149.

Juste, A., Malfliet, S., Waud, M., Crauwels, S., De Cooman, L., Aerts, G., Marsh, T.L., Ruyters, S., Willems, K., Busschaert, P. and Lievens, B., Bacterial community dynamics during industrial malting, with an emphasis on lactic acid bacteria, *Food Microbiology*, 2014, 39, 39–46.

Kaneda, H., Kobayashi, N., Furusho, S., Sahara, H. and Koshino, S., Reducing activity and flavor stability of beer, Master Brewers Association of the Americas, *Technical Quarterly*, 1995, 32, 90–94.

Magnus, C.A., Ingledew, W.M. and Casey, G.P., High-gravity brewing: Influence of high-ethanol beer on the viability of contaminating brewing bacteria, *Journal of the American Society of Brewing Chemists*, 1986, 44, 158–161.

Meilgaard, M.C., Flavor chemistry of beer: Part II: Flavor and threshold of 239 aroma volatiles, Master Brewers Association of the Americas, *Technical Quarterly*, 1975, 12, 151–168.

Moir, M., Hop aromatic compounds, *Proceedings of the European Brewery Convention on Hops*, 1994, 165–180.

Molzon, J.A., The solubility of calcium oxalate as a function of dielectric constant, 1976, MSc thesis.

Muller, R.E., The formation of hydrogen peroxide during oxidation of thiol-containing proteins, *Journal of the Institute of Brewing*, 1997, 103, 307–310.

Narziss, L., Back, W., Miedaner, H. and Lustig, S., Untersuchungen zur Beeinflussung der Geschmacksstabilitat durch Variation technologischer Parameter bei der Bierherstellung, *Monatsschrift fur Brauwissenschaft*, 1999, 52, 192–206.

Nyborg, M., Outtrup, H. and Dreyer, T., Investigations of the protective mechanism of sulfite against beer staling and formation of adducts with *trans*-2-nonenal, *Journal of the American Society of Brewing Chemists*, 1999, 57, 24–28.

Paradh, A.D., Mitchell, W.J. and Hill, A.E., Occurrence of *Pectinatus* and *Megasphaera* in the major UK breweries, *Journal of the Institute of Brewing*, 2011, 117, 498–506.

Peppard, T.L. and Halsey, S.A., Malt flavor—Transformation of carbonyl compounds by yeast during fermentation, *Journal of the Institute of Brewing*, 1981, 87, 386–390.
Priest, F.G. and Campbell, I., *Brewing Microbiology*, 3rd ed., Kluwer, New York, 2003.
Rehmanji, M., Gopal, C. and Mola, A., Beer stabilization technology—Clearly a matter of choice, Master Brewers' Association of the Americas, *Technical Quarterly*, 2005, 42, 332–338.
Rodriguez-Carrasco, Y., Fattore, M., Albrizio, S., Berrada, H. and Manes, J., Occurrence of *Fusarium* mycotoxins and their dietary intake through beer consumption by the European population, *Food Chemistry*, 2015, 178, 149–155.
Sarlin, T., Nakari-Setala, T., Linder, M., Penttila, M. and Haikara, A., Fungal hydrophobins as predictors of gushing activity of malt, *Journal of the Institute of Brewing*, 2005, 111, 105–111.
Schwarz, P.B., Casper, H.H. and Beattie, S.B., Fate and development of naturally occurring *Fusarium* mycotoxins during malting and brewing, *Journal of the American Society of Brewing Chemists*, 1995, 53, 121–127.
Scot, R., *A Perfite Plaforme for a Hoppe Garden*, 1574.
Scott, P.M., Mycotoxins transmitted into beer from contaminated grains during brewing, *Journal of the Association of Official Analytical Chemists International*, 1996, 79, 875–882.
Siebert, K.J. and Lynn, P.Y., On the mechanisms of adsorbent interactions with haze-active proteins and polyphenols, *Journal of the American Society of Brewing Chemists*, 2008, 66, 48–54.
Snow, J., *Mode of Communication of Cholera*, 2nd ed., John Churchill, London, 1855.
Thelen, K., Beimfohr, C. and Snaidr, J., The rapid and easy detection method for beer-spoiling bacteria, Master Brewers' Association of the Americas, *Technical Quarterly*, 2004, 41, 115–119.
Thelen, K., Beimfohr, C. and Snaidr, J., Evaluation study of the frequency of different beer-spoiling bacteria using the VIT analysis, Master Brewers' Association of the Americas, *Technical Quarterly*, 2006, 43, 31–35.
Uchida, M., Suga, S. and Ono, M., Improvement for oxidative flavor stability of beer—Rapid prediction method for beer flavor stability by electron spin resonance spectroscopy, *Journal of the American Society of Brewing Chemists*, 1996, 54, 205–211.
Vanderhaegen, B., Neven, H., Verachtert, H. and Derdelinckx, G., The chemistry of beer aging—A critical review, *Food Chemistry*, 2006, 95, 357–381.
Vesely, P., Lusk, L., Basarova, H., Seabrooks, J. and Ryder, D., Analysis of aldehydes in beer using solid-phase microextraction with on-fiber derivatization and gas chromatography/mass spectrometry, *Journal of Agricultural and Food Chemistry*, 2003, 51, 6941–6944.
Walker, M.D. and Simpson, W.J., Control of sulfury flavors in cask-conditioned beer, *Brewers' Guardian*, 1994, 123(11), 37–38, 40.
Walters, M.T., Heasman, A.P. and Hughes, P.S., Comparison of (+)-catechin and ferulic acid as natural antioxidants and their impact on beer flavor stability. 1. Forced aging, *Journal of the American Society of Brewing Chemists*, 1997a, 55, 83–89.
Walters, M.T., Heasman, A.P. and Hughes, P.S., Comparison of (+)-catechin and ferulic acid as natural antioxidants and their impact on beer flavor stability. 2. Extended storage trials, *Journal of the American Society of Brewing Chemists*, 1997b, 55, 91–98.
Wolf-Hall, C., Mold and mycotoxin problems encountered during malting and brewing, *International Journal of Food Microbiology*, 2007, 119, 89–94.

6 Beer Conditioning, Aging, and Spoilage

Roger Barth

CONTENTS

6.1 DEFINITIONS AND OVERVIEW

From the chemical point of view, beer can be defined as an alcoholic fermented beverage prepared from a source of starch without concentration or distillation. This definition includes a wide range of products, such as sake, chicha, kvass, makgeolli, chibuku, and many others—only a few of which are marketed in the same way as the more familiar barley-based product. We will restrict our attention to European-style malt beer, which has a very wide range of styles, enough to fill a chapter and more.

6.1.1 Definitions

There is no generally accepted consistent terminology for the processes we will discuss, so we will follow, to some extent, the terminology in the encyclopedic *Oxford Companion to Beer* (Oliver 2011). Even so, the postfermentation processes, and even fermentation itself, are not isolated in time and place, so the distinctions are not always perfectly sharp. Many compounds significant to beer have acid/base properties; they exist in two or more forms whose ratios depend on pH. To simplify the notation, we will use the names and formulas of the neutrally charged forms—hence, acetic acid, rather than acetate.

Aging: Storage of beer for an extended period, generally exceeding 100 days, to allow characteristic flavors to develop

Conditioning: Storage of beer in contact with yeast after primary fermentation to improve flavor, allow solids to coagulate and settle, and, in some cases, to achieve desired levels of carbonation

Green beer: Ruh beer

Lagering: A form of cold conditioning applied to lager beer

Maturation: A general term for storing beer to improve its quality; includes conditioning, lagering, secondary fermentation, and/or aging; also refers to the chemical changes that occur in the beer during maturation

Primary fermentation: Process in which yeast consumes sugars in wort and releases ethanol, carbon dioxide, and flavor compounds

Ruh beer: Beer before it is conditioned

Secondary fermentation: A somewhat loosely applied term for fermenter processing of beer after the wort sugars have been depleted and before the yeast is removed

Spoilage: Undesired flavor changes that take place in beer over time

Staling: Spoilage

6.1.2 Conditioning

Beer is seldom served directly from the fermenter. It is usually held for some time in a conditioning vessel, which can be the fermenter itself, to remove off-flavors and their precursors. After this stage of conditioning, the beer is often subjected to filtration or centrifugation to remove yeast and other particulate matter. Prior to conditioning, the beer is called ruh beer or green beer and is not considered saleable.

The most recognized off-flavors in ruh beer are the vicinal diketones (VDKs), including butanedione (known in the trade as diacetyl) and 2,3-pentandione (Boulton and Quain 2006). VDKs impart a buttery aroma sometimes described as "movie popcorn" because of the use of these compounds as an artificial butter flavoring. VDKs will be discussed in detail in Section 6.2. During conditioning, a measured dose of sugar or beer wort may be added to carbonate the beer by fermentation, or food-grade carbon dioxide may be introduced after removal of the yeast. Conditioning also involves clarification of the beer by sedimentation of yeast and other solids. Conditioning is usually accomplished in vessels in the brewery, but there are some beer styles for which conditioning is completed in the packages, either bottles or casks. The duration of conditioning depends on the type of beer. Ale, which is fermented at temperatures above 15°C, is conditioned for a few days. Lager beer, which is fermented at lower temperatures, is conditioned at about 5°C, typically for 20–30 days, but sometimes for more than 100 days.

6.1.3 Aging

The deliberate aging of beer for months or years to develop unique flavors is undertaken commercially in only a few low-volume styles of beer, the best known of which is lambic ale. Controlled studies of the chemistry behind the aging processes are virtually nonexistent. Most of the available information is anecdotal. Nonetheless, there has been a revival of interest in aged beer styles, such as lambic, and also in consumer aging of nonaged commercial styles (Dawson 2014). Aging at the brewery is often undertaken by storing the beer in wooden barrels, which allow access to oxygen and microbes, as well as flavor-active substances leached from the wood.

6.1.4 Spoilage

In contrast to aging, all forms of beer are susceptible to spoilage; it is a constant preoccupation of brewers and brewing scientists. A good deal of research has gone into the issue of making packaged beer as stable as possible. In the past, growth of undesirable microbes was the major concern (Vanderhaegen et al. 2006). In modern beer the key issues are flavor changes and the development of haze during the period from the release of the beer to its consumption. The problem is exacerbated for today's high-volume brands. Many of these are marketed globally and must retain their flavor for an extended shelf-life. In addition, the most popular beer styles are mildly flavored, so flavor changes are readily noticed by consumers.

6.2 LAGER BEER CONDITIONING

Ale and lager are the two major classes of beer. The fermentation temperature for ale is higher than 15°C, which gives rise to esters that lend fruity notes to the flavor. Lager beer is fermented below 15°C, usually below 12°C. The lower temperature and the characteristics of the yeast used result in beer with low ester content whose flavor is described as "clean" (Barth 2013a). The word "lager" means "warehouse" in German, reflecting the importance of extended conditioning.

6.2.1 Lager Yeast

Lager beer is fermented with a cold-tolerant species of yeast called *Saccharomyces pastorianus* (also called *S. carlsbergensis*), named after Louis Pasteur. The species is believed to be a hybrid of *S. cerevisiae* (ale yeast) and *S. eubayanus*, a species discovered in Patagonian beech trees in Argentina (Libkind et al. 2011) and subsequently found in several distant locations. *S. pastorianus* is active for fermentation at temperatures below the range of *S. cerevisiae.* The lower fermentation temperature results in lower production of alcohols of three or more carbons (fusel alcohols), and lower ester content. Because of differences in the compositions of the cell walls, lager yeast tends to sink to the bottom of the fermenter during fermentation, in contrast to ale yeast, which is carried to the top with carbon dioxide bubbles. For this reason, lager yeast was formerly called "bottom yeast" and ale yeast was called "top yeast." The bottom/top distinction is less relevant today because of the widespread use of closed fermenters.

6.2.2 Vicinal Diketones

Alcoholic fermentations yield off-flavor compounds called vicinal diketones (VDKs) (Krogerus and Gibson 2013a), so named because they have two keto (oxo) groups on adjacent carbon atoms. The two major compounds are butanedione (Figure 6.1) and 2,3-pentanedione (Figure 6.2). These compounds impart a buttery aroma that is considered a serious defect in most beer styles. The aroma is most noticeable if the beer is delicately flavored, like most commercial lager beer. Discussion of VDKs is complicated by brewing jargon, which often treats butanedione, 2,3-pentanedione, their precursors, and sometimes even their reduction products as VDKs. Many analytical methods for VDK, including those of the American Society of Brewing Chemists (ASBC 2014) and of the European Brewing Congress (EBC 2008), involve treatments that convert VDK precursors. The expression "total VDK" or "total diacetyl" can always be taken to include both VDKs as well as their precursor compounds: acetolactic acid and acetohydroxybuteric acid (AHBA). It is often necessary to consult the minutiae of the methods section to determine what exactly is meant by the term "diacetyl" or "VDK" in a particular publication.

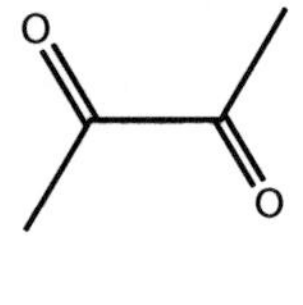

FIGURE 6.1 Butanedione (diacetyl).

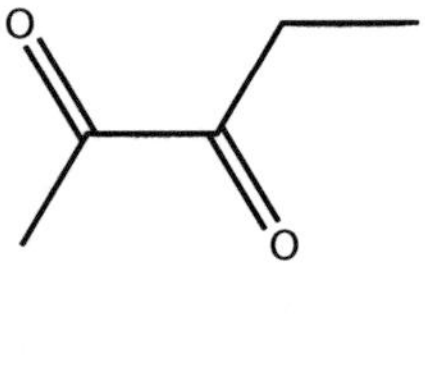

FIGURE 6.2 2,3-Pentanedione.

Vicinal diketones arise from nonenzymatic oxidation of intermediates in the synthesis of amino acids. The process is shown in Figure 6.3 for butanedione and in Figure 6.4 for 2,3-pentanedione. In the case of butanedione, a molecule of pyruvic acid ($C_3H_4O_3$) combines with a molecule of acetaldehyde under the influence of the pH 8 version of acetolactate synthase to give α-acetolactic acid (official name: (S)-2-hydroxy-2-methyl-3-oxobutanoic acid) (Inoue 2008). This compound can undergo a series of enzymatic reactions yielding valine, the amino acid with an isopropyl R-group. The first intermediate, α-acetolactic acid, can undergo nonenzymatic oxidation and decarboxylation to yield butanedione (Inoue and Yamamoto 1970). The route to 2,3-pentanedione, shown in Figure 6.4, is a branch on a pathway that converts threonine to isoleucine. The critical intermediate is α-acetohydroxybutyric acid (official name: (S)-2-ethyl-2-hydroxy-3-oxo-butanoic acid). We will call this compound AHBA. Both butanedione and 2,3-pentanedione are quickly hydrogenated by yeast enzymes to nearly flavorless alcohols (Van den Berg et al. 1983). These enzymatic reductions are rapid compared to the nonenzymatic reduction/decarboxylations that take acetolactic acid and AHBA to VDK. As long as active yeast is present, VDK is reduced as soon as it appears (Boulton and Box 2003). When the beer is taken off the yeast, reduction stops but conversion of acetolactic acid and AHBA to VDK continues. This leads to accumulation of VDK, which can exceed

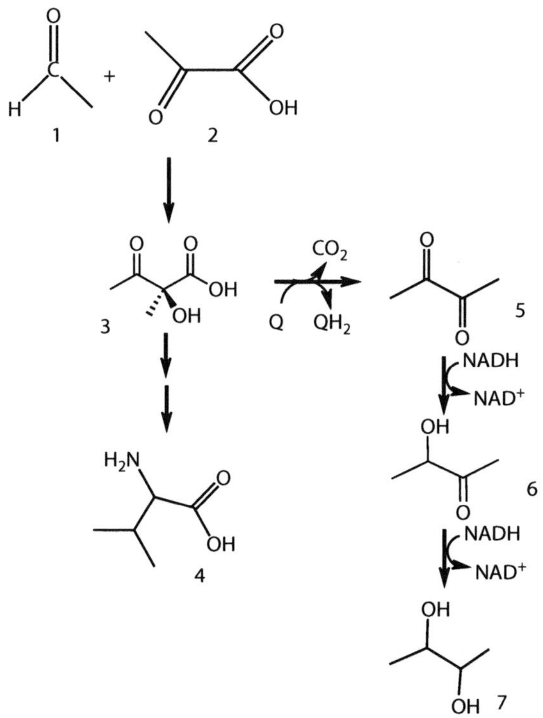

FIGURE 6.3 Butanedione formation. 1: acetaldehyde; 2: pyruvic acid; 3: alpha-acetolactic acid; 4: leucine; 5: butandione; 6: acetoin; 7: 2,3-butandiol.

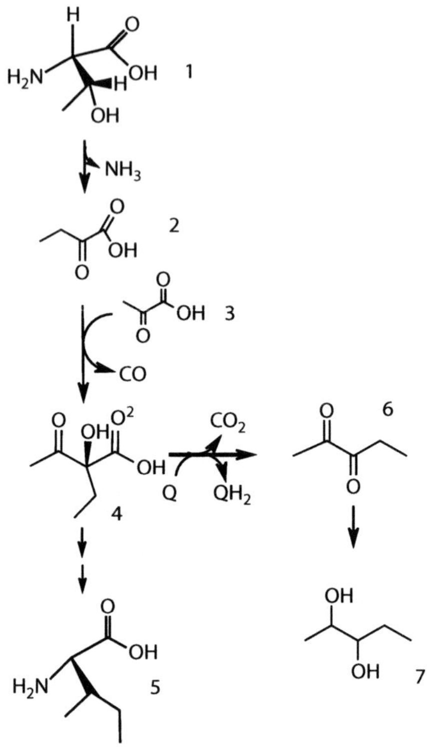

FIGURE 6.4 2,3-Pentanedione formation. 1: threonine; 2: alpha-ketobutyric acid; 3. pyruvic acid; 4: AHBA; 5: isoleucine; 6: 2,3-pentandione; 7: 2,3-pentandiol.

the flavor threshold. In Figure 6.3 butanedione is reduced first to 3-hydroxybutanone (acetoin) and then to 2,3-butanediol.

The flavor thresholds of these compounds increase with reduction, as shown in Table 6.1. This results in a marked decrease in the contribution of butanedione and its reduction products to the beer flavor. For example, if butanedone were present at 0.60 ppm, this would be four times the threshold, or four flavor units, making it a primary flavor compound. After complete reduction to 3-hydroxybutanone, the contribution would be 0.12 flavor units, and after reduction to 2,3-butanediol, it would be

TABLE 6.1
Flavor Thresholds of Vicinal Diketones and Their Reduction Products

Compound	Flavor threshold
Butanedione (diacetyl)	0.15 ppm
3-Hydroxybutanone (acetoin)	50 ppm
2.3-Butanediol	4,500 ppm
2,3-Pentanedione	1 ppm

about 10^{-4} flavor units (Meilgaard 1975). Complete conversion is not attainable, but the flavor contribution of butanone can be brought to an acceptable level. The flavor thresholds of the reduction products of 2,3-pentanedione are not well established, but all agree that they are significantly higher than that of 2,3-pentanedione.

6.2.3 Lager Process

The conditioning method characteristic of lager beer production is the lager process. After primary fermentation, in which nearly all of the fermentable sugar is consumed, the beer is conditioned at about 5°C for 3 weeks or longer. Sometimes cold-conditioning is deferred for as much as a few days to provide a "diacetyl rest." A major objective of the lager process is to convert as much VDK precursor to VDK while the live yeast is available to reduce VDK to diols. Once the yeast is removed, any remaining acetolactic acid or AHBA is liable to be converted to VDK. The chilled vessels for the lager process must accommodate the entire output of the brewery for the lagering period. As such, the lagering cellar typically represents 25% to 40% of the cost of building a brewery; it is the largest capital outlay for a lager brewery (Patino 1999).

Because the proximate objective of the lager process is the removal of acetolactic acid, the duration of lagering depends on the amount of acetolactic acid in the beer. One way to lower production of acetolactate is to avoid depletion of valine from the wort. Acetolactic acid is an intermediate in the synthesis of valine; yeast cells regulate this pathway with a feedback mechanism. When no valine is needed, the production of acetolactic acid is suppressed. Fermentation conditions that favor valine depletion, including the use of low-protein adjuncts, higher fermentation temperature, and high-gravity fermentation, require careful management. It has been shown that supplementation of wort with valine or other branched-chain amino acids results in lower production of VDK precursors during fermentation (Krogerus and Gibson 2013b). Another approach to acetolactic acid removal is to raise the temperature (or defer its lowering) at the end of primary fermentation to accelerate acetolactic acid decomposition to VDK, which is reduced by the yeast. This is called the "diacetyl rest." The advantage of this method is that at this stage, the increased temperature has little effect on the beer flavor profile. The disadvantage is that under the sugar-starved conditions at the end of fermentation, the warmer temperature decreases yeast viability. There is evidence that rehydrated dry lager yeast is less effective at reducing VDK during lagering than the corresponding strain of freshly propagated yeast (Cyr et al. 2007).

6.2.4 Accelerated Methods

A moderately sized regional brewery might brew 3.5×10^5 m^3 (3 million US beer barrels) of lager beer each year. If the beer is lagered for 3 weeks, over 20,000 m^3 of cold storage would be required. Any acceleration of the lager process would yield very large savings in capital outlay and utilities. Efforts to accelerate the lager process have mostly been directed at the use of immobilized yeast cells (Inoue 1988). Immobilized cell technology involves growing an adherent layer of live cells on a support material, often in the form of beads. One process that is in commercial use involves three steps. The fermentation yeast is removed by sedimentation followed by centrifugation. The clear beer is heated to a temperature in the range of 60°C to 90°C for a few seconds, then chilled. This heat treatment converts virtually all of the acetolactic acid and AHBA to the corresponding VDKs. The ruh beer passes through the bed of yeast-bearing beads to allow the yeast to reduce the VDK (Virkajärvi 2006). Support materials for yeast immobilization have been made from alginate (anionic carbohydrate polymer derived from brown algae), ceramic material, porous glass, wood chips, carrageenan (a sulfated carbohydrate polymer derived from seaweed), functionalized cellulose, polyvinyl alcohol, and many others (Leskošek-Čukalović and Nedović 2005).

To improve contact of the beer with the yeast-bearing beads, the flow conditions can be set up to fluidize the beads. Fluidization means that the beads are carried for some distance in the flow of beer before they are separated and returned to their starting point. It can be advantageous to enhance fluidization by introducing a gas at the bottom of the reactor to help drive the beads upward with the beer flow. Immobilized yeast technology has the potential to lower the lagering time from several weeks to as little as 20 hours. Immobilized yeast lagering systems have been in commercial use at some facilities for decades. Nonetheless, adoption has not been widespread despite the clear economic advantages. Some factors that may inhibit adoption are the possibility of flavor changes and issues of consumer acceptance. Although beer prepared by immobilized yeast technology has been shown to be satisfactory to consumers, it is not as clear that the flavor of a particular brand will be identical to the flavor produced by the traditional lager process. Consumers tend be unforgiving of any perceived flavor changes in established brands of beer. A related marketing issue is that brewers often position beer as a product of traditional craftsmanship. Changes in the process that lower cost can introduce a negative perception of flavor even if it were to be unequivocally demonstrated that no perceptible change in the flavor has occurred. A possible third factor is that sales volume for very large breweries has been stagnant as "share of throat" has been shifting to other beverages and to craft beer. This lowers the incentive to get more beer from the facility.

6.3 ALE CONDITIONING

Ale has been around for thousands of years longer than lager beer. The diversity of styles and of yeast strains greatly exceeds those of lager beer. There are pronounced local variations in processing practices. Compared to the lager process, ale conditioning at the brewery is abbreviated. Depending on the style of ale, conditioning can range from zero days to several weeks. Typical for modern processing is 3 to 7 days (Carey and Grossman 2006). Ales with higher initial gravity are typically subjected to longer conditioning. Ale is usually conditioned at higher temperatures than would be typical for lager beer. As a result of the warmer fermentation and conditioning temperatures, VDKs are not usually a major concern unless the fermentation becomes infected with spoilage bacteria (Bamforth 2006).

6.3.1 Cylindricoconical Vessels

Ale can be and often is conditioned in cylindricoconical vessels identical to those used for primary fermentation. One common practice is to chill the ale to about 10°C to cause much of the yeast to settle. The yeast sediment is recovered through a valve at the bottom of the vessel. After 2 or 3 days, the temperature is lowered to about −1°C ("crashing") to begin the process of sedimentation to clarify the ale. If the vessel can take pressure, it may be closed off with a relief valve before fermentation is complete to allow carbon dioxide pressure to build up and carbonate the beer. This is called "spunding" or "bunging" ("Spund" is the German word for bung).

6.3.2 Cask Conditioning

Another type of conditioning vessel is the cask, a type of barrel that serves both as a conditioning and serving vessel. Cask-conditioned ale is popular in the British Isles, where its traditions are safeguarded by a consumer organization called the Campaign for Real Ale (CAMRA). Beer, a source of additional sugar, and finings (agents to clarify the beer) are added to the cask through a bung hole in the side of the cask. The bung is sealed with a plug called a shive, and the cask is shipped to a tavern to undergo final conditioning on site. At the tavern a peg called a spile is driven into a hole in the center of the shive. The spile allows gas to escape to establish the degree of carbonation. When the beer is ready to be served, a tap is driven into an opening (the keystone hole) on the end of the cask for direct service or connection to a dispensing system (Barth 2013a).

6.3.3 Bottle Conditioning

A third type of conditioning vessel is the bottle in which the beer is sold. The beer with a measured concentration of added sugar is bottled and capped. The beer can be added with a low density of yeast cells remaining from fermentation, or the beer can be filtered or centrifuged and a different strain of yeast can be introduced for the bottle conditioning. The yeast ferments the sugar and carbonates the beer. One advantage of this method is that the yeast will consume dissolved oxygen down to a very low level. Some disadvantages are that there will be a sediment of yeast in the bottle that can be a nuisance to the consumer. Also the shelf-life of the beer is limited by potential autolysis of the yeast yielding flavors reminiscent of Marmite, a yeast-based food invented by Justus von Liebig (1803–1873), the founder of organic chemistry.

6.4 BEER CLARIFICATION

6.4.1 About Haze

In the modern age, consumers expect bright, haze-free beer. This is largely a cosmetic issue, but colloidal stability (brewing jargon for resistance to cloudiness) is a major preoccupation of the brewing industry. Haze in beer is the result of light scattered by suspended particles that are nearly as large as or larger than the wavelength of light. Haze can arise from dead or live microbes, from particles of starch, or from residual processing aids (Bamforth 2006), none of which should be present in properly brewed beer.

It has been accepted for a long time that the major source of haze in beer is from particles resulting from the interaction of certain fractions of proteins with certain types of polyphenols (Moll 1987). Specific fractions of proteins seem to be especially haze active, particularly those derived from prolamines (grain storage proteins) that contain a high content of proline and glutamic acid (Asano et al. 1982). Nonetheless, there is evidence that other proteins could also be involved (Evans et al. 2003; Robinson et al. 2007; Iimure et al. 2009). Because the polyphenolic compounds are characterized by a somewhat loose interlocking vocabulary, we will introduce some of the terminology often found in the literature:

- Polyphenol is a general term for a compound with more than one hydroxyl group covalently bound to one or more conjugated rings.
- A tannin is a type of polyphenolic biomolecule that can make a precipitate with proteins.
- A flavanoid is a compound with a ring structure similar to that of flavane, shown in Figure 6.5, but with varying substituents.
- An anthocyanidin is a flavanoid with multiple hydroxide substituents.
- A proanthocyanidin is an oligomer of anthocyanidines.

None of these should be confused with anthocyanins, which are glycosides of anthocyanidines (Wu et al. 2004).

The polyphenols most implicated in haze formation are proanthocyanidins (Siebert 1999). Procyanidin B3, which is a dimer of an anthocyanidin called catechin, is shown in Figure 6.6. If an additional hydroxyl group were to be added to the single

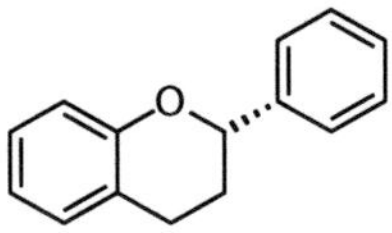

FIGURE 6.5 Flavane.

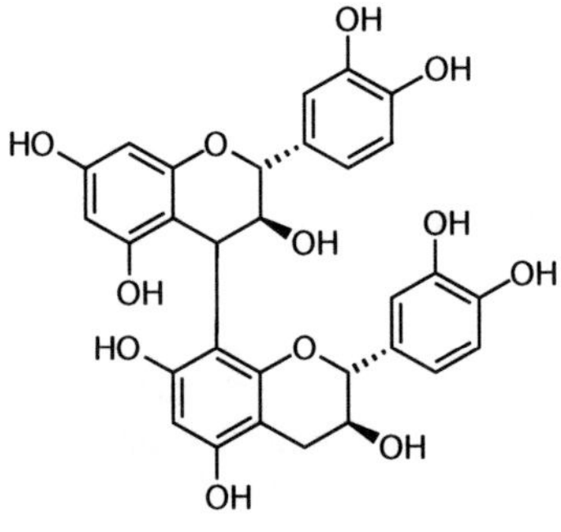

FIGURE 6.6 Procyanidine B3.

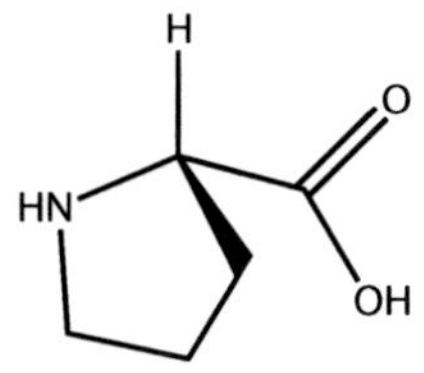

FIGURE 6.7 Proline.

ring of catechin, the compound would be gallocatechin, also known as delphinidin. A third compound is epicatechin, which is a stereoisomer of catechin. These three anthocyanidins and their oligomers are generally present in beer. The structure of the peptide–polyphenol adduct is not known. Structures have been proposed based on the following observations:

1. When polyphenol is limiting (as in beer), the uptake of polyphenol is a linear function of the proline (Figure 6.7) content of the protein or polypeptide (Asano et al. 1982).
2. There is a ratio of polyphenol to polypeptide that gives maximum haze. Excess of either results in less haze (Siebert 1996).
3. Polyphenols with vicinal hydroxides are more haze active than those with isolated hydroxyls (Siebert 1996).
4. Haze increases very slowly or not at all for an induction period, after which it increases much more rapidly (McMurrough and O'Rourke 1997).

The usual explanation for the first three observations is that hydroxides on the polyphenols are involved in noncovalent interactions with proline sites on the polypeptide molecules. Polyphenols can bind to polypeptides, but not with one another. Polypeptides can bind to polyphenols, but not to one another. A haze particle requires that multiple polypeptide molecules be bound together with polyphenols. If polypeptides are in excess, the aggregates will be small because of insufficient polyphenol. If polyphenol is in excess, the polypeptide binding sites will be saturated with polyphenol, so the bound polyphenols will not find available protein sites to form aggregates (Siebert 1999).

The key difficulty with this elegant model can be summed up with a simple question. Why proline? Proline is a hydrogen bond acceptor, but there are other amino acids that are at least as good. Proline can engage in a hydrophobic stacking-type interaction, but there are others that would be expected to be better. So why proline rather than histidine or tryptophan? The answer may have to wait for direct structural information about binding in the haze particles. The fourth observation, that there is an induction period, suggests that the haze precursors must be modified in some way before they can react effectively. Oxidation of the polyphenols and polymerization of the anthocyanidins have been suggested as the preliminary process. It is also possible that haze formation proceeds by a nucleation and growth mechanism.

6.4.2 Beer Clarification

The three general approaches to beer clarification are sedimentation, filtration, and the use of finings. These approaches are often used in combination.

6.4.2.1 Sedimentation

Sedimentation is a method of separation exploiting differences in density between the particles to be separated and the bulk medium. The classic method is to chill the beer to a low temperature and allow undesired particles to fall out of suspension in the field of gravity. This can take weeks; the sediment is likely to entrain a good deal of beer, and the tank bottom will need to be cleaned. The sedimentation process can be greatly expedited with a centrifuge. A beer centrifuge spins a stack of disks. The beer is fed onto the disks, which cause the beer to spin, producing acceleration of up to 8,000 times that of gravity (Patino 1999). The solids, which are denser than the liquid, are driven to the outer edge of the disks and discharged. The centrifuge operates continuously at full speed, producing clarified beer at rates as high as 10 hL/min (1 m^3/min), depending on the size of the device. The beer from the centrifuge is largely clear of yeast cells, but small haze particles are not effectively separated. The solids fraction is mostly yeast and trub. Because the centrifuge does not add anything to the solids, they can be used in the brewery or added to the spent grain for use as cattle feed. The use of a centrifuge reduces beer loss in filtration. Some disadvantages of centrifugation are that it warms the beer, so an additional cooling stage may be needed before further processing or packaging (Briggs et al. 2004; Munroe 2006). Also, centrifuges are noisy enough to make hearing protection necessary. The high speeds and close tolerances of these devices make effective maintenance critical.

6.4.2.2 Filtration

Filtration involves driving the beer through a system of pores or channels that trap undesired particles that are too large to fit through the pores. In practice the filter medium may be supplemented with substances that trap particles or dissolved species on the basis of specific interactions, such as electric charge or intermolecular interactions. Beer filtration is usually effected on a layer of fine inert particles supported on a set of grids or screens. The inert particles are sometimes called a "filter aid," although they are not necessarily aiding any other filtration process. A layer of filter aid is first applied to the screens. This is called the precoat. The precoat may have more than one layer, usually with the finer layer below and a coarser layer above, so the larger particles can be removed nearer to the surface at which the beer first meets the precoat. Often additional filter aid is suspended in the feed to the filter. This is called body feed or admix. The body feed makes the filter cake (material that is retained on the filter) less dense and more permeable.

The most popular filter aid for beer filtration is diatomaceous earth, also called kieselguhr or diatomite. Diatomaceous earth is a mineral composed of silica-bearing skeletons of marine algae. Other filter aids are cellulose and pearlite, which is hydrated volcanic glass (Lindemann 2009). The flow rate through the filter is directly proportional to the filter area and the applied pressure and is inversely proportional to the thickness of the filter cake and the viscosity of the fluid. During filtration the filter cake gets thicker, so either the pressure must increase or the filtration rate slow down. It is also possible to filter through a thin sheet with small pores without the use of filter aids. This is called sheet, surface, or membrane filtration. Sheet filtration can be useful if the filtrate must be recovered without admixture of filter aids.

6.4.2.3 Finings

Finings are materials added to the beer or to the wort to assist with clarification. Finings adsorb or form coagulated precipitates with the undesired component, which can then be removed by sedimentation or filtration. Finings can be added to the beer during transfer to the conditioning vessel, or they can be included with filter aids or even incorporated into sheet filters. There are types of finings that are used during boiling, but these are outside our scope. Several materials are currently or have been historically used as postfermentation finings (Goldhammer 2008):

- **Gelatin**: Gelatin is collagen, a structural protein of animal origin, prepared from the skins of cattle or pigs. The main effect of collagen is to react with polyphenols to make a precipitate. This lowers the polyphenol concentration. Typically, a slurry of gelatin and deaerated water is added at the rate of about 4 g gelatin per hectoliter of beer.
- **Isinglass**: Like gelatin, isinglass is collagen protein, but the source of the collagen is the swim bladders of certain equatorial freshwater fish. Isinglass is usually used to trap yeast cells, forming larger particles that coagulate rapidly. It is added as slurry at the rate of about 2.5 g isinglass per hectoliter of beer. The actual dosage is determined by experiment; too much or too little reduces the effectiveness of the treatment. Both isinglass and gelatin are animal products that may not be acceptable to some consumers.
- **Tannic acid**: Tannic acid is a compound with many polyphenolic rings attached to a central sugar molecule or to one another by ester linkages. It strongly and unselectively precipitates proteins. Tannic acid can enhance colloidal stability of beer at the expense of foam stability. The latter can be made up by the addition of additives, such as propylene glycol alginate (Jackson et al. 1980), but this can be risky in terms of customer acceptance. Tannic acid is derived from plant sources.
- **Bentonite**: Bentonite is a layered clay that is very effective at binding polypeptides. Because it removes both haze-active and foam-active fractions, it is no longer used for beer clarification, but it is used for wine and fruit juices.
- **Silica gel**: Silica gel is prepared by dissolving sand in sodium hydroxide, giving sodium silicate (Na_2SiO_3), also called water glass. Acid as added to precipitate hydrated silica (SiO_2), which is usually kiln dried before shipping. Silica gel is very porous and has a large surface area. The grades used as beer finings are selective adsorbents for haze-active polypeptides, and have minimal affinity for foam-stabilizing polypeptides. It is believed that the pore structure of the gel selectively admits the undesired polypeptides and excludes the desired fraction, leaving them in the beer. Silica gel is

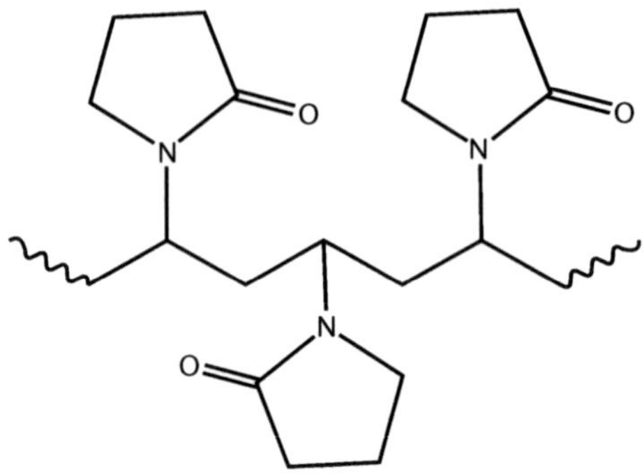

FIGURE 6.8 PVPP.

added at the rate of 40–100 g/hL. It is sometimes included with the body feed of filter aid. The use of silica gel can greatly enhance the colloidal stability of beer by removing one of the components of haze.

- **Polyvinylpolypyrrolidine (PVPP)**: PVPP (Figure 6.8) is an artificial polymer. It is a selective adsorbant for polyphenols. The structure of PVPP is somewhat reminiscent of proline, which may account for its binding. PVPP is added to beer at a rate of about 25 g/hL. The use of PVPP, either alone or in conjunction with silica gel, can greatly enhance the colloidal stability of beer. PVPP is sometimes recovered from the filter, washed with sodium hydroxide solution, and reused. In this case, a coarser grade is used. The PVPP is fed downstream from the main filter, and a dedicated filter is used to recover the PVPP.

6.5 CARBONATION

Modern beer contains dissolved carbon dioxide, which makes a major contribution to the beer flavor. Carbonation is usually measured by the volume of carbon dioxide gas, measured at 1 atm and 0°C, dissolved per volume of beer. One liter of carbon dioxide at 1 atm and 0°C weighs 1.96 g, so a “volume” of carbonation comes to 0.2% by weight. Most beer is carbonated to about 2.5 volumes, with British cask ale styles as low as 1 volume and weizenbier as high as 4 volumes. The saturation carbonation has a simple relationship with the pressure, called Henry’s law: $P = kc$, where c is the saturation carbon dioxide concentration and P is the carbon dioxide absolute (not gauge) pressure. The Henry’s law constant, k, depends on the temperature in a not-so-simple way (Barth 2013b). The numerical value of the Henry’s law constant depends on the units of pressure and concentration. There are two basic methods to carbonate beer. The carbon dioxide from fermentation can be retained in the beer by the application of pressure, or food-grade carbon dioxide can be driven into the beer under pressure. The first method was discussed in Section 6.3. The second method, called forced carbonation, can be done in the bright beer tank or in line as the beer flows to the bottling plant. In either case, the gas is introduced into the liquid in the form of fine bubbles through pores in a diffuser, often called a stone. In-line carbonation is usually controlled by feedback from a carbonation sensor. In-tank carbonation is often controlled by maintaining a set temperature and carbonating until the required pressure, as indicated by Henry's law, is reached. The production of foam during carbonation should be avoided, because the surface of the bubbles can solidify, introducing coarse haze.

6.6 MICROBIAL STABILITY

Beer has some inherent resistance to microbial spoilage as a result of the antimicrobial action of ethanol and of hop bitter compounds. Beer that is intended for use within a short time can be packaged and sold without any special treatment to reduce microbes. Antimicrobial treatments include pasteurization and microbial filtration. Pasteurization can occur either before packaging, which is called flash pasteurization, or afterwards, which is called tunnel pasteurization. Microbial filtration must be before packaging. Flash pasteurization is effected in a heat exchanger in which the beer is brought to a specific temperature for a specific amount of time. The system is designed to control the flow rate to ensure that the beer is at the treatment temperature for the correct amount of time. Microbial filtration involves using a series of filters of decreasing pore size. The final sterilization is carried out with a 0.45 μm membrane filter. Both flash pasteurization and microbial filtration require that package filling be done under sterile conditions. Tunnel pasteurization involves heating the beer in its package. The usual arrangement is for the containers to travel on a conveyer through a spray of hot water for long enough to raise the temperature of the beer to the required temperature for the required time. All of these antimicrobial processes have the potential to change the flavor of the beer. Microbial filtration is thought to have the least effect and tunnel pasteurization the most (Goldhammer 2008).

6.7 BEER AGING

Beer aging is the storage of beer in order to produce unique flavors. Although this practice is common in wine and spirits production, it is unusual for most beer styles. The commercial styles most associated with aging are Belgian lambic and related styles, such as Belgian red-brown ale and American coolship ale (Bokulich et al. 2012). These beers usually include barley malt as well as unmalted wheat. For some styles the chilled wort is exposed to allow it to pick up ambient fungi and bacteria (Guinard 1990). In others the wort is protected and a mixture of yeast and bacteria is introduced. These styles undergo prolonged fermentation involving a variety of yeast species as well as bacteria. Some of the microbes would be regarded as spoilage organisms in standard beer styles. The resulting beers are usually sour as a result of lactic acid or acetic acid. They have flavor notes often described as "barnyard," "horse blanket," and the like. The fermentation process is dominated by several families of organisms in succession: enterobacteria, yeast, and acidification (lactic and acetic) bacteria.

After the fermentation takes hold, the beer is often transferred to wooden casks, where it is likely to encounter organisms retained in the wood. In the case of lambic, the aged beer is often mixed with fruit or with unaged beer. Sometimes a final phase of aging takes place in the bottle (Martens et al. 1997; Spitaels et al. 2014, 2015; Snauwaert et al. 2016). Aging of beer in wooden barrels, especially used bourbon whisky barrels, is increasing in popularity among American craft brewers. The practice is made economically possible in part by US Standards of Identity that require bourbon whisky to be aged in new oak containers that are charred on the inside. Because the bourbon distiller can use the barrel only once, it then becomes available for beer aging. The aging process exposes the beer to a variety of microbes that are not totally eliminated from the wood. The wood itself and its breakdown products provide flavors, such as vanillin (Gonzales 2014). The first time the barrel is used for beer it contributes whisky flavors. A few brewers build large vessels of particular types of wood to get certain wood-aging effects. For example, the Dogfish Head brewery in Milton, Delaware, has 750 hL of aging capacity in two vessels built of Palo Alto (*Bursera graveolens*) wood (Cantwell and Boukaert 2016).

6.8 BEER SPOILAGE

6.8.1 Stale Off-Flavors

The flavor of beer changes with time, mostly for the worse. For standard lager, bitterness decreases, sweetness increases, and a flavor officially called ribes [rye-bees], unromantically known to beer drinkers as "cat pee," increases and then fades away. Eventually a cardboard flavor dominates (Dalgliesh 1977). This general scheme varies a good deal depending on the type of beer and the details of the storage conditions, especially temperature. Reactive oxygen species are implicated in many spoilage processes; if oxygen at packaging is low enough, the ribes flavor will not develop (Vanderhaegen et al. 2006). Conversely, beer will spoil and, in particular, the cardboard flavor will develop even at extremely low levels of oxygen. Some other flavors that have been noticed in stale beer are vinous (winey, or sherry-like), onion, flowery, and smoky (tobacco-like). One special case is hopped beer that has been exposed to light, which gives rise to an aroma called lightstruck, which resembles that of the defensive compounds of skunks (*Mephitis mephitis*).

It is difficult to associate a flavor with a specific molecular structure. The process of isolating a flavor-active fraction that may be present at the microgram per liter level can be complex, tedious, and expensive (Williams and Wagner 1978; Vanderhaegen and Derdelinckx 2005). The contribution of a compound to the flavor of a food or beverage can be estimated in two ways. If the compound is known and available in pure form, its flavor threshold can be determined. The flavor threshold is the lowest concentration of the compound at which the average participant can distinguish presence or absence of the compound. The ratio of the concentration of the compound in a sample to the threshold concentration is the number of flavor units. When the pure compound is unavailable or unknown, the contribution to the flavor is measured by flavor dilution (FD). The FD is the highest ratio of diluent to food/beverage at which the average participant can discern the flavor. There is no comprehensive theory of flavor as there is with color. Even describing flavors in words is difficult. A compound that one person may perceive as banana may be perceived by another person as bubble gum. Some standardization has occurred, but tasters must be trained in the use of the standard terms (Meilgaard et al. 1979). The structure of a molecule is not generally helpful in determining its contribution to flavor. In addition to the compounds that are known to be responsible for stale flavors, there are compounds that are statistically associated with a perception of staleness. These are called staling index compounds.

6.8.2 Spoilage Chemistry

Many, but not all, staling and staling index compounds are aldehydes or ketones, usually with additional functionality. The best known and perhaps most important staling compound is (E)-2-nonenal (also called *trans*-2-nonenal), shown in Figure 6.9. This compound gives a cardboard stale flavor with a flavor threshold of 0.11 μg/L (Meilgaard 1975). The ribes flavor is believed to arise from 3-methyl-3-sulfonylbutyl formate, shown in Figure 6.10. This compound has a flavor threshold in beer of 57 ng/L (57 parts per trillion) (Tran et al. 2013). An onion aroma sometimes found in stale beer seems to arise from dimethyl trisulfide (threshold = 0.1 μg/L), shown in Figure 6.11. There is evidence that this compound is an oxidation product of (3-methylthio)propanal (Gijs et al. 2000). It was reported recently that 4-vinylsyringol imparts a smoky, tobacco-like aroma to stale beer. The magnitude of its contribution to the flavor is similar to that of (E)-2-nonenal (Callemien et al. 2006). The structure of 4-vinylsyringol is shown in Figure 6.12. Sweet or flowery aromas may involve β-damascenone, shown in Figure 6.13 (Callemien et al. 2006). This compound is prominent in the aroma of some flowers, and is often used in the formulation of perfume. Furfural and 5-hydroxymethylfurfural, shown in Figure 6.14,

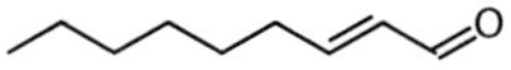

FIGURE 6.9 (E)-2-nonenal.

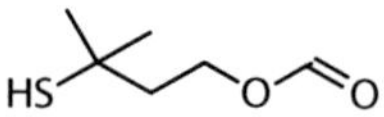

FIGURE 6.10 3-Methyl-3-sufanylbuteryl formate.

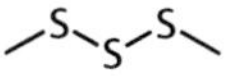

FIGURE 6.11 Dimethyl trisulfide.

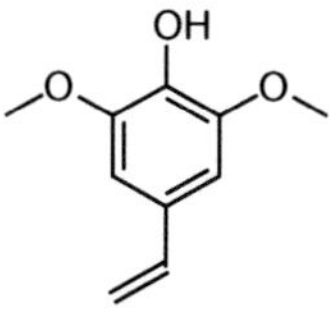

FIGURE 6.12 4-Vinylsyringol.

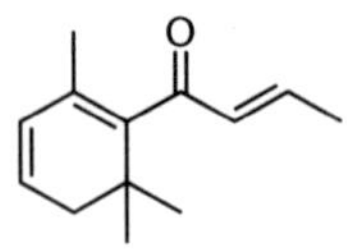

FIGURE 6.13 Beta-damascenone.

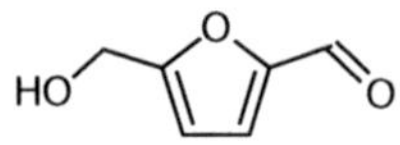

FIGURE 6.14 5-Hydroxymethylfurfural.

are often used as index compounds for beer spoilage. Their own contribution to the stale flavor is not the issue, but their presence correlates with staleness perceived by flavor testers (Shimizu et al. 2001). Because it is more economical to analyze for an index compound than to assemble a panel of tasters, these compounds can be useful, especially in evaluating process or raw material changes for their effects on flavor stability.

Oxygen at various stages of the brewing process is suspected of being involved in the formation of some staling compounds. The oxygen molecule itself is only moderately reactive. Oxygen reactions in solution are mediated by reactive oxygen species, most of which are free radicals; that is, they have an unpaired electron. The first step is $O_2 + e^- \rightarrow {}^{\bullet}O_2$, the superoxide ion, a free radical. This reaction is catalyzed by metal ions that have multiple valencies, such as Fe^{2+}/Fe^{3+} and Cu^{+}/Cu^{2+}. The superoxide ion is a Brønsted base; it abstracts a proton from water: ${}^{\bullet}O_2^- + H_2O \rightarrow OH^- + HO_2^-\,{}^{\bullet}HO_2^-\,{}^{\bullet}$, the hydroperoxide radical, which is extremely reactive; it can accept a hydrogen atom from a C–H bond: $RH + HO_2{}^{\bullet} \rightarrow R^{\bullet} + H_2O_2$. The free radical $R^{\bullet}$, which can be derived from a lipid or a carbonyl compound, can undergo a cascade of reactions yielding staling compounds (De Schutter et al. 2009; Bamforth et al. 1993). Some staling compounds that imply the participation reactive oxygen species can form even when oxygen is rigorously excluded. This has been taken to suggest that species present in beer can be oxidized during the brewing process. These species could then serve as oxidizing agents resulting in beer spoilage. One suspect for the species involved is melanoidins, which are the darkly colored polymeric polyphenols that give beer its color (Anderson et al. 2000). Strategies to inhibit spoilage by reactive oxygen species involve excluding oxygen, complexing metal ions, and trapping free radicals. We will discuss some approaches in the following section.

6.8.3 Spoilage Inhibition

6.8.3.1 Role of Processing

Nearly all commercial brewers take steps to limit the dissolved oxygen in packaged beer to the parts per billion level. From the time the beer is taken off the yeast, it must be protected from air, often by the use of a blanket of inert gas. Packages are purged of air before filling. Some packaging processes introduce a small jet of hot water that makes the beer foam to drive out any air in the head space just before the package is closed. When the beer is brewed in concentrated form (high gravity brewing), the dilution water to bring the beer to the final strength must be deaerated. Some brewers take steps to limit air in earlier brewing steps. Deaerated water can be used for mashing and sparging. These operations can be conducted under a gas blanket.

Not all brewers agree on the desirability of these steps. A number of ideas have been put forward to enhance the flavor stability of beer, but few have found widespread application. Often an improvement in one aspect has undesired side effects. One key factor is the nonacceptance on the part of consumers of flavor-stabilizing additives. If the effective concentrations of catalytic iron and copper ions can be reduced, the rate of production of reactive oxygen species could be decreased, resulting in slower spoilage and longer shelf-life for the beer. Chelating agents that have a high affinity for metal ions, such as phytic acid (Figure 6.15), can bind the metal ions, making them ineffective for catalyzing the production of reactive oxygen species (Graf et al. 1987). Brewers are reluctant to add antioxidants or other preservatives to the beer, but phytic acid is naturally present in the grain. Adjustments to the brewing process that retain phytic acid and similar chelating agents could be expected to enhance beer flavor stability. Some polyphenols can play this role, but their use could compromise the colloidal stability of the beer.

Certain compounds can react with free radicals and make more stable free radicals. These compounds are called free radical traps. The free radical trap most relevant to beer is sulfur dioxide, SO_2, present at beer pH as the hydrogen sulfite (also called bisulfite) ion HSO_3^-. Sulfur dioxide is sometimes added during malt kilning to limit formation of carcinogenic nitrosamines, but this practice has been fading from use (Thomas 2014). Sulfur dioxide is a natural product of fermentation, and its release can be enhanced by choice of yeast strain and other fermentation conditions, all of which have other effects on beer flavor. In addition to its role as a free

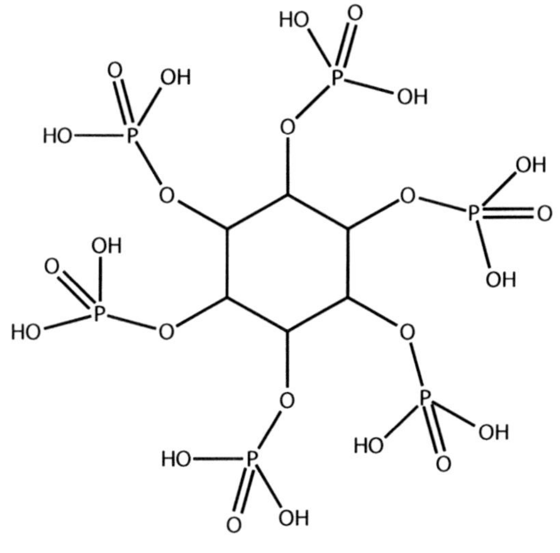

FIGURE 6.15 Phytic acid.

radical trap, the hydrogen sulfite ion can add across the C=O bond in staling carbonyl compounds, or to the C=C bond in unsaturated staling compounds. This can remove some of these compounds, giving the beer a fresher flavor (Nyborg et al. 1999). Conversely, the binding of staling compounds formed before fermentation can protect them from reduction by the yeast. If the hydrogen sulfite then releases the compounds, stale flavors can increase (Baert et al. 2015). The addition of soybean extract as a source of the enzymes superoxide dismutase (SOD) and catalase, as means of removing reactive oxygen species has been investigated (Bamforth and Parsons 1985). The use of diammine resins to remove staling compounds from beer has been undertaken on a pilot scale (Bech and Bravo 2007).

6.8.3.2 Role of Packaging

Beer sold directly to consumers is generally packaged in aluminum cans or glass bottles. Efforts to use plastic packages have been thwarted by permeability to oxygen and by consumer resistance. The aluminum can is impermeable to light and oxygen. The metal itself is corroded by beer, so the cans are lined with epoxy polymer. Glass bottles transmit light to some extent, depending on the color of the glass. The polymer used to make the seal between the crown cap and the bottle itself is slowly permeable to oxygen. The major packaging issue with respect to beer spoilage is exclusion of dissolved oxygen and gas-phase oxygen in the head space. This is a major concern in the design and operation of all aspects of beer processing after fermentation, especially filtration and filling systems. Dissolved oxygen is routinely kept below 100 μg/L (ppb), and some breweries keep it below 25 ppb.

6.8.4 Lightstruck Beer

Virtually all modern beer is flavored with hops, which provide bitterness and aroma. The bittering molecule in hops is isohumulone, shown in Figure 6.16, and its analogs. These compounds make the beer susceptible to a specific form of spoilage. If the beer is exposed to light in the wavelength range of 350 to 500 nm (near ultraviolet to green), the group shown in the box in Figure 6.16 can be cleaved off homolytically. The shared electrons become unshared, so two free radicals are formed. The six-carbon group loses carbon monoxide to yield a five-carbon isobutene free radical. The free radical abstracts S–H, possibly from an amino acid or polypeptide, giving 3-methylbut-2-ene-1-thiol (MBT), shown in Figure 6.17 (De Kuekeleire et al. 2008).

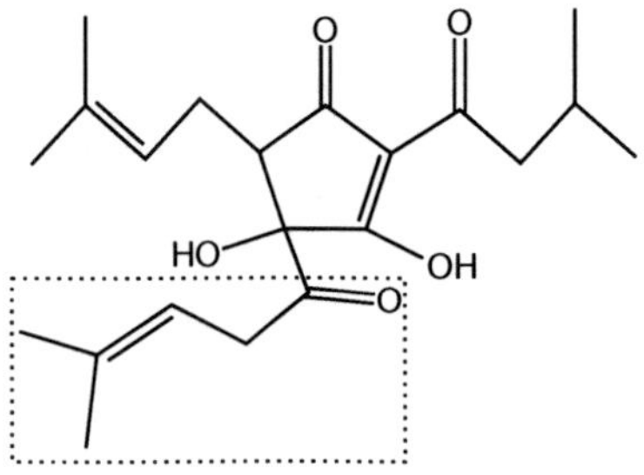

FIGURE 6.16 Isohumulone.

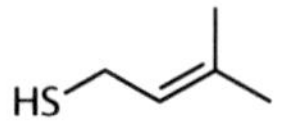

FIGURE 6.17 MBT.

MBT has a distinctive musky odor; beer with the MBT odor is officially called light-struck, but known to American drinkers as "skunked." The flavor threshold of MBT in beer is an astounding 7 ng/L, or seven parts per trillion (Irwin et al. 1993).

The most straightforward way to avoid spoilage of beer by lightstriking is to exclude light. The beer can be packaged in aluminum cans, which are impervious to light, or in brown bottles, which admit very little light in the 350–500 nm range. When, for marketing reasons, the beer is packaged in glass that transmits light, the brewer can use hop preparations whose bittering compounds have been hydrogenated at critical double bonds to lower their susceptibility to light. A variety of hydrogenated hop products is available for this purpose. When the beer is bittered with hydrogenated hop products, even traces of nonhydrogenated bittering compounds must be excluded to maintain the light-proof status of the beer. The use of hydrogenated hop products inevitably results in other changes to the character of the beer.

6.9 CONCLUSION

Beer is subjected to a variety of postfermentation treatments to improve its flavor, clarity, and stability. Most of these have evolved from earlier practices. Some treatments, like sedimentation, have been done for centuries, and maybe millennia. The modern extension, centrifugation, is well established in large breweries, and is now seeing use in small craft breweries. Filtration has evolved from traditional practices like pouring the beer through reed mats to much more efficient and controllable processes. Finings have been in use for centuries, but new fining materials have greatly enhanced beer colloidal stability. Pasteurization was invented over 150 years ago. Although beer brewing was among the first industries to adopt the practices of the industrial revolution, it represents itself as a traditional craft process. Innovations that could change the character of the beer are viewed with suspicion by brewers and by beer consumers. Despite this conservatism, beer is a subject of active, innovative, and productive research that yields incremental improvements in the quality and stability of the beer and the efficiency of its production.

REFERENCES

Anderson, M.L., H. Outtrup, and L.H. Skibsted. 2000. Potential antioxidant in beer assessed by ESR spin trapping. *J. Agric. Food Chem.* 48:3106–3111.

Asano, K., K. Shangawa, and N. Hashimoto. 1982. Characteristics of haze-forming proteins of beer and their roles in chill haze formation. *J. Am. Soc. Brew. Chem.* 40:147–154.

ASBC. Methods of Analysis, online. 2014. Beer-25. Diacetyl. St. Paul, MN: American Society of Brewing Chemists. doi: 10.1094/ASBCMOA-Beer-25.

Baert, J.J., J. De Clippeleer, L. De Cooman, and G. Aerts. 2015. Exploring the binding behavior of beer staling aldehydes in model systems. *J. Am. Soc. Brew. Chem.* 731(1):100–108.

Bamforth, C.W. 2006. *Scientific Principles of Malting and Brewing*. St. Paul MN: American Society of Brewing Chemists, p. 171.

Bamforth, C.W., R.E. Muller, and M.D. Walker. 1993. Oxygen and oxygen radicals in malting and brewing: A review. *J. Am. Soc. Brew. Chem.* 51(3):79–88.

Bamforth, C.W., and R. Parsons. 1985. New procedures to improve flavor stability of beer. *J. Am. Soc. Brew. Chem.* 43(4):197–202.

Barth, R. 2013a. *The Chemistry of Beer: The Science in the Suds*. Hoboken, NJ: Wiley, pp. 211–223.

Barth, R. 2013b. The science of carbonation. *Zymurgy*. 36(6):65–69.

Bech, L.M., and A. Bravo. 2007. Use of a 1,3-diammine scavenger resin to reduce the concentration of staling markers in beer. *Proc. 31st Congr. Eur. Brew. Conv.* Amsterdam. Paper 97.

Bokulich, N.A., Bamforth, C.W., and Mills, D.A. 2012. Brewhouse-resident microbiota are responsible for multi-stage fermentation of American coolship ale. *PLoS One* 7(4):e35507.

Boulton, C., and W. Box. 2003. Formation and disappearance of diacetyl during lager fermentation. In *Brewing Yeast Fermentation Performance*, 2nd ed., ed. K. Smart. Oxford, UK: Blackwell.

Boulton, C. and D. Quain. 2006. *Brewing Yeast and Fermentation*. Hoboken NJ: Blackwell, p. 447.

Briggs, D.E., C.A. Boulton, P.A. Brookes, and R. Stevens. 2004. *Brewing Science and Practice*. Boca Raton, FL: CRC, pp. 569–572.

Callemien, D., S. Dasnoy, and S. Collin. 2006. Identification of a stale-beer-like odorant in extracts of naturally aged beer. *J. Agric. Food Chem.* 54:1409–1413.

Cantwell, D. and P. Boukaert. 2016. *Wood and Beer: A Brewer's Guide*. Boulder, CO: Brewers Publications.

Carey, D., and K. Grossman. 2006. Fermentation and cellar operations. In *Fermentation, Cellaring and Packaging Operations*, ed. K. Ockert. St. Paul, MN: Master Brewers Association, pp. 119–130.

Cyr, N., M. Blanchette, S.P. Price, and J.D. Sheppard. 2007. Vicinal diketone production and amino acid uptake by two active dry lager yeasts during beer fermentation. *J. Am. Soc. Brew. Chem.* 65(3):138–144.

Dalgliesh, C.E. 1977. Flavour stability. *Proc. 16th Congr. Eur. Brew. Conv. Amsterdam*, pp. 623-659.

Dawson, P. 2014. *Vintage Beer*. North Adams, MA: Storey.

De Kuekeleire, D., A. Heyerick, K. Huvaere, and L.H. Skibsted. 2008. Beer light struck flavor: The full story. *Cerevisia*. 33(3):133–144.

De Schutter, D.P., D. Saison, F. Delvaux, G. Derdelinckx, and F.R. Delvaux. 2009. The chemistry of beer aging. In *Beer in Health and Disease Prevention*, ed. V.R. Preedy. Burlington, MA: Academic Press.

EBC. European Brewery Convention. 2008. *Analytica-EBC*, 7th ed. P. 680. Nürnberg: Hans Carl Getränke-Fachverlag.

Evans, D.E., L.H. Robinson, M.C. Sheehan, A. Hill, J. S. Skerritt, and A.R. Barr. 2003. Application of immunological methods to differentiate between foam-positive and haze active proteins originating from malt. *J. Am. Soc. Brew. Chem.* 61(2):55–62.

Gijs, L., P. Perpète, A. Timmermans, and S. Collin. 2000. 2-Methylthiopropionaldehyde as a precursor of dimethyl trisulfide in aged beers. *J. Agric. Food Chem.* 48:6196–6199.

Goldhammer, T. 2008. *The Brewer's Handbook*, 2nd ed. Clifton, VA: Apex, pp. 275–293.

Gonzales, S. 2014. Bourbon barrel aging optimization. *MBAA Tech. Q.* 51(4):106–109.

Graf, E., K.L. Empson, and J.W. Eaton. 1987. Phytic acid: A natural antioxidant. *J. Biol. Chem.* 262(24):11647–11650.

Guinard, J-X. 1990. *Lambic*. Boulder, CO: Brewers Publications.

Iimure, T., N. Nankaku, M. Watanabe-Sugimoto, N. Hirota, Z. Tiansu, M. Kihara, K. Hayashi, K. Ito, and K. Sato. 2009. Identification of novel haze-active beer proteins by proteome analysis. *J. Cereal Sci.* 49:141–147.

Inoue, T. 1988. Immobilized cell biotechnology—A new possibility for brewing? *J. Am. Soc. Brew. Chem.* 46:64–66.

Inoue, T. 2008. Diacetyl formation by yeast and bacteria. Chapter 3 in *Diacetyl in Fermented Food and Beverages.*

Inoue, T. and Y. Yamamoto. 1970. Diacetyl and beer fermentation. *Proc. Amer. Soc. Brew. Chem.* 1970:198–208.

Irwin, A.J., L. Bordeleau, and R.L. Barker. 1993. Model studies and flavor threshold determination of 3-methyl-2-butene-1-thiol in beer. *J. Am. Soc. Brew. Chem.* 51(1):1–3.

Jackson, T., R.T. Roberts, and T. Wainwright. 1980. Mechanism of beer foam stabilization by propylene glycol alginate. *J. Inst. Brew.* 86:34–37.

Krogerus, K., and B.R. Gibson. 2013a. 125th Anniversary review: Diacetyl and its control during brewery fermentation. *J. Inst. Brew.* 119:86–97.

Krogerus, K., and B.R. Gibson. 2013b. Influence of valine and other amino acids on total diacetyl and 2,3-pentanedione levels during fermentation of brewer's wort. *Appl. Microbiol. Biotechnol.* 97:6919–6930.

Leskošek-Čukalović, I.J. and V.A. Nedović. 2005. Immobilized cell technology in beer brewing. *Proc. Nat. Sci. Matica Srpska Novi Sad.* 109:129–141.

Libkind, D., C.T. Hittinger, E. Valério, C. Gonçalves, J. Dover, M. Johnston, P. Gonçalves, and J.P. Sampaio. 2011. Microbe domestication and the identification of the wild genetic stock of lager-brewing yeast. *Proc. Natl. Acad. Sci. U. S. A.* 108(35): 14539–14544.

Lindemann, B. 2009. Filtration and stabilization. In *Handbook of Brewing*, ed. H.M. Eßlinger. Weinheim, Germany: Wiley-VCH, pp. 518–520.

Martens, H., D. Iserentant, and H. Verachtert. 1997. Microbiological aspects of a mixed yeast—Bacterial fermentation in the production of a special Belgian acidic ale. *J. Inst. Brew.* 103:85–91.

McMurrough, I., and T. O'Rourke. 1997. New Insight Into the mechanism of achieving colloidal stability. *MBAA Tech. Q.* 34(1): 271–277.

Meilgaard, M.C. 1975. Flavor chemistry of beer: Part II: Flavor and threshold of 239 aroma volatiles, *MBAA Tech. Q.* 12(3):151–167.

Meilgaard, M.C., C.E. Dalgliesh, and J.F. Clapperton. 1979. Beer flavor terminology. *J. Am. Soc. Brew. Chem.* 37(1):47–52.

Moll, M. 1987. Colloidal stability of beer. In *Brewing Science*, ed. J.R.A. Pollack. London: Academic Press, pp. 1–327.

Munroe, J.H. 2006. Fermentation. In *Handbook of Brewing*, eds. F.G. Priest and G.G. Stewart. Boca Raton, FL: CRC, pp. 518–520.

Nyborg, M., H. Outtrup, and T. Dreyer. 1999. Investigations of the protective mechanism of sulfite against beer staling and formation of adducts with trans-2-nonenal. *J. Am. Soc. Brew. Chem.* 57(1):24–28.

Oliver, G. 2011. Aging. In *Oxford Companion to Beer*, ed. G. Oliver. Oxford, UK: Oxford University Press.

Patino, H. 1999. Overview of cellar operations. In *The Practical Brewer*, 3rd ed., ed. J.T. McCabe. p. 299. Wauwatosa, WI: Master Brewers Association.

Robinson, L.H., P. Healy, D.C. Stewart, J.K. Eglinton, C.M. Ford, and D.E. Evans. 2007. The identification of a barley haze active protein that influences beer haze stability. *J. Cereal. Sci.* 45:335–342.

Shimizu, C., Y. Nakamura, K. Miyai, S. Araki, M. Takashio, and K. Shinotsuka. 2001. Factor affecting 5-hydroxymethyl furfural formation and stale flavor formation in beer. *J. Am. Soc. Brew. Chem.* 59(2):51–58.

Siebert, K.J. 1999. Effects of protein-polyphenol interactions on beverage haze, stabilization, and analysis. *J. Agric. Food Chem.* 47(2):353–362.

Siebert, K.J., A. Carrasco, and P.Y. Lynn. 1996. Formation of protein-polyphenol haze in beverages. *J. Agric. Food Chem.* 44:1997–2005.

Snauwaert, I., S.P. Roels, F. Van Nieuwerburg, A. Van Landschoot, L. De Vuyst, and P. Vandamme. 2016. Microbial diversity and metabolite composition of Belgian red-brown acidic ales. *Int. J. Food. Microbiol.* 221:1–11.

Spitaels, F., A.D. Wieme, M. Janssens, M. Aerts, H-M. Daniel, A. Van Landschoot, L. De Vuyst, and P. Vandamme. 2014. The microbial diversity of traditional spontaneously fermented lambic beer. *PLoS One.* 9(4):e95384.

Spitaels, F.S., S. Van Kerrebroeck, A.D. Wieme, I. Snauwaert, M. Aerts, A. Van Landschoot, L. De Vuyst, and P. Vandamme. 2015. Microbiota and metabolites of aged bottled gueuze beers converge to the same composition. *Food Microbiol.* 47:1–11.

Thomas, D. 2014. *The Craft Maltsters' Handbook.* Hayward, CA: White Mule, p. 170.

Tran, T.T.H., S. Nizet, J. Gros, and S. Collin. 2013. Occurrence of the ribes odorant 3-sulfanyl-3-methylbutyl formate in aged beers. *Flavour Fragr. J.* 28(3):174–179.

Van den Berg, R., P.A. Harteveld, and F. B. Martens. 1983. Diacetyl reducing activity. *Proc. Congr. Eur. Brew. Conv.* 19:497–504.

Vanderhaegen, B. and G. Derdelinckx. 2005. Characterization of flavor compounds in fresh and aged beer by purge and trap-gas chromatography-mass spectrometry. *Monatsschr. Brauwiss.* Nov/Dec:1–9.

Vanderhaegen, B., H. Neven, H. Verachtert, and G. Derdelinckx. 2006. The chemistry of beer aging—A critical review. *Food Chem.* 95:357–381.

Virkajärvi, I. 2006. Accelerated processing of beer. In *Brewing New Technologies*, ed. C. Bamforth, Chapter 12. Cambridge, UK: Woodhead.

Williams, R.S., and H. Wagner. 1978. The isolation and identification of new staling related compounds from beer. *J. Am. Soc. Brew. Chem.* 36:27–31.

Wu, X., L. Gu, R.L. Prior, and S. McKay. 2004. Characterization of anthocyanines and proanthocyanidines in some cultivars of *Ribes, Aronia*, and *Sambucus* and their antioxidant capacity. *J. Agric. Food Chem.* 52:7846–7856.

7 Vinegars: Process, Aging, and Spoilage

Paolo Giudici, Tommaso Bonciani, Federico Lemmetti, and Giuseppe Corradini

CONTENTS

7.1 VINEGAR

The production of vinegar can be presumably dated back to the dawn of agriculture and, although once conceived as a poor product, these days its use is widespread as a condiment and food preservative. Vinegar can be defined as a varied ensemble of condiments and/or beverages characterized by a set of specific features:

- They result from the onset of at least two successive events of fermentation starting from agricultural raw materials, usually consisting of an alcoholic fermentation followed by an acetic fermentation.

- They contain acetic acid in significant amounts, which gives them a pungent smell. Low amounts of residual ethanol are also allowed.

The UN's Food and Agriculture Organization (FAO) and the World Health Organization (WHO) define vinegars as liquids fit for human consumption, exclusively obtained starting from suitable substrates containing starch and/or sugars by the process of double fermentation: alcoholic fermentation followed by acetous fermentation. The maximum residual ethanol content in wine vinegar and other vinegars must be 1.5% and 1%, respectively. In the European Union the regulations for vinegar are regional and entail both acetic acid and ethanol contents. In addition, these contents are defined type-wise: A generic vinegar contains a minimum acetic acid concentration of 5% (w/v) and a maximum ethanol concentration of 0.5% (v/v), while vinegars made starting from different raw materials feature varying thresholds for the two metabolites. For instance, wine vinegar is defined (EC 1999) by a minimum 6% (w/v) acetic acid and a maximum ethanol of 1.5% (v/v). In the United States, the Food and Drug Administration (FDA) requires a minimum acidity of 4% for vinegars, while not addressing alcohol content (FDA 1995). Also, Chinese legislation concerning ethanol and acetic acid content in vinegar relies on regional regulations; however, the term "vinegar" is used for both fermented and artificial vinegars, which means vinegars made by directly blending acetic acid with other ingredients (CNS 2005).

In Asia and Africa, vinegar is used as a beverage and is thus characterized by a less sour taste. Therefore, in China and Southeast Asia, sweetened fruit vinegars with low acidity and aromatic flavors can be found. China is also a huge producer of cereal vinegars, such as rice, sorghum, and barley vinegars (Chen et al. 2009). Japanese vinegars are conceived both as condiment and as beverage; the two varieties of rice vinegar (kurosu [unpolished rice vinegar] and komesu [polished rice vinegar] address the two purposes, respectively). Whereas komesu is clear and low in acidity, kurosu is black and richer in amino acids and vitamins (Murooka et al. 2009). In Africa, the production of vinegar starts from raw materials such as palm sap and cocoa pod. Western vinegars are not conceived as mere condiments, but rather also have a role as traditional cleaning or health-related products; they are obtained starting with raw materials that already contain high amounts of simple chemical compounds (both sugars and ethanol). Wine vinegar is a typical product of EU grape-producing countries and has numerous applications: condiment, food preservative, and cleaning product. On the other hand, the traditional balsamic vinegar of Modena and Reggio Emilia, produced in the Emilia region of Italy, is its nobler counterpart. It is obtained starting from concentrated musts and, according to different production modes, can be strictly industrial or more traditional (Lemmetti and Giudici 2011). Apple vinegar is typical, on the other hand, of central-northern countries of the EU, where apples constitute a more available raw material. Although its main usage is as preservative or flavor agent for the food industry, it is also used in folk medicine as a claimed remedy against arthritis, asthma, cough, diarrhea, colitis, and eczema (a little bit too much, in our opinion).

7.1.1 Raw Materials

Many botanical species can be employed for vinegar production as far as they have two basic attributes: (1) safe for human and animal consumption and (2) composed of either a directly or indirectly fermentable sugar source. Raw materials comprise fruits, cereal seeds, legumes, roots, tubers, exudates, sap, and even whey and honey The nature of the raw material significantly affects the process, the sensory properties, and the physical–chemical stability of each vinegar (Solieri and Giudici 2009). The pH of the raw material is crucial in determining the fermentability and in controlling the potential spoilage by microbiota. For instance, juices obtained from fruits and vegetables, characterized by high pH values, are more easily altered by microorganisms, which can affect either the sensory attributes of the vinegar or, in some more extreme cases, even health-related attributes. General considerations can be made based on the chemical composition of the edible parts and their fermentation attitude. Generally, raw materials can be grouped according to their pH and their carbohydrate composition:

- Acidic and easily fermentable: pH lower than 3.5; glucose, fructose, and sucrose as main carbon sources (e.g., berries and grape)
- Moderately acidic and easily fermentable: pH between 3.5 and 4.5 (e.g., tomato, apple, and plums)
- Scarcely acidic and easily fermentable: pH greater than 4.5 (e.g., figs, dates, and palm sap)
- Nonfermentable: hydrolysis required prior to fermentation (e.g., rice, sorghum, and barley)

Alcohol oxidation by acetic acid bacteria (AAB) is the main event of acetic fermentation and is necessarily preceded by a stage of alcoholic fermentation, which is usually performed by yeasts. This means that the availability of readily fermentable sugars in the raw material is pivotal for the process. This is especially true if we want to use complex polysaccharides, such as starch and inulin, which need to be broken down to oligomers or monomers, which are more easily fermented by the yeasts. This goal will be achieved via chemical, enzymatic, or biological processes (Giudici and Gullo 2012).

A very important raw material is distilled alcohol, obtained by fermentation of fruits and/or cereals, especially from corn starch, which leads to the production of the so-called "spirit vinegars" or "white vinegars," widely used in both Northern Europe and United States. Distilled alcohol has many advantages: It is easily stored, thus delivering vinegar production from the cycle of seasons; it allows the overall reduction of volumes; it allows obtaining vinegars with a higher titer of acetic acid, even higher than 20% v/v if submerged fermentation is applied; and even second-choice and/or altered fruits and vegetables can be fermented, as distillation removes any possible mycotoxin present on decayed fruits. Finally, spirit vinegar is very distinguishable from the other types of vinegar, due to both the peculiar pungency of the smell and the total lack of sensory-active molecules related to the starting raw material.

7.1.2 Process Overview

Although the proper acetous fermentation implies the conversion of ethanol to acetic acid, all types of vinegar are actually produced starting from unfermented sources of sugars such as glucose or starch. Sugars are converted to alcohol by yeasts before the acetous fermentation can begin, while starchy raw materials require treatments to break down the long-chain sugar polymers before the alcoholic fermentation. They are eventually yielded in mono- or oligomeric form, thus being more easily converted to ethanol by yeast. In the case of Chinese cereal vinegar, the raw materials are roughly milled to enhance the contact surface with water, humidified, and mixed with qu (koji in Japanese), a natural starter containing molds, yeasts, and bacteria (Figure 7.1). Qu is both a saccharification agent and a starter. It usually comes in the form of pressed bricks and is produced starting from cereals (wheat, barley). The cereals act as growth medium for molds, yeasts, and bacteria; these microorganisms produce high amounts of hydrolythic enzymes (amylases, glucoamylases, proteases, and lipases) that remain associated with the cereal bricks (Chen et al. 2009). The bricks are powdered and added to the fermentation mash.

FIGURE 7.1 Bricks of qu. The bricks are a mixture of a previously fermented batch of qu with milled cereals. They are kept in a room under strictly controlled conditions of humidity and temperature.

The beginning of saccharification, triggered by the hydrolytic enzymes, immediately pushes the beginning of alcoholic fermentation. In the traditional way, a mixture of qu and starchy materials is contained in ceramic urns, and it gives the cereal wine as a final product, known as pei in Chinese. Then the pei is transferred and mixed with rice husks, which work as a physical support for AAB (see later discussion). The acetous fermentation requires a proper oxygenation of the medium, as the AAB will act by oxidizing ethanol. The acetification process is highly exergonic and can be performed in three different operational modes:

- Surface culture fermentation (SCF)
- Submerged fermentation (SF)
- Solid-state fermentation (SSF)

The three operational modes differ not only for the shape of the bioreactor and the control over the involved variables but also for the main microbiological actors. Under a more practical point of view, the differences among SCF, SF, and SSF can be seen in terms of acetic acid yield and final titer and of the overall sensory properties of the vinegar. In SF, microorganisms grow and ferment in the liquid bulk; all the "spirit vinegar" and the greatest part of wine vinegar on an industrial scale are produced by SF. This mode gives the highest yields and final concentrations. The acetous fermentation is carried out in mechanically stirred bioreactors; the fan/agitator causes the homogeneous dispersion of air bubbles and maximizes the contact of nutrients and microorganisms. The bioreactors need a fine control on temperature and online monitoring of alcohol, acidity (García-García et al. 2009), and oxygen availability (dissolved oxygen tension [DOT]). Concerning sensory properties, these vinegars have more polished flavors and contain just limited amounts of esters (Callejón et al. 2009).

In SCF, AAB develop in a biofilm with variable thickness and texture at the interface between air and liquid; the totality of home-made production and the greatest part of the small-scale and traditional production plants rely on SCF. Fermentation is performed in open vessels (barrels or others) made of different materials. These vessels are not filled to the top, as oxygen is needed for the fermentation to be carried out; in fact, the transfer of O_2 through the inner surface of the vessel is limited in the case of materials such as wood and otherwise is completely absent if the process is carried on in metal tanks. Vinegar obtained through SCF has a lower titer in acetic acid and rarely reaches values higher than 10% v/v. Yields are low and the process is longer. In addition, it can even lead to the consumption of acetate in the long term, which results in an even lower titer of acetic acid. However, the resulting vinegars are more complex under the sensory point of view, especially if compared to the ones obtained by SF, featuring a higher concentration of esters and of metabolites associated to the fermentation (Callejón et al. 2009).

SSF refers to the acetous fermentation performed by AAB growing on solid substrates (Figure 7.2) and is used to produce Eastern cereal vinegars such as Chinese cereal vinegars. The solid substrate can be totally inert (i.e., a mere solid support) or act also as a source of nutrients. The former case is the one of the Schutzenbach method, where wood shavings are used to provide a large contact surface for AAB

FIGURE 7.2 Urn containing rice husks and rice wine (pei) for traditional solid state fermentation (SSF) of cereal vinegars.

with both wine and air, which are forced to flow counter currently to one another. The second case is the one of Chinese cereal vinegar: Bacteria grow on cereal bran (husks) at the interface with wine. The tanks (urns, in this case) are shuffled at regular intervals, either manually or by the application of a mechanized procedure. This shuffling has a triple aim: the aeration, the maintenance of a constant moisture in the mass, and the dissipation of heat.

7.2 FERMENTATIONS

The spontaneous fermentation of vinegar is the result of a microbial activity comprising both bacteria (AAB and lactic acid bacteria) and yeasts (ascomycetes). The final step is always the oxidation of ethanol by AAB, which happens after the alcoholic fermentation. The presence and/or dominance of a given species is strictly dependent on the physical–chemical parameters of the fermentation such as pH, temperature, aeration, alcohol, acetic acid, and sugar content. In other words, these variables determine the ecological landscape of the fermentation and select only those species that can cope with it.

7.2.1 Microorganisms

The AAB applied to vinegar production belong to the genera *Acetobacter*, *Gluconacetobacter*, and *Komagataeibacter*. AAB are obligate aerobic bacteria that oxidize a variety of carbohydrates, thus releasing corresponding metabolites (organic acids, aldehydes, and ketones) into the surrounding medium. They can produce different types of exopolysaccharides, such as cellulose (Sievers and Swings 2005). The production of these molecules can be seen either as a production defect in the context of vinegar production or as innovative in the context of microbial polymers for material engineering (Gullo et al. 2017). From the industrial point of view, taxonomy is not as important as the phenotypic traits displayed by the single strains. Strains of AAB to be applied in the industry need to feature at least two traits: (1) no production of a thick "mother" of vinegar, which affects the yield of acetic acid and the quality of the final product and (2) high tolerance to ethanol and the ability to grow in wines with more than 12% ethanol. Apart from these two compulsory traits, an AAB strain to be applied in the industrial production of vinegar should feature high acetic acid resistance as the most basic feature, as it needs to resist progressively rising concentrations of this metabolite (Giudici et al. 2016).

The other relevant traits include the ability to oxidize ethanol as a preferred carbon source and to produce acetic acid in a rapid manner (Giudici et al. 2016). The latter feature becomes even more handy if seen in the context of large-scale productivity by SF, as will be shown later in this chapter. The combination of acetic acid tolerance and high production rate maximizes acetic acid yield and, thanks to the antimicrobial action of acetic acid, prevents the growth of spoilage microorganisms that could affect the overall quality of the product. Remarkably, neither big nor the small industry relies on selected strains of AAB; the most common practice is the back-sloping from previous fermentation batches. The first reason is technological: The success of the fermentation is strictly dependent on how well the culture responds to the environmental changes during the acetic fermentation and not on the application of a specific strain (Gullo et al. 2012). The second is an economical reason: The practice of back-sloping is indeed simple and inexpensive, still preserving the reproducibility of the production process.

As for yeasts, genera and species that are involved in vinegar production vary greatly according to the fermented raw material; for instance, the genus *Kluyveromyces* is involved in the production of whey vinegar, and more osmophilic/osmotolerant genera such as *Zygosaccharomyces*, *Saccharomycodes*, *Hanseniaspora*, and *Candida* (Solieri and Giudici 2008) are involved in the production of traditional balsamic vinegar. Although the importance of specific yeast strains is not proved for vinegar production because detailed studies about the metabolic and sensory effects of specific strains are not available, some exceptions exist. In fact, similarly to wine production, variations in secondary metabolites of alcoholic fermentation have been observed (Gobbi et al. 2014)—for instance, succinic and malic acids, glycerol, acetoin, and higher alcohols. Given this premise, we could argue that different yeast strains might likely influence the taste and flavor of the intermediate wine and, consequently, of the final vinegar.

7.2.2 Substrates and Final Product

Acetic acid bacteria can oxidize many different compounds. The most straightforward transformation of AAB is the conversion of ethanol to acetic acid. However, some AAB species are prone to oxidize many different chemical species. By microbial metabolism, glucose can be converted to gluconic acid, glycerol to dihydroxyacetone, and D-sorbitol to L-sorbose (Asai 1998). One should also consider that the available substrates vary according to the specific vinegar; different vegetal sources have different proportions of sugars, acids, and alcohols.

A varied range of compounds are also yielded by the other microbiological agents, especially for the less controlled processes. In fact, every time we deal with complex microbial activity (acetic acid bacteria, yeasts, molds, lactic acid bacteria), we always witness a complex ensemble of metabolites. This statement is even more verified if we take into account the more traditional types of vinegar, where less control is exerted over the process; as a direct consequence, the microbial contributions are stronger and more varied.

7.2.3 Yield

Theoretically, the molar ratio of ethanol conversion to acetic acid is 1:1 according to the following exergonic reaction:

$$1\ C_2H_5OH + 1\ O_2 - 1\ CH_3COOH + 1\ H_2O;\ (\Delta G^\circ = -455\,\text{kJ/mol})$$

The best industrial fermentation yield is around 95%–98%, because of the cell metabolism and multiplication. Furthermore, a part of ethanol and produced acetate can either evaporate or be depleted via entrance into the TCA and subsequent respiration. In general, acetic acid yield for the SCF is much lower than the one obtained by SF. However, the vinegar industry still measures yields based on ethanol volume and acetic acid concentration, expressed in grams per liter. This is mainly related to the way the winemaking industry refers to alcohol concentration, which is volume percentage. Remarkably, as vinegar is valued on the basis of acetic acid content, a hybrid and somewhat inaccurate definition of the yield arises. This makes for an apparent rise of the values above the theoretical maxima; in fact, the maximum apparent yield is shifted from 100% (mol/mol), from a purely chemical perspective, to 102% (w/v) in industrial terms. In addition, the industry of vinegar also expresses concentrations of ethanol and acetic acid in a conjoined way, by means of the so-called "total concentration," which stands for the sum of ethanol (% v/v) and acetic acid (grams per 100 mL). This means that a vinegar featuring 5 g/100 mL of acetic acid and a residual ethanol value of 2% has a total concentration value of 7. In this case, the yield is calculated as the ratio between the total concentration at the beginning and at the end of fermentation (Ebner et al. 1996).

7.3 AGING

Age is generally perceived as a mere time-related attribute, but its definition is not as easy as it seems; in fact, time does not act in the same manner for each separate situation. Without turning to Einstein's relativity principle, the aging of vinegar can be described through two distinct parameters:

- *Residence time* (RT) is purely physical and is expressed with the dimension of time (weeks, months, years). It simply describes the permanence of the vinegar inside the aging vessel.
- *Age* is conceived in a more complex manner and is partially dependent on the RT. In fact, age can be seen as a physical–chemical variation of the product. It depends on the initial composition of the vinegar, the material of the aging vessel, the temperature, and, of course, the RT (Lemmetti and Giudici 2011).

During the long aging period, vinegar undergoes significant chemical and physical changes; we have significant data on Jerez, cereal, and balsamic vinegars. In the case study of traditional balsamic vinegar, the following variables undergo these changes: water content, specific weight, viscosity, color, Brix degree, titratable acidity, and pH. Under a more general point of view, an indefinite number of chemical compounds are accumulated according to the complexity of the matrix and of the specific process. The variations of these variables and the formation of new compounds contribute to the establishment of the peculiarities that differentiate the many types of vinegar. These variations are strictly dependent on the vessel applied to the aging phase and are affected by the nature of the constituent materials.

7.3.1 Vats and Barrels

The aging of vinegar, as for beverages and liquid foods in general, is carried out in containers made of different materials: wood, steel, glass, pottery (either plain or glazed), concrete, and, more recently, plastic polymers. Other metals used in the past should absolutely be avoided; the reasons are evident and do not require any further explanation. Concerning the more specific case of vinegar, wood barrels, steel vats, and ceramic urns are the containers used for aging. Vessels and materials can be grouped according to their effect on solutes, solvents, dissolved gases, and vapors of vinegar:

- Totally impermeable: this is the case of stainless steel, glazed pottery, and glass. The vessels made with these types of materials, if hermetically closed, do not allow any molecule to leave the vinegar. Minor changes in volume could only be due to differences in molar volumes between reagents and products of aging-related chemical reactions.
- Permeable to small molecules: This is the case of all wooden barrels and vats (Figure 7.3). Wood acts as a semipermeable membrane for molecules according to their steric hindrance. It retains all bigger molecules (alcohols, esters, acetic acid) inside the container, while smaller molecules such as water are allowed to flow out by evaporation. This event gives rise to the concentration effect of solutes inside the barrel. Remarkably, all higher molecular weight compounds like the taste- and flavor-active ones are retained, making for greater sensory complexity of the product.

FIGURE 7.3 Solera system for the aging of sherry vinegar in the Jerez region. The vinegar is withdrawn from the lowest barrel set (solera), and then the barrels are refilled with the vinegar coming from the set that is directly above (first criadera), which are in turn filled with the vinegar of the third barrel set (second criadera). New base vinegar is then used to refill the top barrel set (third criadera).

- Open vats: This operational mode is typical of certain Chinese cereal vinegars (Figure 7.4). In this case, evaporation proceeds at the highest rate, as the molecules are left to pass directly from the liquid to the gaseous phase. The concentration effect is maximized for nonvolatile molecules, while both small and large volatile molecules pass to the gaseous phase.

7.3.2 Water Transfer in Wooden Barrels

Beverages and liquid foods preserved in wooden vessels for long periods of time, such as those associated with aging, always undergo a certain loss of water due to transpiration. The volume loss in a barrel is typical also of the aging processes of distilled beverages and is known as the "angels' share" in the English-speaking world. The finest vinegars undergo similar aging processes and the physicochemical modifications of vinegar can be understood in the terms of a diffusion model. The water losses can happen either by plain evaporation, if some direct contact between vinegar and external air exists (as in the case of open-vat fermentations), or by vapor transpiration through the staves of the barrels. The two different types of evaporation have great implications in the overall characteristics of the final product. As already mentioned, the pores of the wood act as size-exclusion filters that allow the passage of small molecules, such as water and molecular oxygen, but retain larger ones (Siau 1984), such as alcohols, aldehydes, esters, and organic acids (including acetic acid), which would otherwise be lost by plain evaporation. The extent of water transpiration through the casks can be significant: the yearly volume loss for wine

FIGURE 7.4 Open ceramic vats used for the aging of Chinese cereal vinegar in the region of Jiangxi, China.

in sealed wooden barrels ranges from 1% to 9% (Canas et al. 2002). In the southern hemisphere, in areas with low relative humidity, volume losses reach a value of 15% every 2 years.

The variables affecting the transfer of molecules from the inner to the outer part of the barrel are strictly related to the chemical nature of the material and thus on the type of wood. Secondly, one should also consider the shape of the barrel, as expressed by the surface/volume ratio and the environmental conditions such as relative humidity, temperature, and air flow. Finally, the chemical composition of the product (vinegars, distilled alcoholic beverages, wine) also affects the way the process proceeds. The permeability and the diffusion layer are both related to the nature of the solution and the porosity of the material; their behavior can be described with Darcy's law. On the other hand, the mass transfer through the wood follows Fick's model of diffusion (Ruiz de Adana et al. 2005). It is easily understood how the transfer of molecules through the structure of the wood is also dependent on the different molecular sizes of the solutes. Small molecules such as water and ethanol easily pass through the wooden staves (Guymon and Crowell 1972), while many higher molecular weight compounds are retained inside the barrel. Water penetrates the wood even more quickly if the environmental values of relative humidity are low. As a matter of fact, at low values of relative humidity, the evaporation of water increases. Interestingly, the opposite happens for ethanol (Guymon and Crowell 1972; Singleton 1995).

Two topologically detached phenomena are responsible for the transfer of the water from the inside to the outside of the barrel. Firstly, the liquid impregnates the wood. Secondly, at the interface with nonimpregnated wood, also known as "front of impregnation," evaporation occurs with change of state from liquid to gas. The front of impregnation, in some environmental conditions, can be so extended as to reach

FIGURE 7.5 Barrel with a thriving fungal mycelium on the external wooden surface. This is due to the water impregnation front that reaches the outer surface of the barrel.

the outer surface. The higher the relative humidity is, the greater is the impregnation layer that will reach the outer surface of the barrel. Figure 7.5 shows what happens when the front of impregnation reaches the outer surface of the barrel: Due to the moisture of the wood surface, fungal mycelia can grow. Of course, the outflow of water causes the concentration of solutes; this was also observed in the barrel aging of both wine and brandy, where the concentration of ethanol was observed (Onishi et al. 1977).

The selective transfer of molecules can be described on a more rigorous basis by considering the steric hindrance of the involved molecules. Table 7.1 shows the molecular size of some vinegar-related compounds, calculated by the method of Connolly. The solvent is simulated by a sphere that is run upon the van der Waals surface of the molecule. The sphere maps the points of contact with the atomic radius of van der Waals of the molecule. The set of these points constitutes the "solvent-accessible surface" and the volume enclosed within is the "solvent-excluded volume" that approximates very well the volume of the molecule. The simulation accounts for the minimum-energy conformation, which that minimizes the electronic repulsions between nonbonded atoms. It should be noticed that the vapor pressure is not related to the size or the molecular weight; indeed, it depends on the intermolecular van der Waals forces and hydrogen bonds. For example, the ester ethyl acetate has a strong volatility despite its volume, which is eightfold greater than that of water and is thus retained inside the barrel by the structure of the wood. The loss of solvent also has a role in the activation of the chemical processes of aging and contributes to the overall increase of specific gravity. When seen in the specific case of traditional balsamic vinegar, the concentrations of solutes can rise up to values of 70% (w/v) with a two- to threefold increase of the concentration of sugars.

7.3.3 Residence Time

Residence time (RT) is the first parameter to be considered; this is defined as the actual persistence of the vinegar in the vessels. RT is a variable that is easily

TABLE 7.1
Molecular Size of Some Constituents of the Traditional Balsamic Vinegar of Modena and Reggio Emilia, Calculated by the Method of Connolly

Compound	Molecular Weight (g/mol)	Relative Density $d_{20/20}$	Vapor Pressure at 25°C (KPa)	Connolly Surface (Å²)	Connolly Molecular Volume (Å³)
Water	18.0	1.00	3.17	23.3	10.3
Acetic acid	60.0	1.04	2.07	73.1	46.9
Butanoic acid	88.1	0.95	0.221	111.9	83.2
Ethanol	46.1	0.79	5.73	74.1	48.6
2-Methyl-1-propanol	74.1	0.80	2.39	109.2	84.3
2,3-Butandiol	90.1	1.00	0.02	114.3	90.8
Acetaldehyde	44.0	0.78	120	65.4	40.7
Furfural	96.1	1.16	0.29	100.1	70.3
Ethylacetate	88.1	0.90	12.6	114.6	81.8
Ethylbutanoate	116.2	0.87	2.01	152.4	116.5

Notes: The relative densities and vapor pressures are from the *CRC Handbook of Chemistry and Physics*, 88th ed. Surface and volume of molecules are calculated on the minimum-energy conformation (determined with the MM2 algorithm) with the tool "Connolly" in ChemB.

determined if a specific vinegar is aged in a single vessel. However, some types of vinegars have more complex definitions for the residence time. This is mainly due to the usage of blending procedures, which prevent the straightforward measurement of the aging time. This is the case of Jerez vinegars, aged with the solera system, and traditional balsamic vinegar of Modena and Reggio Emilia. In the solera system a set of barrels is disposed in parallel rows. The lowest row, the one directly above the floor, is the solera. The ones above are called first, second, and third criadera. Every year a certain amount of end-product is taken from the solera row, which is then promptly refilled with the vinegar coming from the first criadera. Then, in the same manner, the first criadera is refilled starting from the second one, which is in turn refilled starting from the third one. The missing liquid volume of the third criadera is replaced with new vinegar (Tesfaye et al. 2002b).

In the case of traditional balsamic vinegar, an analogous process is applied, which goes by the name of Rincalzo (Figure 7.6). This type of vinegar is produced by a series of barrels (five or more) of decreasing sizes, which is called by the Italian term *batteria*; the younger the vinegar is, the bigger the barrel is. Every year a certain amount of vinegar is collected, as end-product, from the smallest, and thus oldest, barrel. The same barrel is then refilled with the vinegar coming from the barrel of the next size up and so on for each barrel of the set until the biggest, which is topped with new base vinegar (Giudici et al. 2015). In such a case, the age of the product in each barrel can be calculated by applying the following equation, according to Giudici and Rinaldi (2007):

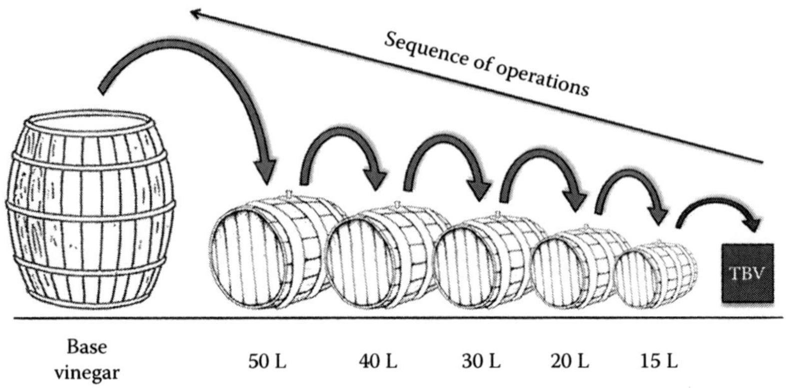

FIGURE 7.6 General scheme for the Rincalzo procedure in the production of traditional balsamic vinegar of Modena and Reggio Emilia. (Adapted from Lemmetti, F. and P. Giudici. 2011. *Ind. delle Bevande* XL. 235:12–24.)

$$E(i,n)=\frac{R(i)}{V(i)}\cdot\left[E(i+1,n-1)+1\right]+\frac{V(i)-R(i)}{V(i)}\cdot\left[E(i,n-1)+1\right] \tag{7.1}$$

where

i,j indexes refer to the barrel number
n index is the number of years from the start-up of the batteria
V and R refer to the volumes of the barrels and the refilling
$E(i,n)$ is the residence time of the vinegar on the barrel i after n years
t is the number of barrels

The theoretical model was applied to a typical batteria of five completely filled barrels (50, 40, 30, 20, and 15 L), for a total volume of 155 L and without structural losses (Vp = 0). We defined the range of the amount of traditional balsamic vinegar and cooked must as well as the evaporation rate. The maximum volume of traditional balsamic vinegar collected every year is supposed to be 6 L (that is, 40% of the volume of the smaller barrel or about 3.9% of the total volume of the batteria). The maximum volume of cooked must is equal to the capacity of the largest barrel (50 L) and the yield does not exceed 40%. These limits impose a maximum evaporation rate of about 31%. In such an equation (Equation 7.1), the actual age of vinegar is calculated by weighing the contribution of each addition. In fact, the result is a blending of vinegar of different ages in each barrel.

$$\lim_{n\to+\infty} E(i,n)=\sum_{j=i}^{t}\frac{V(j)}{R(j)}-1 \tag{7.2}$$

where

i,j indexes refer to the barrel number
n index is the number of years from the start-up of the batteria

V and R refer to the volumes of the barrels and the refilling
$E(i,n)$ is the residence time of the vinegar on the barrel i after n years
t is the number of barrels.

The application of such a conceived blending procedure leads to a finite limit (Equation 7.2) for the actual aging of vinegar, which strictly depends on the volumes of vinegar transferred from one barrel to the next (Giudici and Rinaldi 2007); the greater the transferred volume is, the lower is the actual age of the vinegar. Concerning the chemical changes associated with aging, differences can be detected for the different types of vinegar and for the different materials of the storing vessels.

7.3.4 Age

The aging of vinegar comprises changes to its chemical composition alongside its sensorial attributes. We already defined the age of vinegar as the combination of many physical–chemical parameters rather than the mere time count of its permanence in the aging vessel. The greatest part of published studies concern the chemical modifications undergone by Jerez vinegars, where a major contribution by wood-released phenols was acknowledged (Tesfaye et al. 2002c). However, some basic principles and considerations have also been assessed for balsamic vinegar, where measures of pH, total acidity, sugar content, and rheological properties were efficient at discriminating between aged balsamic vinegar and its industrial counterpart (Lemmetti and Giudici 2011).

When vinegars are aged in barrels, the long contact time between the vinegar and the wood makes for the exchange of molecules and the formation of new ones. This brings a great contribution to the final product, especially to color and sensory complexity: vinegar extracts from wood aromatic aldehydes, lactones, and phenols. This aspect has been extensively studied for Spanish vinegars. During wood-aging, an increase of the total polyphenol index (TPI) is usually detected over time (Tesfaye et al. 2004). This comes as an effect of the release of compounds such as gallic acid, hydroxymethylfuraldehyde, and coumaroyltartaric acid glycoside, which have been detected in amounts proportional to the residence time of barrel-aged vinegar (Garcia-Parrilla et al. 1997). An increase in aromatic aldehydes such as 5-(hydroxymethyl)-2-furaldehyde, 2-furfuraldehyde, protocatechualdehyde, and vanillin was also found to be related to wood aging (Tesfaye et al. 2002c). In addition, in both regular and accelerated aging, some molecules can be used as markers for the specific types of wood. This is the case of taxifolin, naringenin, and kaempferol, which are only present in vinegars aged in cherry wood, while ellagic acid is typical of chestnut and oak. Similarily, robinetin and fustin were associated to acacia wood and isovanillin to oak wood (Cerezo et al. 2010).

Many of these compounds have deep effects on the sensory attributes of vinegars. For instance, furfural (2-furancarboxyaldehyde) is a heterocyclic aldehyde that originates from the degradation of hemicellulose-derived monosaccharides and is important in the definition of some flavor attributes of aged vinegar (dry fruit, almond). Eugenol (2-methoxy-4-(2-propenyl)-phenol) and vanillin (4-hydroxy-3-methoxybenzaldehyde), on the other hand, both derive from the breakdown of lignin and are both involved in the onset of pleasant flavors in the aged vinegar.

Lactones are also an important class of wood-released compounds, such as *cis*-β-methyl-γ-octalactone, which determine the flavor-related features of vinegar. Lactones are associated to the barrel aging of whisky and bourbon and are therefore also known as whisky lactones or, alternatively, oak lactones. They usually impart coconut-like flavors. Another class of compounds associated to aging are the products of Maillard reactions. These reactions are typical of all foods containing sugars and species with amine and amide groups, when they are cooked or roasted. The heat gives rise to a nonenzymatic browning reaction during which a wide range of compounds is produced as AGEs (advanced glycation end-products) and especially melanoidins. Melanoidins are mainly composed of a skeleton of carbohydrate moieties with small amounts of nitrogen (Falcone and Giudici 2008; Lemmetti 2013).

Most interestingly, melanoidins play a role in the definition of the aroma, color, flavor, and rheology of foods. Furthermore, they have antioxidant, antimicrobial, antihypertensive, and probiotic properties. Concerning the physical changes, one can notice a concentration effect for solutes for vinegars stored in semipermeable vessels, as in the case of traditional balsamic vinegar. Remarkably, starting from a 30%–40% sugar concentration of the cooked must (raw material), concentrations of up to 50% can be obtained. What is more interesting is that the percentage of solutes detected through the usage of a refractometer is even higher than 70°Brix in most cases, which would account for other soluble species interfering with the measurement of sugars, mostly acquired during the aging phase from the wood or newly formed by condensation reactions. The presence of species other than sugars would also account for the otherwise unexplainable values of specific weight in traditional balsamic vinegar.

7.4 PACKAGING

Vinegar containers must satisfy some requirements in order to be applied to the packaging of such an aggressive product. This is due to the nature of acetic acid itself, especially in its gaseous form. For instance, in Figures 7.7 and 7.8 it can be seen how an iron net and a stone (used in the past to close the hole of the barrels of traditional balsamic vinegar) have been corroded and degraded. Another fundamental attribute for the packaging material is the price, which should be low, especially for low-added-value vinegars: After all, vinegar should be *bueno, bonito, y barato* [good, fine, and cheap], according to an old Spanish saying or cheap at twice the price for the Anglo-Saxon world. For these vinegars, the value of the bottle might be, in most cases, higher than that of the product itself. However, the market segmentation also requires more costly products, with a high evocative power. For this price target, more refined materials and models are usually found. For instance, the bottles and boxes for the traditional balsamic vinegar of Modena and Reggio Emilia were designed by the famous designer Giorgetto Giugiaro (Figure 7.9). The concept of ease of use should not be neglected. Examples for this concept are well known and widespread in the catering industry—for example, the top-down or squeeze bottles, used for many different condiments (ketchup, mustard, mayonnaise), be they served by street-food vendors or by proper restaurants. To realize how important this is, one should observe the variety of packaging models for vinegars and vinegar-based

FIGURE 7.7 The figure shows the corrosion exerted by acetic acid on an iron net, used to cover the holes of the barrels during the aging of vinegar.

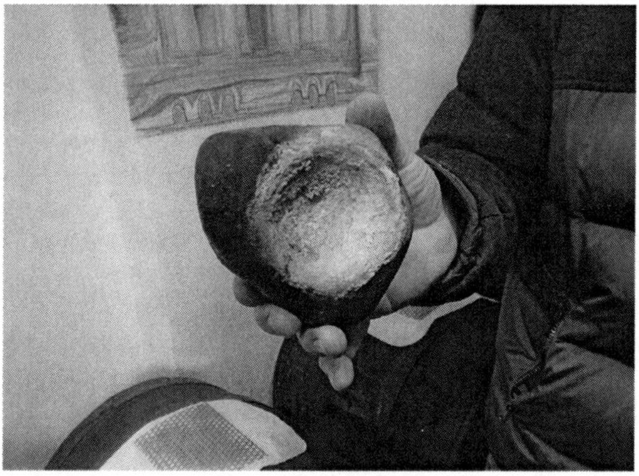

FIGURE 7.8 The figure shows the corrosion exerted by acetic acid on a stone, used to cover the holes of the barrels during the aging of vinegar.

condiments that can be found in a supermarket: Customers will be able to find regular glass bottles, top-down glass bottles, hard-plastic bottles, squeeze bottles, bottles designed to aid the decoration of plates, and so on.

Another fundamental aspect in the choice of a material is avoiding the release of potentially harmful molecules or elements from the material itself. There is a gargantuan amount of scientific literature about the packaging-to-food transfer of chemicals. Many parameters are usually considered: for example, the temperature, the contact time, the surface/volume ratio, and the chemical nature of the both the food and container matrices. However, the available literature concerning vinegar is more limited and more oriented towards process-related contaminations (see the next section). Obviously, acetic acid does not exert the same aggressiveness on all

FIGURE 7.9 The bottle for the traditional balsamic vinegar of Modena designed by Giorgetto Giugiaro.

materials. Wood is a really good material to contain vinegar (see the paragraph about aging), as are high-quality steel and glass. A more complex behavior can be expected for the many types of plastic polymers. Actually, the most used materials for the packaging of vinegars are glass and polyethylene terephthalate (PET), while pottery, steel, and wood are employed to a minor extent. Although ceramic packaging is now only applied to eastern cereal vinegars, it was once used also in the West, at least until the recent past; older people will certainly remember the local wine being served in pottery. This practice is fortunately being abandoned, due to the many risks derived from the release of metals into the liquid.

Although glass is always perceived as an inert material, this is actually not so straightforward. Furthermore, there are many types of glass, just to begin with. Secondly, one should always consider that even good glass could be contaminated with residuals coming from the manufacturing process (e.g., earth oxides). To our knowledge, there are currently no data about the transfer of elements or molecules from glass to vinegar. PET is the most widespread polymer for vinegar packaging, both for small, consumer-oriented volumes (up to 1 L) and for bigger volumes that are not necessarily destined to mere food applications. Usually, the permanence in the vessels is limited; however, since longer periods cannot be excluded, the container-to-vinegar transfer is a crucial factor that must be considered. Also in this case, residuals of the production process such as antimony trioxide (Sb_2O_3), used as a catalyst in the polymerization reaction, can easily migrate from the PET containers to the food product (Sánchez-Martínez et al. 2013).

In the case of vinegar, packaging does not retain the same importance it has for wine or oil. Regular vinegar is usually sold in clear bottles; transparency allows the consumer to better appreciate the color and turbidity of the product, especially for red vinegars or vinegars with peculiar color tones. Turbidity is a common attribute of vinegar and derives from both biological and physicochemical events. Dark bottles are used for vinegars and condiments with two purposes. Firstly, the dark color is unconsciously associated by consumers with aged products. Secondly, under a more practical point of view, a dark bottle masks the turbidity of the products. For instance, balsamic vinegars of low added value (Giudici et al. 2015) are obtained by blending many ingredients (wine vinegar, concentrated must, caramel); therefore, they frequently yield unstable and turbid solutions, which are promptly masked by the usage of dark glass bottles. A general and very simplified assumption is that the good or best balsamic vinegars are bottled in transparent glass bottles.

7.4.1 Safety and Spoilage

Although vinegar can be considered as a safe product from a general point of view, it is not devoid of risks associated with the raw materials, the production process, and the packaging. The hazards are conventionally grouped into three categories: chemical, physical, and biological. The risks related to chemical hazards are derived from the raw materials and are also spread over all the steps of the process: residuals of phytochemicals, antimicrobials, cleaning products, and chemicals released by the materials in contact with vinegar. The biological risks are mainly due to the raw material and to the initial steps of the process. Aflatoxins are a common hazard of raw materials, especially decaying fruits and cereals. Furthermore, if the process is not performed properly, the vinegars obtained by high-pH or starchy raw materials can suffer from the same risk. The physical risks are mainly concentrated at the packaging level and are restricted to the presence of pieces of glass and other solids. The spoilage of vinegar sounds improbable, since it is a food preservative thanks to its high level of acetic acid and its low pH; vinegar actually undergoes frequent alterations, both during the process and after the packaging.

7.4.1.1 Raw Materials

The health-related risks coming from the raw materials are of both chemical and biological natures. The physical risks are completely removed during the process, only to reappear at the packaging level. Some hazards are related to specific raw materials, while others are more general; for example, the aflatoxin patulin is found frequently in decaying apples (Wilson and Nuovo 1973). The chemical hazards of raw materials are mainly due to the agronomical practices, in particular the usage of agrochemicals such as weed-killers and pesticides. However, pollutants cannot be excluded, especially for raw materials coming from strongly anthropic environments. Agrochemicals are frequently detected even in fermented products (Sala et al. 1996). The biological alteration of raw materials is mainly due to the growth of plant pathogens (fungi, usually) on fruits and seeds. The list of these pathogenic fungi and of their toxins is far from the aim of this chapter. The literature on potential hazards of food raw material is very conspicuous and only a few comprehensive reviews are mentioned: Lima Tribst et al. (2009); Tournas (2005); Barth et al. (2009). Raw materials contaminated with aflatoxin cannot be fermented as such, because there are no stages in the process of acetification that can eliminate the risks associated with them. However, there is the possibility to use decayed fruit and contaminated cereals for the production of distilled alcohol, because the distillation, which follows the alcoholic fermentation, allows complete elimination of the risk of the presence of aflatoxins. In fact, there are no concerns about the biological safety of raw materials applied to the production of spirit vinegars.

7.4.1.2 Production Process

The basic production process is the same for each vinegar: alcoholic and acetic fermentation. However, there are many differences in the details of both the technology and the raw materials (see previous paragraphs). The type of raw material needs a specific technological procedure, such as the storage of the raw material and its handling. For example, the easily perishable fruits need to be transformed quickly or stored to avoid the development of potentially toxigenic and spoilage microorganisms. These operations are to be conducted in order to completely eliminate the risk of toxic compounds. The risk of manifestation of biological hazards exists until the completion of the alcoholic fermentation and sometimes even of the acetification. The cereal vinegars are more susceptible to biological risks during the process, because the hydrolysis of starch requires the intervention of fungi (Chen 1999; Miao and Ma 2006; Wu 2004) in a process that is, at least in the case of the traditional one, not simple to govern or control. Humidity and temperature are the two most important process variables to be controlled to enable the development of the desired fungi. Furthermore, some fungal species, such as *Monascus* spp., though appreciated for the production of certain bioactive metabolites such as monacholins (cholesterol-reducing molecules), might cause the by-production of other harmful secondary metabolites or toxins, such as citrinin (Wang and Lin 2007).

Finally, as the alcoholic fermentation by yeasts is also involved in the overall process of vinegar production, one should not exclude the possible presence of biogenic amines. Concerning chemical hazards, operators should also care about the many types of vessels used from the harvest to the bottling phases. As already proved for wood, vinegar interacts with different materials. This could easily lead to the presence of undesired or harmful solutes. The aggressiveness of acetic acid can even spoil metal vessels, which could lead to the release of metal ions (lead, for instance) into the solution. Another hazard that cannot be neglected comes from the usage of ceramic vessels for aging or packaging. In fact, if the proper food-grade materials are not employed, many heavy metals, which are the natural components of clay, can be transferred to the vinegar (Lin et al. 2014). This aspect, as already stated, has important consequences on the choice of the packaging.

7.4.1.3 Spoilage and Shelf-Life

Vinegar spoilage seems a contradiction in terms, because vinegar is perceived as a very stable product. Vinegar undergoes alterations both during the production process and during storage and distribution. The main alterations that occur are of biological origin and due to different groups of organisms, both eukaryotes and prokaryotes. Film-forming yeasts belonging to the genus *Candida* and *Pichia* are particularly dangerous, especially for vinegars produced in the traditional way of surface fermentation. These yeasts can completely colonize the surface of the liquid in the barrel and respire both alcohol and acetic acid, with the production of CO_2 and water. In the end, no more vinegar is in the barrel, but only a solution of something that is completely useless. To avoid the growth of these yeasts, it is necessary to follow carefully the first steps of acetification: The wine has to be properly inoculated with a culture of active AAB. In addition, phagic infections can cause the interruption of acetous fermentation, which usually happens in the production of high-acidity vinegars such as those obtained by SF. This is due to the extremely harsh environmental conditions, which select only one or a few strains of the same species (usually *K. europaeus*), thus resulting in low biological variability. Processes at lower acidity have a greater biological diversity, with many species and strains featuring varying degrees of phage sensitivity. Thus, a built microbial consortium is more likely to resist the infection, with a consequent positive impact on the safety and reliability of the process.

However, the industry minimizes the risk of phage infection by pasteurizing the substrate, by using sterilized air, and by keeping high acidity bioreactors separated from low acidity tanks, which could constitute the source of the infection. Vinegars can also undergo overoxidation, which happens when the AAB strains involved in the fermentation are resistant to high concentrations of acetic acid. Overoxidation is usually determined by high DOT values and by the absence of ethanol. At the industrial level, the online measurement and control of these parameters helps to keep overoxidation at bay. Conversely, overoxidation is a more concrete risk for less controlled fermentation systems. Biological alteration of vinegar is not just strictly related to microbes, such as in the case of cider vinegar; nematodes belonging to the species *Anguillula aceti*, also known as vinegar eels, can occur at any stage of the fermentation. They are usually responsible for a lower acid content in the final product. The spoilage by vinegar eels is usually treated by applying thermal treatments at 40°C–45°C in order to kill the nematodes without affecting the vitality of AAB. Alternatively, they can be removed by pasteurization after the end of the fermentation.

The shelf-life of regular vinegar does not have any particular problems, and it is fairly long. The only known biological hazard is the development of AAB that produce exopolysaccharides (Gullo and Giudici 2008), known as vinegar mother. Therefore, vinegar shows the presence of gelatinous materials that are not dangerous at all, but are particularly unappreciated by consumers. This kind of spoilage can be particularly evident when the bottles are opened and preserved a long time before consuming. To avoid the risk of "mother," vinegar is filtered and/or pasteurized immediately before or after bottling.

The shelf-life of vinegar can also be limited by chemical modifications at the expense of the phenolic constituents: wine and apple vinegars are particularly susceptible. The visual appearance of the vinegar changes in terms of color and turbidity; the color hue tends towards brick-red for red vinegars and brown-yellow for white vinegars. To avoid this alteration, the adopted strategy is to induce polymerization of phenols as much as possible before bottling. Polyphenols are expected to polymerize and therefore precipitate to the bottom of the vat. However, if this phase is not brought to completion and vinegar is bottled anyway, polymerization will continue inside the bottle, though at a slower rate. Haze will arise when the bottle is opened; oxygen will enter the bottle and catalyze the polymerization at a higher rate, with the formation of sediments.

Some stabilization methods are sufficient to avoid the onset of long-term damage. For instance, a longer maturation time readily prevents the formation of hazes. Otherwise, hazes are prevented with methods similar to those applied in the winemaking industry. SO_2 can be added as an antioxidant, thereby preventing the oxidative polymerization of polyphenols, or vinegar can be treated with fining agents such as gelatin, kieselsol, or bentonite. Some special vinegars, such as traditional balsamic vinegar, are subjected to physical instability because of the high concentration of solutes. The shelf-life of these vinegars is often shortened by the formation of crystals inside the bottles. In some cases, the vinegar becomes completely solid; we studied the problem in the past and formulated a tentative explanation: unstable vinegars are characterized by a glucose/fructose ratio in favor of the first. Moreover, they have, with the same total solutes, a greater amount of sugars.

7.4.2 Sensory Analysis

Sensory analysis of vinegar must be approached with caution because the high content of acetic acid is enough to numb the taste buds even after a single sample. This fact can easily interfere with the proper perception of aromas and taste; for this reason, specific tasting procedure have been set up (Tesfaye et al. 2010). Moreover, until recently, there was no such thing as a shared vocabulary for the sensory evaluation of all the different types of vinegars. The literature on the sensorial attributes of vinegars is mainly associated to sherry vinegar, traditional balsamic vinegars, and, more recently, Chinese cereal vinegars (Corradini et al. 2016; Giudici et al. 2017; Callejón et al. 2008; Tesfaye et al. 2002a). Many olfactory descriptors have been reported; some are shared by all the kinds of vinegar, and others are more peculiar to specific vinegars. However, work is still required in order to reduce the redundancy of terms and obtain a minimal set of descriptors to be universally applied.

The flavor of vinegar involves all the five tastes: "sweet," "bitter," "sour," "salty," and, for cereal vinegars, "umami" (pleasant savory taste) (Giudici et al. 2017). Also, the trigeminal sensations play an important role. They can be of two different types: tactile sensu stricto (physical) and chemical ones. The first category comprises "consistency," "viscosity," "granularity," "unctuosity," and "texture." The second comprises the sensations related to pain and irritation: "astringency," "pungency," "spiciness," "effervescence," "metallic," "pseudo-heat," and "pseudo-freshness." In the end, a visual evaluation of the samples can also be performed; this is aimed at judging the color and viscosity of the sample (Corradini et al. 2016).

7.4.2.1 Vinegar Vocabulary for Flavor and Taste

In the literature, the most common attribute or vinegar descriptor is pungency, which is common to all vinegars worldwide. It can be regarded as both an olfactive and tactile sensation and it is due to the presence of acetic acid. Other recurring descriptors are "glue," "alcohol," "winy," "raisins," "woody," "clove," "citrus," "red fruits," "vanilla," "sweet aroma," "bitter almond," "medicine," "apple," and "coconut" (Tesfaye et al. 2010; Corradini et al. 2016). Concerning fruity and floral flavors, if we can rely on a thorough subdivision of the single aromas for wine, the detection of single fruits in vinegar is made more difficult due to the extreme sensory aggressiveness of this product. When it comes to balsamic vinegars, great importance is acquired by caramel-like descriptors: "cooked wine," "mustard," "prune," "carob," "cocoa," and "madeira." These can be attributed to the products of the Maillard reaction such as furfurals and melanoidins.

Other descriptors are regarded as negative; this is the case of "seasoned cheese," which is also often termed "acid sweat" and "foot odor." "Spicy" is a commonly detected descriptor and should not be confused with "pungent," which is due to the effect exerted by acetic acid on the olfactory mucosa. It should be noted that, in contrast, the molecules responsible for the spicy sensation are mainly volatile phenols derived from guaiacol. The same sensory-active molecules, if present at different ratios, are also responsible for "smoky" flavor. However, other descriptors are important if you consider the varied landscape of eastern cereal vinegars, such as the different types of Chinese cereal vinegar. Descriptors for these products detach from the preceding list. This is obviously due to the radically different raw material. In fact, recent studies have detected the occurrence of descriptors such as "walnut," "licorice," and "toasted" (Giudici et al. 2017). However, it should be pointed out that the term "Chinese cereal vinegar" refers to an extremely varied ensemble of products that differ from one another for raw materials (bran, wheat, rice, sesame, barley, sorghum), as well as production and aging processes.

Taste sensations are somewhat less refined than olfactory ones. However, they can roughly be classified according to four different descriptors for vinegar: "sweetness," "bitterness," "acidity," and "saltiness." Many compounds can affect the different tasting sensations of vinegar. If sugars—most importantly, glucose and fructose—are obviously related to the sample's sweetness, the many different organic acids can have varied effects ranging from "acidity," to "bitterness" and "sweetness." "Saltiness" is mainly related to both organic and inorganic sodium salts. Remarkably, all these descriptors are present and are not, as one could argue, mutually exclusive (e.g., "bitterness" and "sweetness"). In fact, the four descriptors arise during different moments of the tasting procedure (Lemmetti et al. 2014).

7.4.2.2 Olfactive Perception and Chemical Composition

Odor is a complex sensory perception; in most cases, a descriptor has no straightforward correspondence to a single compound. Conversely, it is dependent on the concentration ratios and on many other factors (Frijters 1987; Helson 1964; Laffort 1989). However, the research of flavor determinants in different chemical compounds is not devoid of meaning. Ethyl acetate is always present in vinegars in great amounts; it is a sensorally relevant molecule and it is associated with a glue-like, solvent-like flavor. Acetoin is produced during the alcoholic fermentation and it can be accounted for the olfactive perception of "almond" and "butter." This compound is strictly related to diacetyl, a diketone contributing to the tactile sensation of "fluidity" in alcoholic beverages and is also used to impart caramel-like and butter-like odor in butter production. Diacetyl derives from the oxidation of acetoin and therefore its concentration rises during the aging process.

Isovaleric acid is characterized by an unpleasant smell of "sweat" and "cheese"; on the other hand, its volatile esters impart to vinegar a pleasant, fruity smell.

Diacetyl acetaldehyde is always present in vinegars; it is also typical of malt whisky and of sherry wine, where it rises during the aging as a result of ethanol reacting with acetaldehyde produced by Flor-yeasts (Zea et al. 2015). This particular type of yeast grows in a biofilm on the surface of wine and yields peculiar flavor characteristics thanks to its oxidative metabolism. Diacetyl acetaldehyde is an extremely important flavor-active molecule that yields sensory notes such as "balsamic," "fruit," and "wood" (Muñoz et al. 2007; Zea et al. 2008). As already stated, the derivatives of the Maillard reaction have a role in the definition of the final aroma of vinegar. Particularly, 2-furfural, the aromatic aldehyde derived from furan, has got a sweet aroma similar to "wood," "bread," and "caramel," with a vaguely phenolic feel. The 5-hydroxymethyl-2-furfural (HMF) has a pleasant aroma of "caramel," "toasted bread," and "coffee." This aroma is typical of many bakery products and is always present in concentrated and cooked must, which are both applied in the production of balsamic vinegars. More generally, it is present in all sugar-containing foods undergoing thermal treatments such as drying or cooking. Its oxidized form, 5-methyl-2-furfural, has a sweet smell of "coffee," "caramel," and "maple syrup."

Other similar compounds are 2-furanoic acid ("caramel") and 5-acetoxymethyl-2-furfural, which is produced by the esterification of HMF with acetic acid. It should be noted that the toasted aroma is also due to enolic derivatives such as cyclotene ("caramel," "maple syrup," "burned," "coffee," "bread") and maltol, which has a flavor similar to cotton candy and caramel (Zeppa et al. 2002). Vinegars that undergo the aging phase in wooden barrels, generally oaken barrels, also have a sensory contribution from lactones. These lactones are also known as "quercus lactones" or "whisky lactoni," as they are typical of whisky and, more generally, of aged alcoholic beverages. They impart aromas of "coconut," "celery," and "fresh wood" (Masson et al. 2000). The β-methyl-γ-octalactone increases during aging (Zeppa et al. 2002), while other lactones such as γ-decalactone result from the cyclization of fatty acids, which are hydrolyzed during alcoholic fermentation (Dufossé et al. 1994). γ-Butyrolactone is naturally present in wine in small amounts; it has a cheese-like, creamy flavor, with an after-flavor of "peach." Other compounds that should not be neglected are pantolactone, which smells like cotton candy, and sotolon, which has an aroma of hazelnuts at low concentration and of curry at high concentration (Callejon et al. 2008). Sotolon is a typical aging-related compound of sherry vinegar; of certain white wines such as Jura, Porto, and Tokay; and of wines obtained from botrytized grapes (Sousa Câmara et al. 2004).

Other compounds related to barrel-aging are vanillin, 4-ethylphenol, guaiacol, and its derivative 4-vinylguaiacol. Guaiacol is especially present in balsamic vinegar and its aroma can be described as "phenolic," "smoky," or "spicy" and sometimes is associated to "medicine," "vanilla," and "smoked bacon." 4-Vinylguaiacol's aroma, on the other hand, is described as "fresh," "wood," "resin," "citron," and "toasted peanuts." 4-Ethylphenol has a smoky and phenolic smell, similar to that of creosote. Eugenol, an allyl chain-substituted guaiacol, is always present in samples of both vinegar and red wines and is the molecule responsible for the smell of cloves and of some medicinal herbs, often used as both antiseptics and anesthetics. Still concerning aromatic compounds, 1,1,6-trimethyl-1,2-dihydronaphtalene (TDN) is conversely responsible for unpleasant, petrol-like smells if present above the threshold value of 0.20 μg/L (Sacks et al. 2012). However, it seems that the concentration of TDN decreases during the aging phase (Callejon et al. 2008).

Vinegar esters are very active from the sensory point of view and, if considered by themselves, they are perceived as fruit aromas. However, just monoethylsuccinate is the significant molecule, present in the oldest barrels (Zeppa et al. 2002). Ethyl-isobutyrate has an ethereal, pungent, and fruity smell, which also leans towards "egg" and "rum." Ethyl butyrate smells like "orange," "cherry," "pineapple," "peach," "apricot," "fig," and "prune." Butyl acetate has a "sweet banana" and "apple" flavor. Ethyl isovalerate has a sweet and spicy aroma that resembles apple and pineapple. Ethyl propanoate has an ethereal and sweet flavor of pineapple. Isoamyl acetate is linked to the descriptors "banana" and "pear." Ethyl octanoate's aroma is "sweet," "waxy," and "fruity," with some notes of "cream," "fat," "mushrooms," and "cognac." Ethyl-2-methylbutirrate has an aroma of "red fruits," "grapes," "pineapple," "mango," and "cherries." Amyl acetate has a fruity-fresh flavor reminiscent of "pear," "banana," and "apple." Ethyl exanoate smells like "pineapple," "wax," and "green banana." Exyl acetate gives a green and fresh aroma of fruits similar to "banana peel," "apple," and "pear."

7.4.2.3 Olfactive Descriptors and Reference Standards

It has to be noticed how, in sensory analysis, the same descriptor can be given by many different compounds, and many examples have been provided in the previous paragraphs. Therefore, reference standards become essential. Ideally, each descriptor should correspond to a single chemical compound or, alternatively, to specific and reproducible mixes. Natural compounds and essences could also be used. Reference standards are extremely useful in the formation of the tasters, as they allow the construction of a shared vocabulary of sensory evaluation. The amount of scientific literature on sensory analysis is gargantuan for many consumer goods; this is proved by the impact of sensory analysis on both the description and commercialization of wine and beer. For vinegar, we have less available literature; however, the advantage yielded by reference standards is that they are reliable and applicable regardless of the analyzed food. For instance, diacetyl will always be the reference standard for butter, whether we are talking about wine, butter, or vinegar. The main goal is to achieve the correspondence between a descriptor and a standard; this easily leads to a common misconception. It is believed that the presence of a chemical compound in a given sample can simply be assessed by the mere perception of the corresponding descriptor. However, since the same descriptor can be given by many different molecules, this statement is de facto incorrect. Just rarely, there is a straightforward correspondence between molecule and perception. More frequently, a perceived descriptor results from the combined and sometimes synergic effect of many molecules (Frijters et al. 1987; Helson 1964; Laffort 1989).

REFERENCES

Asai, T. 1968. *Acetic Acid Bacteria*. Tokyo University Press, Tokyo.

Barth, M., T.R. Hankinson, H. Zhuang and F. Breidt. 2009. Microbiological spoilage of fruits and vegetables. In: *Compendium of the Microbiological Spoilage of Foods and Beverages*. Springer, New York. 135–183.

Callejón, R.M., M.L. Morales, A.C. Silva Ferreira and A.M. Troncoso. 2008. Defining the typical aroma of sherry vinegar: Sensory and chemical approach. *J. Agric. Food Chem.* 56:8086–8095.

Callejón, R.M., W. Tesfaye, M.J. Torija, A. Mas, A.M. Troncoso and M.L. Morales. 2009. Volatile compounds in red wine vinegars obtained by submerged and surface acetification in different woods. *Food Chem.* 113:1252–1259.

Canas, S., P. Belchior, A.M. Mateus, M.I. Spranger and R. Bruno-de-Sousa. 2002. Kinetics of impregnation/evaporation and release of phenolic compounds from wood to brandy in experimental model. *Ciência Téc. Vitic.* 17:1–14.

Cerezo, A.B., W. Tesfaye, M.E. Soria-Díaz, M.J. Torija, E. Mateo, M. C. Garcia-Parrilla and A.M. Troncoso. 2010. Effect of wood on the phenolic profile and sensory properties of wine vinegars during ageing. *J. Food Compos. Anal.* 23:175–184.

Chen, F., L. Li, J. Qu and C. Chen. 2009. Cereal vinegars made by solid-state fermentation in China. In: *Vinegars of the World*. Milan: Springer, 243–259.

Chen, Z.F. 1999. Vinegar. In: Wang F.Y. (ed) *Modern food fermentation technology*. China Light Industry Press, Beijing, 506–566.

CNS (Chinese National Standard) (2005). Edible vinegar. No.14834, N5239. Ministry of Economic Affairs, Taiwan, Republic of China.

Corradini, G., F. Lemmetti and P. Giudici. 2016. Wine and balsamic vinegars: Olfactory and chemesthetic attributes; sensory active substances. *Ind. Aliment.* 56:17–23.

Dufosse, L., A. Latrasse, and H.E. Spinnler. 1994. Importance of lactones in food flavours: Structure, distribution, sensory properties and biosynthesis. *Sci. Aliment.* 14:17.

Ebner, H., S. Sellmer, and H. Follmann. 1996. Acetic acid. In: Rehm, H.J. and G. Reed (eds.), *Biotechnology*, 2nd ed., 6:381–401.

EC (Council Regulation) (1999). No 1493/1999 of May 17 1999 on the common organization of the market in wine. *Official Journal of the European Communities* L 179:1–84.

Falcone, P.M. and P. Giudici. 2008. Molecular size and molecular weight distribution impacting traditional balsamic vinegar ageing. *J. Agr. Food Chem.* 56:7057–7066.

FDA (Food and Drug Administration) (1995). Compliance policy guides (CPG) Sec. 525.825. Vinegar, Definitions—Adulteration with vinegar eels. Available from http://www.fda.gov/ICECI/ComplianceManuals/CompliancePolicyGuidanceManual/ucm074471.htm (accessed: April 2016).

Frijters, J.E. 1987. Psychophysical models for mixtures of tastants and mixtures of odorants. *Ann. N.Y. Acad. Sci.* 510:67–78.

García-García, I., I.M. Santos-Dueñas, C. Jiménez-Ot, J.E. Jiménez-Hornero and J.L. Bonilla-Venceslada. 2009. Vinegar engineering. In: Solieri, L. and P. Giudici (eds.), *Vinegars of the World*. Milan: Springer, 97–120.

García-Parrilla, M.C., G.A. González, F.J. Heredia and A.M. Troncoso. 1997. Differentiation of wine vinegars based on phenolic composition. *J. Agr.Food Chem.* 45:3487–3492.

Giudici, P., G. Corradini, T. Bonciani, J. Wu and F. Chen. 2017. The flavor and taste of cereal Chinese vinegars. *Acetic Acid Bacteria*. 6:1–6.

Giudici, P., L. De Vero, M. Gullo, L. Solieri and F. Lemmetti. 2016. Fermentation strategy to produce high gluconate vinegar. *Acetic Acid Bacteria*. 5:1–6.

Giudici, P. and M. Gullo. 2012. Aceto. In: Farris, G.A., M. Gobbetti, E. Neviani and M. Vincenzini (eds). *Microbiologia dei Prodotti Alimentari*. Cea edizioni. 305–323.

Giudici, P., F. Lemmetti and S. Mazza. 2015. *Balsamic Vinegars: Tradition, Technology, Trade*. Milan: Springer.

Giudici, P. and G. Rinaldi. 2007. A theoretical model to predict the age of traditional balsamic vinegar. *J. Food Eng.* 81:121–127.

Gobbi, M., L. De Vero, L. Solieri, F. Comitini, L. Oro, P. Giudici, and M. Ciani. 2014. Fermentative aptitude of non-*Saccharomyces* wine yeast for reduction in the ethanol content in wine. *Eur. Food Res. Technol.* 239:41–48.

Gullo, M. and P. Giudici. 2008. Acetic acid bacteria in traditional balsamic vinegar: Phenotypic traits relevant for starter cultures selection. *Int. J. Food Microbiol.* 125: 46–53.

Gullo, M., D. Mamlouk, L. De Vero and P. Giudici. 2012. *Acetobacter pasteurianus* strain AB0220: Cultivability and phenotypic stability over 9 years of preservation. *Curr. Microbiol.* 64:576–580.

Gullo, M., A. Sola, G. Zanichelli, M. Montorsi, M. Messori and P. Giudici. 2017. Development and characterization of bacterial cellulose produced by a selected *Komagataeibacter xylinus* strain in glucose and ethanol broths. *Applied. Microbiol. Biotechnol.* 101(13): 1–13.

Guymon, J.F. and E.A. Crowell. 1972. GC-separated brandy components derived from French and American oaks. *Am. J. Enol. Viticult.* 23:114–120.

Helson, H. 1964. *Adaptation-Level Theory.* Harper & Row, Oxford.

Laffort, P. 1989. Models for describing intensity interactions in odor mixtures: A reappraisal. In: Laing, D.G., W.S. Cain, R.L McBride and B.W. Ache (eds.). 1989. *Perception of Complex Smells and Tastes.* Academic Press, Sidney. 173–188.

Lemmetti, F. 2013. Systemic approach for improving balsamic vinegars. PhD dissertation, University of Modena and Reggio Emilia.

Lemmetti, F. and P. Giudici. 2011. Simple chemical and physical parameters to identify true aged balsamic vinegars. *Ind. delle Bevande* XL. 235:12–24.

Lemmetti, F., L. Solieri, T. Bonciani, G. Zanichelli and P. Giudici. 2014. Sensory analysis of traditional balsamic vinegars: Current state and future perspectives. *Acetic Acid Bacteria.* 3:2240–2845.

Lima Tribst, A.A., A. de Souza Sant'Ana and P.R. de Massaguer. 2009. Review: Microbiological quality and safety of fruit juices—Past, present and future perspectives. *Cr. Rev. Microbiol.* 35:310–339.

Lin, Q.B., Y. Chen, H. Song, H.J. Wu and X.Y. Wang. 2014. Kinetic migration of chemical elements from ceramic packaging into simulated foods and mature vinegar. *Packag. Technol. Sci.* 27:59–67.

Masson, E., R. Baumes, C. le Guernevé, and J.L. Puech. 2000. Identification of a precursor of β-methyl-γ-octalactone in the wood of sessile oak (*Quercus petraea* (Matt.) Liebl.). *J. Agric. Food Chem.* 48:4306–4309.

Miao, J. and H.X. Ma. 2006. Function and harm of microorganism in producing soy sauce and vinegar. *Chinese Condiment.* 2:9–12.

Muñoz, D., R.A. Peinado, M. Medina and J. Moreno. 2007. Biological aging of sherry wines under periodic and controlled microaerations with *Saccharomyces cerevisiae* var. *capensis*: Effect on odorant series. *Food Chem.* 100:1188–1195.

Murooka, Y., K. Nanda, and M. Yamashita. 2009. Rice vinegars. In: *Vinegars of the World.* Milan: Springer, 121–133.

Onishi, M., J.F. Guymon and E.A. Crowell. 1977. Changes in some volatile constituents of brandy during aging. *Am. J. Enol. Viticult.* 28:152–158.

Ruiz de Adana, M., L.M. Lopez and J.M. Sala. 2005. A Fickian model for calculating wine losses from oak casks depending on conditions in ageing facilities. *Appl. Therm. Eng.* 25:709–718.

Sacks, G.L., M.J. Gates, F.X. Ferry, E.H. Lavin, A.J. Kurtz and T.E. Acree. 2012. Sensory threshold of 1,1,6-trimethyl-1,2-dihydronaphthalene (TDN) and concentrations in young Riesling and non-Riesling wines. *J. Agric. Food Chem.* 60:2998–3004.

Sala, C., F. Fort, O. Busto, F. Zamora, L. Arola, and J. Guasch. 1996. Fate of some common pesticides during vinification process. *J. Agr. Food Chem.* 44:3668–3671.

Sánchez-Martínez, M., T. Pérez-Corona, C. Cámara and Y. Madrid. 2013. Migration of antimony from PET containers into regulated EU food simulants. *Food Chem.* 141:816–822.

Siau J.E. 1984 *Transport Process in Wood.* Berlin, Heidelberg, New York: Springer.

Sievers, M. and J. Swings. 2005. Family Acetobacteraceae. In: Garrity G.M. (ed.). *Bergey's Manual of Systematic Bacteriology*. New York: Springer, 41–95.

Singleton, V.L. 1995. Maturation of wines and spirits: Comparisons, facts, and hypotheses. *Am. J. Enol. Viticult.* 46:98–114.

Solieri, L. and P. Giudici. 2008. Yeasts associated to traditional balsamic vinegar: Ecological and technological features. *Int. J. Food Microbiol.* 125:36–45.

Solieri, L. and P. Giudici. 2009. Vinegars of the world. In: Solieri, L. and P. Giudici (eds). *Vinegars of the World*. Milan: Springer, 1–16.

Sousa Câmara, J., J.C. Marques, M.A. Alves, A.C. Silva Ferreira. 2004. 3-Hydroxy-4, 5-dimethyl-2(5H)-furanone levels in fortified Madeira wines: Relationship to sugar content. *J. Agric. Food Chem.* 52:6765–6769.

Tesfaye, W., M.C. García-Parrilla and A.M. Troncoso. 2002a. Sensory evaluation of sherry wine vinegar. *J. Sens. Stud.* 17:133–144.

Tesfaye, W., M.L. Morales, B. Benıtez, M.C. Garcıa-Parrilla and A.M. Troncoso. 2004. Evolution of wine vinegar composition during accelerated aging with oak chips. *Anal. Chim. Acta.* 513:239–245.

Tesfaye, W., M.L. Morales, R.M. Callejón, B. Cerezo Ana, A.G. González, M.C. García-Parrilla and A.M. Troncoso. 2010. Descriptive sensory analysis of wine vinegar: Tasting procedure and reliability of new attributes. *J. Sens. Stud.* 25:216–230.

Tesfaye, W., M.L. Morales, M.C. Garcıa-Parrilla and A.M. Troncoso. 2002b. Wine vinegar: Technology, authenticity and quality evaluation. *Trends Food Sci. Tech.* 13:12–21.

Tesfaye, W., M.L. Morales, M.C. García-Parrilla and A.M. Troncoso. 2002c. Evolution of phenolic compounds during an experimental aging in wood of sherry vinegar. *J. Agr. Food Chem.* 50:7053–7061.

Tournas, V.H. 2005. Spoilage of vegetable crops by bacteria and fungi and related health hazards. *Cr. Rev. Microbiol.* 31:33–44.

Wang, T.H. and T.F. Lin. 2007. Monascus rice products. *Adv. Food Nutrition Res.* 53:123–159.

Wilson, D.M. and G.J. Nuovo. 1973. Patulin production in apples decayed by *Penicillium expansum*. *Appl. Microbiol.* 26:124–125.

Wu, J.H. 2004. The ecological distribution and productivity of microbe. In: Da Qu for Shanxi aged vinegar. Dissertation, Shanxi Agricultural University.

Zea, L., L. Moyano, M. Medina. 2008. Odorant active compounds in Amontillado wines obtained by combination of two consecutive ageing processes. *Eur. Food Res. Technol.* 227:1687–1692.

Zea, L., M.P. Serratosa, J. Merida and L. Moyan. 2015. Acetaldehyde as key compound for the authenticity of sherry wines: A study covering 5 decades. *Compr. Rev. Food Sci. F.* 14:681–693.

Zeppa, G., M. Giordano, V. Gerbi and G. Meglioli. 2002. Characterization of volatile compounds in three acetification batteries used for the production of Aceto Balsamico di Reggio Emilia. *Ital. J. Food Sci.* 14:247–266.

8 Stabilization of Distilled Spirits

Paul S. Hughes

CONTENTS

8.1 INTRODUCTION

The category of drinks that comes under the broad banner of distilled spirits is highly diverse. The alcohol base must be derived from carbohydrates that come ultimately from an agricultural source (Table 8.1). In the case of complex carbohydrates, such as starch (e.g., from cereals) and inulin (from *Agave tequiliana*, used in the production of tequilas), there is a requirement to first break down these carbohydrates, to release simple sugars that can then be converted into alcohol (ethanol) by the action of yeasts or a broad range of fermentative bacteria. In other instances, where simple sugars are already present in the raw materials (e.g., sugar cane, fruits), the sugars are generally extracted by crushing and/or pressing and then fermented directly into alcohol. The production of distilled spirits can therefore be considered to consist of three broad operations (Figure 8.1):

- Preparation of a fermentable extract (either directly after crushing/pressing or following hydrolysis of complex carbohydrates), and subsequent fermentation
- Concentration of the alcohol in the broth to at least 15% alcohol by volume (ABV) by distillation, but often up to considerably higher alcohol concentrations (i.e., 60% ABV or more)
- Post-stillhouse product elaboration (e.g., compounding, maturation in the presence of wood, dilution to sales strength, filtration) to create the final product for sale

The implication here is that the raw materials used for the production of alcohol, and the requirements for downstream processing, drive much of the diversity of the

TABLE 8.1
Summary of Production Features of Common Distilled Spirits Drinks

Category	Sugar Source	Batch or Continuous?	Maturation
Whiskies	Cereals[a]	Both	Mandatory
Brandies	Grapes, fruit	Both	Mandatory
Gins	Various[b]	Batch[c]	Usually not
Vodkas	Various[b]	Both[c]	Usually not
Rums, cachaça	Sugar cane, molasses	Both	Optional
Tequila	Agave	Both	Optional
Liqueurs	Various[b]	Both[c]	Optional

[a] Malted barley, raw barley, maize and wheat. Molasses is a common source of carbohydrate for Indian whiskies.
[b] Most, but not all, of these products are produced from a neutral spirit and redistilled to create the desired product.
[c] Alcohol production is often separate from the creation of the product.

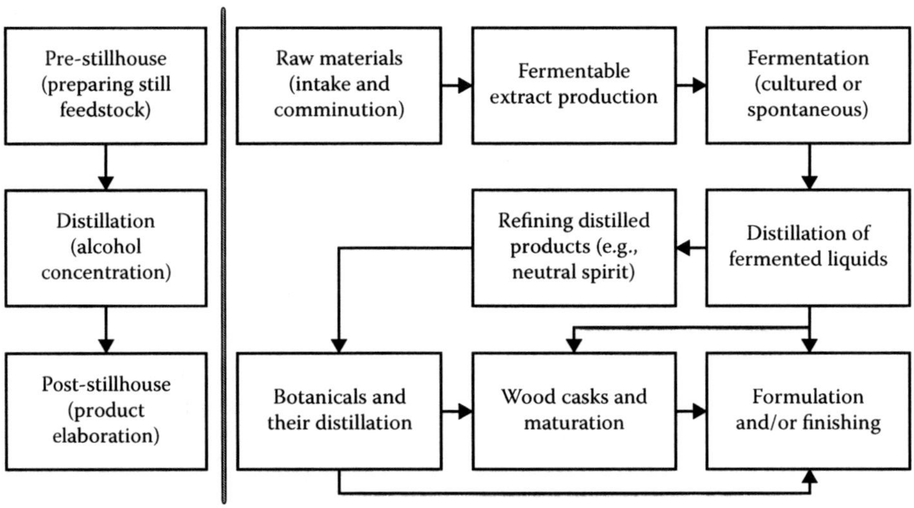

FIGURE 8.1 Scheme indicating the broad production operations employed in the manufacture of distilled spirits.

distilled spirits category, both in terms of process design and the final product attributes. For most distilled spirits producers, it is important that their products are sufficiently stable to minimize the risk of nonconformance both after leaving the plant and up to the point of sale, as well as in the hands of the consumer. Given the sheer range of drinks based on distilled spirits, it is hardly surprising that some categories are inherently more resistant to change after leaving the factory gate, while others require more consideration to ensure that they are acceptable for consumption at point of sale and for a reasonable amount of time afterwards. Additionally, a recurring theme here

TABLE 8.2
Sources of Instability for Various Major Spirit Drinks Categories and Estimate of Their Likelihood

	Cause of Instability			
	Biological	**Physical**	**Chemical**	**Light**
Whiskies	x	xx	xx	x
Brandies	x	xx	xx	x
Unaged rums/cachacas	x	xx	xx	x
Aged rums/cachacas	x	xx	xx	xx
Unaged tequilas	x	xx	xx	x
Aged tequilas	x	xx	xx	xx
Gins/botanical spirits	x	xx	xxx	x
Vodkas	x	xx	xx	x
Botanical liqueurs	x	xxx	xxx	x
Cream liqueurs	xx	xxx	xx	xxx
Advocaats	xx	xx	xx	xx

Notes: x: a minor or insignificant problem; xx: a possible shelf-life problem; xxx: a probable shelf-life problem that requires process and/or package modification.

is the assumption that the bottle closure is intact and, if not, how an opened product might demonstrate accelerated deterioration. Here, we outline the four potential stability issues that can compromise quality in the market—microbiological stability, flavor stability, physical stability, color/light stability—and, taking examples from the production of the more commonly produced spirits (Table 8.2), discuss the solutions employed to minimize or eliminate specific stability risks. Finally, we consider potentially new approaches to control product quality more closely.

8.2 MICROBIOLOGICAL STABILITY

The microbiological stability of the final products is often of relatively little concern to the distilled spirits producer, as the final alcohol concentration of more than 15% ABV essentially inhibits microbiological growth. However, alcohol is generally not considered to be bactericidal unless the alcohol concentration is at least 60% ABV (Centers for Disease Control and Prevention, 2008). Nonetheless, in areas of the distillery where alcohol concentrations are sufficiently low, fungal and microbiological contamination can occur. For instance, casks used for the maturation of spirits, such as Scotch whisky, when left unfilled, can develop growths of mold that can produce off-flavors, such as 2,4,6-trichloroanisole (TCA), a compound also responsible for the "corked" taint in wines. This compound has flavor descriptors such as "musty" and "moldy" and is typically detectable by sensory evaluation at concentrations of less than 1 μg/L (Aylott, 2003), although Conner and Reid (2008) report that TCA has a relatively high threshold of around 1.4 μg/L in blended Scotch whisky diluted to 20% ABV.

Historically, empty casks were filled with a citric acid/sulfite mixture to prevent fungal growth and to keep the cask wood swollen (oak staves contract as they dry, loosening the joins between staves and providing sites for liquid losses), but this has fallen into disuse. It is, though, generally good practice to sniff or "nose" casks before filling and also after the spirit maturation process has been completed, to help minimize the risk of tainting more than just the faulty cask. It is also prudent to ensure that any bottle closures based on natural cork are demonstrated to be clear of TCA and other flavor-active species. The formation of TCA is thought to be due to *O*-methyltransferase activity in a range of fungi (Alvarez-Rodriguez et al., 2002), resulting in the methylation (and subsequent detoxification) of the weakly acidic phenolic moiety (Figure 8.2). Other flavor-active compounds also appear to contribute to corked flavor taint, including geosmin, 2-methylisoborneol, guaiacol, 3-octenol, and 3-octenone. The presence of mycotoxin-forming fungi on cereals, such as barley and maize, has rightly attracted a lot of attention from the food and beer production supply chains. In the case of distilled spirits the presence of mycotoxin-forming fungi is of less concern as the mycotoxins themselves are insufficiently volatile to survive the distillation process (Priest and Campbell, 2003). However, the risk remains that mycotoxins are present in the distillery by-products, which is of particular relevance if the by-products enter the food chain—for instance as cattle feed (Gonzalez Pereyra et al., 2011).

Another area of potential concern is the prevalence of sooty molds, exemplified by *Baudoinia compniacensis*. These molds occur both in the wild and in the vicinity of warehouses in which maturing spirit is stored. They have a propensity to utilize ethanol as a carbon source when other carbon sources are depleted (Ewaze et al., 2008). However, there appears to be little cause for concern as they do not appear to impact the sensory quality of the warehoused spirit, the molds having much more impact on local aesthetics (Figure 8.3). The potential for microbial spoilage of cream liqueurs is dependent on whether they are pasteurized or not. Thus, products such as Baileys Irish Cream, the best-selling liqueur in the world, are made with fresh cream (<2 hours old) and pasteurized before packaging. However, there is a small microbiological risk associated if pasteurization is not carried out, although it has been stated that the combination of sufficiently high alcohol concentrations and sugar levels effectively precludes microbial spoilage of cream liqueurs (Deosarkar et al., 2016).

There is also a potential microbiological risk with advocaat products. This is a liqueur based on egg yolks, which in themselves present a microbiological hazard

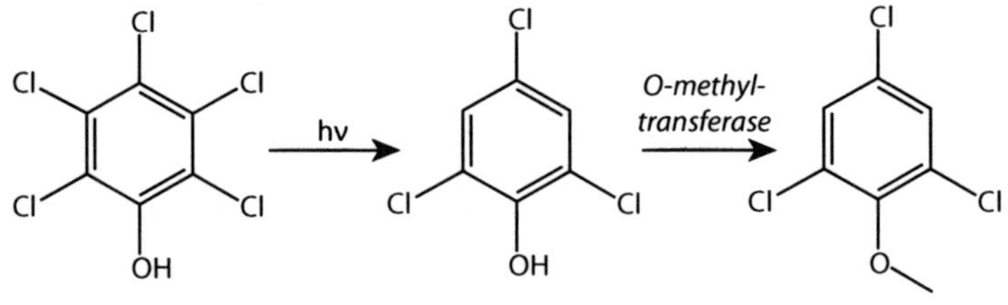

FIGURE 8.2 Mechanism of 2,4,6-trichloroanisole (TCA) formation. Its flavor potency means that even low concentrations of starting material (i.e., trichlorophenol) can result in the development of flavor-active levels of TCA.

FIGURE 8.3 Infestation of *Baudoinia compniacensis* on the outside of a whisky warehouse in Scotland, UK.

as eggs are frequently the cause of foodborne diseases, particularly salmonellosis. Unfortunately, advocaat quality suffers if yolks are heated prior to their use, so manufacturers are reluctant to pasteurize egg yolks prior to production. Various authors have noted, though, that 20% ABV alcohol in advocaats is as effective as a typical pasteurization regime and inactivates *Salmonella typhimurium* (Beckers and Van Erne, 1983; Kuschfeldt and Thiel, 1976). Nevertheless Beckers and Van Erne (1983) did observe that spores of *Bacillus cereus* could survive in advocaats at 17.2% ABV, a potential stability issue as *B. cereus* has lecithinase (i.e., phospholipase C) activity that can negatively affect the emulsion stability of advocaat. The authors concluded that adventitious contamination of advocaats with *B. cereus* during production was a possibility, but should be avoided not only so that shelf-life is not unduly compromised but also because of the pathogenicity of *B. cereus*.

8.3 FLAVOR STABILITY

Given the complexity of many alcoholic drinks, it is perhaps unsurprising that at least some display some form of flavor instability. In this section we will restrict the discussion to the stability of spirits in unopened bottles (i.e., true shelf-life) of products. There are three groups of mechanisms by which the flavor of spirits can change in-pack:

- Oxidation related
- Interaction with the packaging
- Changes in the physical structure of the drink

Light can also induce flavor changes, which are discussed later. In terms of oxidation, we can consider the oxidation products of ethanol itself and those of the phenols and polyphenols that are extracted during wood maturation. In principle, ethanol can undergo oxidation to acetaldehyde (green apples, emulsion paint) and even to acetic acid (vinegar; Figure 8.4), although conditions for the formation of acetic acid are such that this step is unlikely to be significant. However, ethanol can react with acetaldehyde to form diethyl acetal (heavy green notes), and with acetic acid present to form ethyl acetate. However, it is worth pointing out that the flavor thresholds of these various species mean that perhaps only acetaldehyde and diethyl acetal can potentially influence the resulting flavor of the spirit-based drink, and this is more likely in partly filled bottles. As ethanol is present in such relatively high concentrations, this chemistry will have a negligible effect on ethanol content.

The presence of phenols and polyphenols in spirits has been associated with antioxidant potential. Gallic acid, extracted from wood either as is or as ellagic acid or conjugated with other tannins, has been shown to be an effective scavenger of reactive oxygen species such as hydroxyl and superoxide/hydroperoxyl radicals (Marino et al., 2014). McPhail et al. (1999) showed that there is a correlation between the antioxidant potential of Scotch whiskies and the presence of simple phenolics. Similarly, Schwarz et al. (2009) and Canas et al. (2008) have observed similar phenomena in Spanish and Portuguese brandies, respectively. However, it is worth pointing out that here the "antioxidant" activity is merely diverting the destination of any reactive oxygen species that might be formed, and that there may be sensorially relevant changes induced in the phenolic compounds by their reactivity. It remains to be seen whether this is the case. The presence of relatively nonpolar packaging materials such as bottle cap seals can have an impact on the more hydrophobic flavor compounds present in many spirit-based drinks. Perhaps the most complex example is

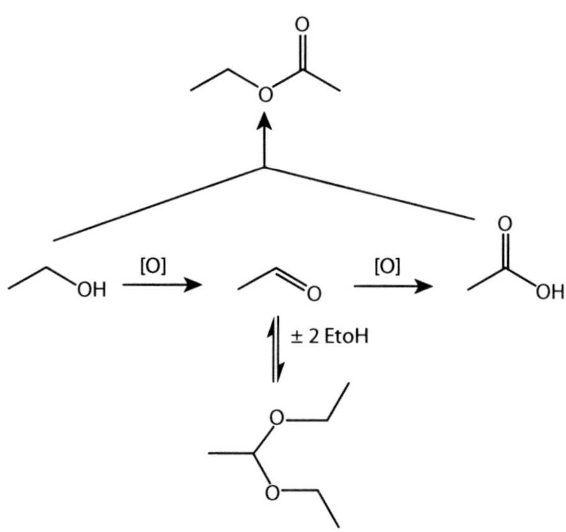

FIGURE 8.4 Proposed scheme for oxidative changes in spirit-based drinks with headspace oxygen.

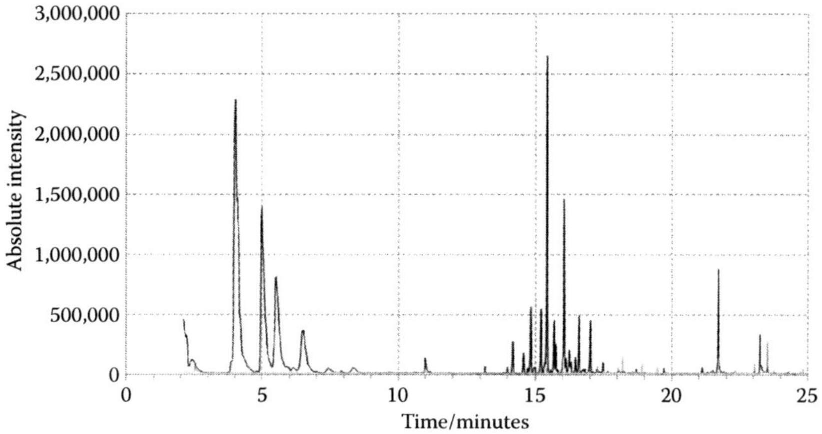

FIGURE 8.5 Total ion current chromatogram of a typical juniper distillation. (Courtesy of G. Evans.) The earlier eluting peaks are typically monoterpenes (C10) while the complex cluster midchromatogram are mainly sesquiterpenes (C15).

the case of spirits containing botanically derived flavors, such as gins, absinthe, and aquavits (Figure 8.5). Some generic features of such drinks include the presence of mono- and sesquiterpene hydrocarbons, esters, and higher alcohols. Nonvolatile compounds will also be present if at least some of the flavors in the botanical spirit are introduced by infusion rather than completely by distillation. These relatively nonpolar compounds can, in principle, react while in the bottle, but can also be adsorbed into the polymeric bottle cap seal, a phenomenon known as flavor scalping. Such considerations are likely to be even more pertinent for spirits packaged in polymeric (e.g., PET) bottles.

8.4 PHYSICAL STABILITY

The chemical processes that result in physical instability are highly dependent on the spirit category under consideration. Sharp et al. (1986) contended that the main culprits of haze formation in whiskies are ethyl and isoamyl esters of long-chain fatty acids (>C8, but especially C16 and C18), as well as phenylethanol, ethyl lactate, and, to a lesser extent, long-chain fatty acids. They also report that haze formation occurs increasingly below 48% ABV, and especially when diluting to popular sales strengths (40%–43% ABV). They also note that, after dilution, most haze is formed within 24 hours. They state that around twice as much haze forms on storage at 5°C compared with storage at 20°C. This suggests that cooling (typically anywhere between −10°C and 10°C), followed by filtration, is a useful strategy to reduce the risk of haze formation in-pack, and indeed it is a popular way of stabilizing whisky against haze formation. There is a perceived drawback in that chill filtration is commonly thought to have an impact on the flavor and mouthfeel of the final product, which becomes more apparent when chill filtration is performed at colder temperatures (Sharp et al., 1986; Booth et al., 1989). Bruichladdich makes a point of bottling its whiskies at 46% ABV, "Bruichladdich is bottled at the higher strength of 46%—any lower can result in cloudiness as a result of this insolubility" (www.bruichladdich.com/)—thus obviating the risk of flavor and mouthfeel changes by chill filtration. Puskas et al. (2013) have shown that chill filtration effectively prevented haze formation in apricot brandies.

It is worth bearing in mind, though, that drinks that are not sufficiently chill filtered, if mixed with ice, water, or tonic by the consumer, are likely to develop clear signs of haze formation in the glass. Although completely harmless, there is a risk that the consumer will react negatively to a cloudy drink. This can be commonly observed in some whiskies, although gins can be prone to this effect, known as "louching," if the oil content of the final product is sufficiently high. Of course, for drinks such as absinthe and pastis, this louching is demanded or even required by the consumer familiar with these drinks categories. Historically, spirits have suffered from the development of "bloom," which is the development of white particles or films on the internal surface of the glass. This is due to the presence of sodium ions in the glass and those on the surface undergoing a series of reactions to form insoluble sodium carbonate. Today, a "dealkalization" (i.e., sodium ion removal) process is applied, for instance, using a mixture of air and 1,1-difluoroethane (HFC-152a) or 1,1,1,2-tetrafluoroethane (HFC-134a). Both compounds degrade at the high processing temperatures required to form the glass bottle to yield small quantities of hydrogen fluoride, which affects the dealkalization, a process known as the Ball I. T. process, named after the corporation that filed the initial patented process. This process typically reduces the surface sodium ion concentrations by more than an order of magnitude (Sharp et al., 1986).

The maturation of spirits in oak casks results in the extraction of various sterols from the wood, mainly β-sitosterol and stigmasterol (Sharp et al., 1986), which are sparingly soluble in aqueous media. Repeated use of casks reduces the quantities of the sterols extracted, so it is primarily an issue for bourbons, for which new wood is mandatory for their maturation process. Chill filtration also reduces the level of sterols in whisky. With the exception of other debris in spirits (e.g., filter aids; Gomez et al., 2014, 2015; carbon dust, glass particles), another potential instability is the formation of crystals of calcium oxalate. Oxalic acid is extracted from wood, not newly made spirit, and reacts with calcium or magnesium to form insoluble oxalate salt crystals that do not redissolve. These metal ions can be introduced with water used for "deproofing" spirit to sales strength, but also ironically from filter sheets, so there is a need to ensure that the filter is scaled appropriately for the volume of spirit volume to reduce the risk of introducing excessive quantities of calcium and magnesium ions into the final product.

Emulsified alcoholic drinks, specifically those containing egg or cream, are produced in such a way as to minimize or prevent phase separation of aqueous and lipid phases. There are several mechanisms by which lipid and aqueous phases separate (Figure 8.6). Firstly, because the lipid phases are less dense than aqueous phases, there is a tendency for the lipid droplets to rise to the surface, a process known as creaming. Lipid droplets can also flocculate, forming clusters of discrete droplets. Droplets may also coalesce, increasing the volume and therefore the diameter of the

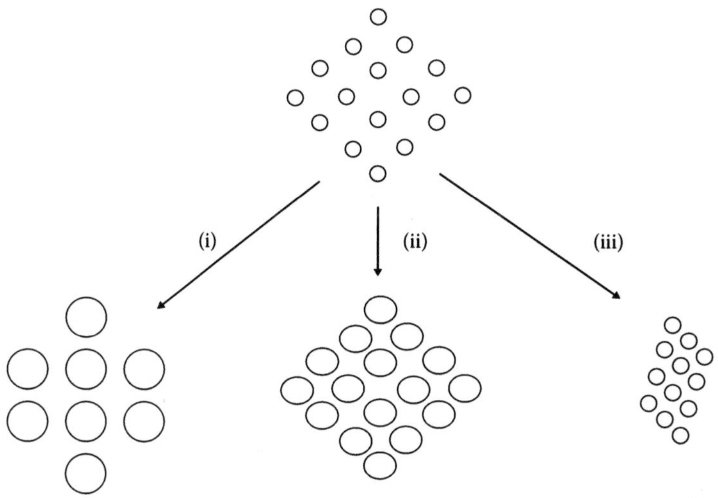

FIGURE 8.6 Schematic outlining the main sources of emulsion instability.

droplets with age. Finally, partial mixing of an aqueous phase with the lipid droplets can result in droplet growth, a process known as physical ripening. While no single mechanism accounts entirely for the instability of oil-in-water emulsions, it is helpful to consider these four processes separately, not least in terms of aiding in the identification of potential palliative options. A Polish study on advocaat stability (Tarko, 2006) suggested that, over time, the median size of fat droplets increases (Table 8.3; Figure 8.7). An interesting additional observation was that the conductivity of the advocaat samples increased over a 7-day period, and that the stability of advocaat samples is higher for those samples with lower conductivity values. The rationale here is that the droplets are surrounded by a layer of anions (the so-called double electric layer; Evans and Wennerstrom, 1994) and that, as droplet size increases, the overall droplet surface area decreases, thereby releasing conducting anions into free solution.

TABLE 8.3
Changes in Median Fat Droplet Sizes for Four Commercial Advocaat Samples

	Median Droplet Diameter (μm)	
Sample	**Fresh**	**7-Day Aged**
I	16.4	19.5
II	17.5	20.0
III	18.5	22.6
IV	17.5	18.9

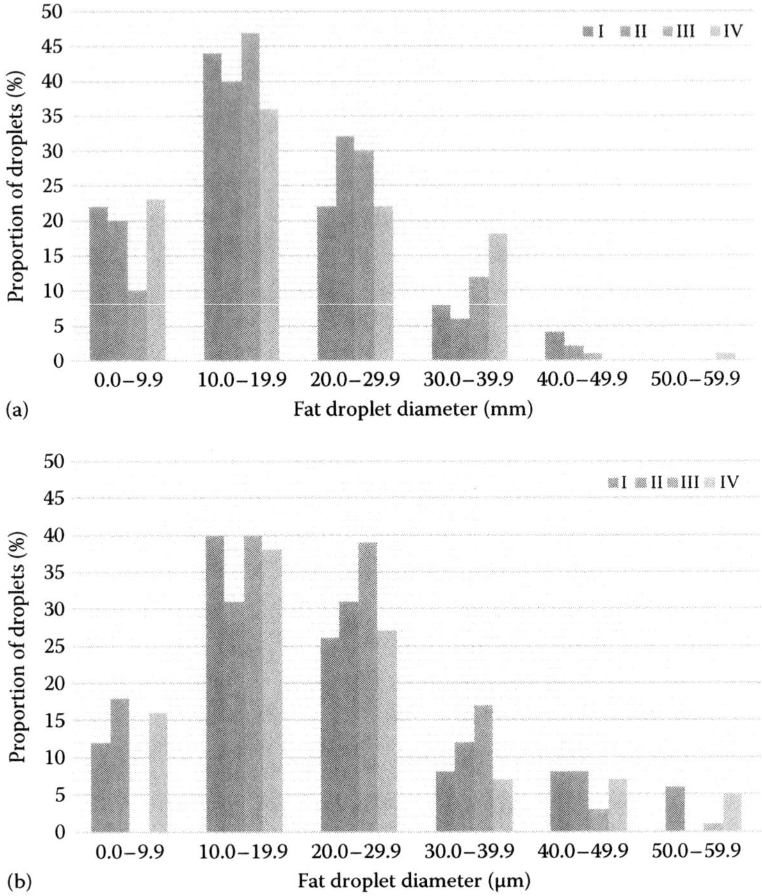

FIGURE 8.7 Changes in fat droplet sizes in four industrial advocaat samples. (Based on Tarko, T., *Acta Scientarium Polonorum, Technol. Aliment.*, 2006, 5, 37–45.) (a) Droplet size distribution in fresh advocaat samples; (b) dropet size distribution in advocaat samples stored for 7 days.

A significant amount of research has been carried out on the stability of cream liqueurs, motivated not least by the popularity of products such as Baileys and Amarula. The development of cream liqueurs was probably motivated, at least in part, in an effort to commercialize the historic concoction Atholl Brose, a Scottish drink made of oatmeal brose (i.e., oatmeal soaked in water and strained), milk, whisky, and honey. However, the stability requirements for the commercial market were an effective barrier to exploiting the concept; Baileys was the first successful product, launched in 1974. While there is no "standard" cream liqueur, typically these liqueurs contain around 160 g/kg milk fat, 195 g/kg sucrose, and are around 17% ABV (Banks and Muir, 1988). The alcohol concentration influences the stability of cream liqueur emulsions and, in particular, it appears that the alcohol content of the liquid phase, rather than total alcohol content, determines cream liqueur emulsion stability (Banks and Muir, 1985). As with advocaat, the critical parameter is to ensure that the lipid globules are sufficiently small, with a typical figure quoted requiring at least 98% of all lipid globules to be no larger than 0.8 μm diameter (Deosarkar et al., 2016). Developments in homogenization technologies have helped manufacturers of cream-based spirits to create more stable products (Heffernan et al., 2011).

Two main faults arise in cream liqueurs. Firstly, cream plugs can form at the bottle neck that do not redisperse (Dickinson et al., 1989a). The ability to form a cream plug appears to correlate with the proportion of large lipid globules present. Increasing the cream homogenization pressures helps to prevent cream plug formation, as it helps to convert larger globules into smaller ones. Low molecular weight surfactants, avoidance of excess acidity (pH > 6.0), control of calcium concentrations, and avoidance of temperature cycling of the product also help cream liqueurs resist cream plug formation. A second fault is a phenomenon known as syneresis. Here the cream starts to form a gel, which expels liquid (or serum) that accumulates on top of the gel. Syneresis can be suppressed by higher pH values, using anhydrous fat cream, and adding in calcium chelators such as trisodium citrate. Dickinson et al. (1989b) suggested that 0.4–0.5 g/100 g of either glycerol monostearate or sodium stearoyl lactylate and pH values > 6.4 enhance the stability of cream emulsions in the presence of ethanol. Other palliatives include replacing caseinate with whey protein and partial replacement of carbohydrate with sorbitol. The quantities of trisodium citrate used should be managed judiciously. While it is an effective calcium chelator, the complex can form a granular precipitate, so each product in which the chelator is used needs to be evaluated for prevention of syneresis versus precipitation of the chelated complex.

8.5 COLOR AND LIGHT STABILITY

Alongside the visual impacts of microbial growth and physical instability, the stability of products towards light is an important consideration for many spirit categories. There are two instances that are recognized light-stability issues, specifically light-induced decolorization. Firstly, spirit caramel (E150A), decolorizes when bottles are exposed to light. For darker products this is less of an issue, because small color changes are less apparent to the eye in stronger colored liquids. For those products where small quantities of spirit caramel are added to provide product consistency (e.g., whiskies, brandies), the decolorization can become evident to the consumer. Refsgaard et al. (1993) have shown that caramel can lose color even when exposed to visible light, although the rates of degradation are generally slower than when exposed to the UV wavelengths they tested (336 and 366 nm).

Another light-unstable pigment, found mainly in some absinthes, is chlorophyll, extracted from botanicals after distillation. Chlorophyll, or rather the range of compounds that are collectively known as the chlorophylls, readily degrade in the presence of heat, light, and acidic pH. Thus, it is essential to keep any naturally pigmented absinthes in the dark to retard the degradation of chlorophylls. Synthetic, stable chlorophyll derivatives are approved for food use (E141 (i) and (ii)), which are stable copper(II) complexes (Figure 8.8). The latter, E141 (ii), is water soluble and may have application in the formulation of a more color-fast absinthe. It is also worth bearing in mind that any product formulated with significant proportions of cream, such as cream liqueurs, is susceptible to light-induced flavor changes. Specifically, it has been reported that riboflavin (vitamin B2) can catalyze the degradation of the amino acid methionine, releasing various compounds including methanethiol, a highly aroma-active compound that is often described as having a smell like rotting vegetables, resulting in "light taste" (Belitz et al., 2009). However, more recent evidence points to trace concentrations of chlorophyll, and its degradation products are more likely to be the catalytic culprit (Wold et al., 2015). Keeping such products in opaque containers can help to alleviate this light sensitivity.

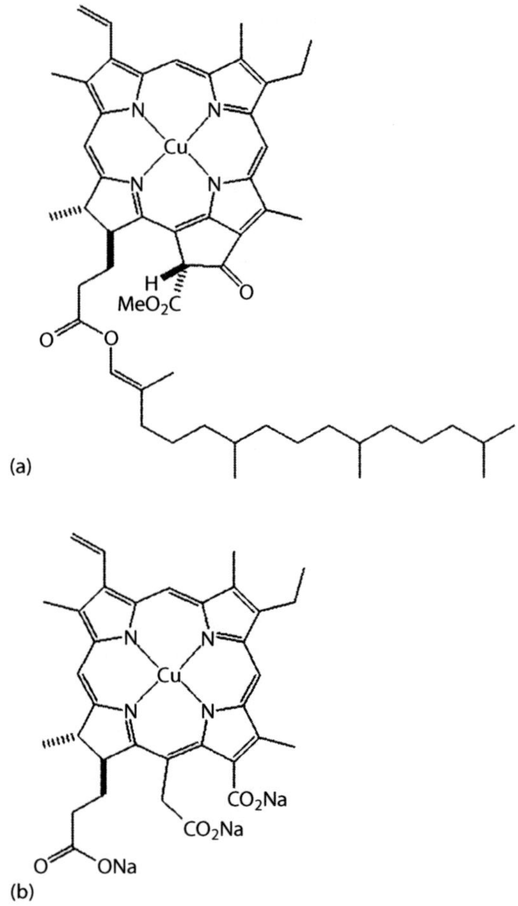

FIGURE 8.8 Colorfast synthetic chlorophyll analogs. (a) Copper(II) pheophytin (E141(i)); (b) sodium copper chlorophyllin (E141(ii)). Hydrolysis of the phytyl ester in (a) results in a more water-soluble green pigment.

Some trace metals can cause discoloration in susceptible spirits. Lead, from some bottle closures, and iron, from pipework, dissolve relatively quickly in acidic whisky (Sharp et al., 1986). Iron can chelate to phenolic compounds that have been extracted from wood, such as gallic acid, resulting in a black discoloration, while lead results in deposits. Copper, which can leach from a copper condenser or transfer pipework, can result in a greenish or bluish tinge in brown and white spirits, respectively. Quality control of filter aids such as kieselguhr is essential to ensure that they are sufficiently low in copper and iron to minimize the risk of discoloration or flavor changes due to oxidation.

8.6 WHAT HAPPENS WHEN THE BOTTLE IS OPENED?

The breaching of the package introduces oxygen into the headspace, the volume of which increases as the drink is consumed. Unlike most other drinks, it is not uncommon to store partially consumed bottles of distilled spirits and liqueurs, which can compromise the quality of the contained drink. Germain-Robin (2016) suggests three ways to circumvent these potentially damaging processes:

- Adding glass balls/marbles as an inert filler to displace air in the bottle
- Decanting spirit into smaller bottles
- Using a nitrogen gas can to flush out the oxygen from the headspace

He also points out that while spirits (or at least French brandies) do not age in glass, they "become softer, rounder and more harmonious" (Germain-Robin, 2016), a positive perspective on age-related changes, at least for French brandies. He follows this up by stating that a bottle three-quarters full will last for around a year and a half-full bottle around 6 months, and he advises that a quarter-filled bottle should be consumed in 3–4 months. We have already discussed the oxidative changes that can occur in spirits, both in terms of ethanol oxidation and the reactivity of phenols and polyphenols, which are likely to be exacerbated in partly filled bottles. In addition, the possible effects that packaging materials, particularly where hydrophobic polymers are present, can have on the portfolio of flavors in a spirit-based drink over the time that the consumer keeps it may be enhanced if the product is kept in a partly filled bottle for an extended period of time.

8.7 SUMMARY

Distilled spirits are a diverse category of alcoholic beverages. The diversity in raw materials, process, and final formulation of product brings specific challenges to each product category (Table 8.2). Product changes are classified here as chemical, microbiological, physical, and light-induced/color changes. For all well-established commercial spirit-based products, shelf-lives are generally at least 2 years, with those products at the lower end of the stability range tending to be emulsion based. No product is infinitely stable, although storing products in intact packaging, in the dark, and away from heat sources all help to keep them in good order.

REFERENCES

Alvarez-Rodriguez, M.L., Lopez-Ocana, L., Lopez-Coronado, J.M., Rodriguez, E., Martiez, M.J., Larriba, G. and Coque, J.-J.R., Cork taints of wines: Role of filamentous fungi isolated from cork in the formation of 2,4,6-trichloroanisole by *O*-methylation of 2,4,6-trichlorophenol, *Appl. Environ. Microbiol.*, 2002, 68, 5860–5869.

Aylott, R., Whisky analysis, in *Whisky—Technology, Production and Marketing*, ed. Russell, I., Academic Press, London, 2003, p. 299.

Banks, W. and Muir, D.D., Effect of alcohol content on emulsion stability of cream liqueurs, *Food Chem.*, 1985, 18, 139–152.

Banks, W. and Muir, D.D., Stability of alcohol-containing emulsions, in *Advances in Food Emulsions and Foams*, ed. Dickinson, E., Elsevier, London, 1988, pp. 257–283.

Beckers, H.J. and Van Erne, E.H.W., Survival of *Salmonella* spp., *Staphylococcus aureus* and *Bacillus cereus* in advocaat, *J. Food Protect.*, 1983, 46, 98–99.

Belitz, H.-D., Grosch, W. and Schieberle, P., *Food Chemistry*, 4th ed., Springer, Berlin, 2009, p. 543.

Booth, M., Shaw, W. and Morhalo, L. Blending and bottling, in *The Science and Technology of Whiskies*, eds. Piggott, J.R., Sharp, R and Duncan, R.E.B., Longman, Harlow, UK, 1989, p. 310.

Bruichladdich (www.bruichladdich.com/article/bruichladdichs-guide-to-chill-filtration), accessed May 8, 2016.

Canas, S., Casanova, V. and Belchior, A.P., Antioxidant activity and phenolic content of Portuguese wine aged brandies, *J. Food Compos. Anal.*, 2008, 21, 626–633.

Centers for Disease Control and Prevention (www.cdc.gov/hicpac/Disinfection_Sterilization/6_0 disinfection.html), accessed July 10, 2016.

Conner, J.M. and Reid, K.J.G., Packaging materials as a source of taints, in *Distilled Spirits: Production, Technology and Innovation*, eds. Bryce, J.H., Piggott, J.R. and Stewart, G.G., Nottingham University Press, Nottingham, UK, 2008, pp. 269–275.

Deosarkar, S.S., Khedkar, C.D., Kalyankar, S.D. and Sarode, A.R., Cream: Types of cream, in *Encyclopedia of Food and Health*, Academic Press, London, 2016, p. 336.

Dickinson, E., Narhan, S.K. and Stainsby, G., Stability of alcohol-containing emulsions in relation to neck-plug formation in commercial cream liqueurs, *Food Hydrocolloids*, 1989a, 3, 85–100.

Dickinson, E., Narhan, S.K. and Stainsby, G., Stability of cream liqueurs containing low molecular weight surfactants, *J. Food Sci.*, 1989b, 54, 77–81.

Evans, D.F. and Wennerstrom, H., *The Colloidal Domain: Where Physics, Chemistry, Biology and Technology Meet*, VCH Publishers, New York, 1994.

Ewaze, J.O., Summerbell, R.C. and Scott, J.A., Ethanol physiology in the warehouse-staining fungus *Baudoinia compniacensis*, *Mycological Res.*, 2008, 112, 1373–1380.

Germain-Robin, H., *The Maturation of Distilled Spirits, Vision & Patience*, White Mule Press, 2016, pp. 48–49.

Gomez, J., Gil, M.L.A., de la Rosa-Fox, N. and Alguacil, M., Diatomite releases silica during spirit filtration, *Food Chem.*, 2014, 159, 381–387.

Gomez, J., Gil, M.L.A., de la Rosa-Fox, N. and Alguacil, M., Formation of siliceous sediments in brandy after diatomite filtration, *Food Chem.*, 2015, 170, 84–89.

Gonzalez Pereyra, M.L., Rosa, C.A., Dalcero, A.M. and Cavaglieri, L.R., Mycobiota and mycotoxins in malted barley and brewer's spent grain from Argentinean breweries, *Lett. Appl. Microbiol.*, 2011, 53, 649–655.

Heffernan, S.P., Kelly, A.L., Mulvihill, D.M., Lambrich, U. and Schuchmann, H.P., Efficiency of a range of homogenisation technologies in the emulsification and stability of cream liqueurs, *Innovative Food Sci. Emerging Technol.*, 2011, 12, 628–634.

Kuschfeldt, D. and Thiel, W., Zur Frage der ausreichende Vorbehandlung von Eiproduckten in besondere von Eigelb, *Archiv fur Lebensmittelhygiene*, 1976, 27, 134–135.

Marino, T., Galano, A. and Russo, N., Radical scavenging ability of gallic acid towards OH and OOH radicals. Reaction mechanism and rate constants from the density functional theory, *J. Phys. Chem. B*, 2014, 118, 10380–10389.

McPhail, D.B., Gardner, P.T., Duthie, G.G., Steele, G.M. and Reid, K., Assessment of the antioxidant potential of Scotch whiskeys by electron spin resonance spectroscopy: Relationship to hydroxyl-containing aromatic components, *J. Agric. Food Chem.*, 1999, 47, 1937–1941.

Priest, F.G. and Campbell, I., *Brewing Microbiology*, 3rd ed., Kluwer, New York, 2003. pp. 165–166.

Puskas, V., Miljic, U., Vasic, V., Jokic, A. and Manovic, M., Influence of cold stabilization and chill membrane filtration on volatile compounds of apricot brandy, *Food Bioproducts Process.*, 2013, 91, 348–351.

Refsgaard, H.H.F., Rasmussen, M. and Skibsted, L.H., Light sensitivity of colourants used in alcoholic beverages, *Z. Lebensm. Unters. Forsch.*, 1993, 197, 517–521.

Schwarz, M., Rodriguez, M.C., Martinez, C., Bosquet, V., Guillen, D. and Barroso, C.G., Antioxidant activity of Brandy de Jerez and other aged distillates, and correlation with their polyphenolic potential, *Food Chem.*, 2009, 116, 29–33.

Sharp, R., Pert, D.M.K. and Canaway, P.R., Some aspects of the shelf life stability of whisky, in *The Shelf Life of Foods and Beverages*, ed. Charalambous, G., Elsevier, Amsterdam, 1986, pp. 205–224.

Tarko, T., Predicting the stability of advocaat type egg liqueurs on a base of selected evaluation indicators, *Acta Scientarium Polonorum, Technol. Aliment.*, 2006, 5, 37–45.

Wold, J.P., Skaret, J., and Dalsgaard, T.K., Assessment of the action spectrum for photooxidation in full fat bovine milk, *Food Chem.*, 2015, 179, 68–75.

9 Distillates Cask Aging

Thomas Collins

CONTENTS

9.1 OVERVIEW

While the composition of freshly distilled spirits consists primarily of compounds volatile enough to be carried over during the distillation process, the composition of cask-aged spirits is more complex. These spirits contain a range of volatile and nonvolatile compounds either extracted directly from the cask or derived from these compounds. While a range of woods have been used for casks for spirits aging, oak remains the most widely used wood for wine and spirits casks. Oak wood is well adapted for use for barrel making, as it is relatively easy to work, bends well, is durable, and, when milled properly, is relatively leak proof, as most of the oak species used for coopering have tyloses that plug the springwood vessels in the heartwood. Some European species used for coopering have fewer tyloses than American oak, requiring those species to be split rather than sawn to produce the staves.

A range of oak sources and cask sizes are in regular use for the aging of spirits. By law, American bourbon whisky must be aged in new charred oak casks, usually of American white oak, *Quercus alba*, or other American white oak species, typically for periods of 2 years ("straight whisky") or more and up to 12 years or more. This restriction on the reuse of these casks for making bourbon has meant the ready availability of once used bourbon casks for the aging of other spirits. Used bourbon barrels are commonly broken down and shipped as staves to Scotland, where they are recoopered into 300 L hogshead casks for the maturation of Scotch whiskies. These casks will be used for two or three fills for maturing single malt whiskies, before being refurbished for use in aging grain whiskies. Some percentage of Scotch whiskies are aged in used sherry butts, historically to put the casks used to ship sherry to England and Scotland to good use, but now sherry casks often arrive in Scotland as staves, just as for ex-bourbon casks. Some of these sherry butts are made of Spanish oak, *Quercus pyrenaica*, but many are American white oak. Cognac and Armagnac brandies are commonly aged in barrels of French oak, usually *Quercus robur* from the nearby Limousin forest. Armagnac is also sometimes aged in the local Gascony black oak. Given the range of oak types used for cask production, as well as the widespread use of previously filled barrels, it is important to understand the aroma and flavor chemistry of oak wood, as well as the processes that are used to convert sawn wood into usable casks for maturing spirits.

9.2 CHEMICAL COMPOSITION OF OAK WOOD

The aroma and flavor chemistry of oak wood and oak casks has been extensively studied. These studies can be divided into three broad areas of investigation—namely, the chemistry of the oak wood used in cask production, the changes arising from the processes applied to the wood during the production of the cask (e.g., seasoning of the wood, toasting or charring processes, etc.), and, finally, the extraction of oak-related compounds from the wood during the aging of the wines or spirits. Investigations into the chemistry of wood have evaluated the impact of the geographic origin of the wood (Towey and Waterhouse 1996; Doussot et al. 2002; Cadahia et al. 2003; Prida and Puech 2006b) and differences among the botanic species of oak (Miller et al. 1992; Masson et al. 1995; Chatonnet and Dubourdieu 1998; Mosedale et al. 1998; Doussot et al. 2002; De Simon et al. 2006; Prida et al. 2006a, 2007), as well as variability in the chemistry of wood from different trees within a specified geographic area or species (Mosedale et al. 1998; Prida et al. 2006a, 2007) and the variability in the chemistry of wood sourced from different sections of a single tree (Masson et al. 1994). Studies have also evaluated the impact of certain characteristics of the wood, such as grain tightness or age of the trees (Prida et al. 2007; Bosso et al. 2008).

In many of these studies, the focus has been on ellagitannins or *cis*- and *trans*-oak lactone content and how they vary by geographic origin, species, location in a tree, etc., in large part because these are primary components present in green wood that persist through the seasoning and toasting/charring processes to be extracted into wines or spirits. Most of the volatile components that constitute the basis for oak aroma in wine or spirits, however, are created during the toasting/charring process and will be discussed later in this chapter in the section on the coopering process. Mosedale et al. (1998), for example, reported on the variability of extractives in mixed stands of *Q. robur* and *Q. petraea* in the Citeaux forest in eastern France. They showed substantial differences in the concentrations of the oak lactones between the two species, with levels in *Q. petraea* two or more times greater than the levels found in *Q. robur* (3.9 μg/g vs. 1.50 μg/g for *cis*-oak lactone and 1.06 μg/g vs. 0.56 μg/g for *trans*-oak lactone). Mosedale et al. also evaluated the ellagitannin content for the samples from the two species, where the average total ellagitannin content of *Q. robur* was 33% greater than for *Q. petraea* (37.9 mg/g vs. 28.5 mg/g). The predominant ellagitannins in both species were castalagin (~33.0% of total ellagitannins), followed by vescalagin (17.0%–19.0% of total ellagitannins) and grandinin (10.7%–11.3% of total ellagitannins). Their study is also remarkable for the level of detail provided regarding the methodology and for the number of trees included in their study.

Similarly, Prida and Puech (2006b) evaluated the ellagitannin content as well as the content for several volatile compounds in freshly cut French (both *Q. robur* and *Q. petraea*) and American (*Q. alba*) oak. French oak wood had significantly higher total ellagitannin content (42.4 mg/g vs. 8.4 mg/g) than American oak, although castalagin and vescalagin were the two most abundant ellagitannins in each oak type. Of the volatile compounds measured in the study, only *cis*-oak lactone (14.2 µg/g in American oak vs. 3.1 µg/g in French oak) and eugenol (2.2 µg/g in American oak vs. 0.4 µg/g in French oak) were significantly different between the wood origins. The *trans*-oak lactone content for both sources was similar (approximately 2.0 µg/g), resulting in a high *cis*/*trans* ratio for the American oak, a feature that is sometimes used to distinguish among American oak and oak from other origins.

Another significant finding from this study was the extent of tree-to-tree variability in the content of these compounds. The standard deviations ranged from 48.0% to 165% of the mean values. Cadahia et al. (2001) measured the content of several ellagitannins in Spanish *Q. robur* and *Q. pyrenaica*, French *Q. robur*, and American *Q. alba*, all seasoned for 36 months. Spanish and French *Q. robur* had similar concentrations of castalagin (6.68 mg/L for Spanish and 6.11 mg/L for French), vescalagin (4.62 mg/L for Spanish and 5.00 mg/L for French), and roburin A (1.09 mg/L for Spanish and 1.27 mg/L for French), while the Spanish *Q. robur* had somewhat higher concentrations of grandinin (3.69 mg/L vs. 2.03 mg/L) than for the French *Q. robur*. Spanish oak, *Q. pyrenaica*, had lower concentrations of castalagin (4.54 mg/L vs. 6.68 mg/L), vescalagin (1.68 mg/L vs. 4.62 mg/L), and roburin A (0.73 mg/L vs. 3.69 mg/L) than for Spanish grown. American oak, *Q. alba*, had the lowest ellagitannin concentrations of any of the wood types, with only 2.86 mg/L of castalagin, 1.14 mg/L of vescalagin, and 0.23 mg/L of roburin A. Cadahia et al. (2003) also measured the content of 55 volatile compounds, many of which were present in only trace amounts, in the same 36-month seasoned oaks. American *Q. alba* had the highest levels of *cis*-oak lactone of any of the woods (32.5 µg/g for *Q. alba* vs. 17.5 µg/g for *Q. pyrenaica* and only 2.8 µg/g for French *Q. robur*). *Trans*-oak lactone content was more similar across the species, ranging from 3.4 µg/g in French *Q. robur* to 4.6 µg/g in Spanish *Q. robur* and 3.5 µg/g in *Q. alba*. Eugenol was also higher in *Q. alba* than in the other woods (5.7 µg/g for *Q. alba* vs. 1.6 µg/g in Spanish *Q. robur* and 2.3 µg/g in *Q. pyrenaica*).

9.3 SEASONING AND TOASTING/CHARRING EFFECTS

The investigations into the impact of processing on wood aroma and flavor chemistry have focused on two primary areas: the seasoning of the wood and the toasting of the wood. Studies of the seasoning process have looked at the impact of the time duration of seasoning as well as the location at which the seasoning was carried out (Doussot et al. 2002). The impact of kiln drying has been studied as well (Masson et al. 2000). Numerous studies have compared toasting regimes (Gimenez Martinez et al. 1996; Chatonnet et al. 1999; Hale et al. 1999; Matricardi and Waterhouse 1999; Cadahia et al. 2003; Campbell et al. 2005; Bosso et al. 2008), often including wood of various geographic origins or species in the design of the experiment (Guichard et al. 1995; De Simon et al. 1999; Sanz et al. 2010, 2011). The volatile content of untoasted wood is relatively low compared to the content of toasted wood. Many of the compounds of interest in the toasted wood arise from the degradation of lignin, hemicellulose, and other components of the untoasted wood. Of the compounds that give rise to the aromas of oak in wines and spirits, only the oak lactones, eugenol, and vanillin are present in appreciable concentrations in the untoasted wood. A general model of the toasting process, proposed by Chatonnet et al. (1999), shows that with heating, compounds formed from the thermal degradation of lignin, hemicellulose, or cellulose will increase in concentration initially, but if heating continues, the concentrations will begin to decrease. The toasting level at which that occurs varies by the chemical nature of the compound in question (Chatonnet and Boidron 1989). Gimenez Martinez et al. (1996), for example, showed that vanillin and syringaldehyde levels in American oak increased with heating, reaching maximum concentrations between 195°C and 215°C; by 250°C, both compounds had been completely degraded.

Most of the volatile compounds in toasted wood of *Q. petraea* reported by Cadahia et al. (2003) are at greater concentration in the toasted wood than in the untoasted wood, often at many-fold higher levels. Vanillin, for example, increases from 14 μg/g of dried wood to 370 μg/g, a 26-fold increase. Furfural and 5-methyl furfural increase from 19.9 μg/g to 58.9 μg/g and 1.73 μg/g to 30.2 μg/g, respectively; both are thermal degradation products of hemicellulose (Guchu et al. 2006). Guaiacol and 4-methyl guaiacol increase from 0.12 μg/g to 0.41 μg/g and 0.37 μg/g to 2.12 μg/g, respectively; these compounds are produced by the breakdown of lignin during the toasting process. In addition to *Q. petraea*, Cadahia et al. (2003) also reported the concentration of a range of volatiles in the 36-month seasoned Spanish, French, and American oak described before, after the wood had been toasted. The seasoned wood was used to make casks at the Demptos cooperage (Bordeaux, France), using the cooperage's protocol for medium toast casks (35 minutes toasting, temperature on the wood surface of 160°C–170°C, over a wood fire). Concentrations of a range of volatile compounds were measured in these same casks after toasting to 160°C–170°C.

Compounds derived from the thermal degradation of lignin increased during the toasting process. Vanillin, which is present in low concentrations in untoasted wood, increased from 7.9 μg/g to 39.7 μg/g in *Q. alba*, from 9.3 μg/g to 172 μg/g in French *Q. robur*, and from 10.6 μg/g to 210 μg/g in *Q. pyrenaica*. Similarly, coniferyl aldehyde increased from 3.1 μg/g to 169 μg/g in *Q. alba*, from 4.0 μg/g to 127 μg/g in French *Q. robur*, and from 4.1 μg/g to 307 μg/g in *Q. pyrenaica*. Cadahia et al. (2003) reported greater increases in monomethoxy and dimethoxy phenols rather than simple phenols, reflecting the predominance of the guaiacyl and syringyl subunits in oak lignin. Additionally, they evaluated the furanic and cyclic ketone compounds, which arise from the degradation of hemicellulose and cellulose structure components in wood. The furanic aldehyde furfural increased from 4.0 μg/g to 346 μg/g in *Q. alba*, from 8.9 μg/g to 340 μg/g in French *Q. robur*, and from 19.8 μg/g to 170 μg/g in *Q. pyrenaica*, while 5-methyl furfural increased from 0.3 μg/g to 85.0 μg/g, from 1.1 μg/g to 80.9 μg/g, and from 1.5 μg/g to 30.1 μg/g in the same woods, respectively. Other sugar degradation showed similar increases in concentration between the untoasted, seasoned woods and the toasted wood. There was no clear pattern, however, for eugenol or the oak lactones, concentrations of which were not greatly changed by the toasting process used for these casks.

Cadahia et al. (2001), in a separate paper, also looked at the ellagitannin content in these barrels as well. Ellagitannins present in the untoasted wood were degraded in the toasting process to ellagic acid and gallic acid, resulting in a reduction in ellagitannin concentrations for all woods and all ellagitannins measured. Castalagin, for example, was reduced from 6.68 mg/L in the untoasted Spanish *Q. robur* to 3.60 mg/L in the toasted wood and from 4.54 mg/L in the untoasted *Q. pyrenaica* to 2.77 mg/L in the toasted wood. Similar degradation was measured for the other ellagitannins. The ellagitannins were completed degraded in the American *Q. alba*, which had lower concentrations than the other wood types of ellagitannins in the untoasted wood. Other comparisons of the impact of degree of toasting have been made as well. As an example, in a study from 1999, Chatonnet et al. compared the composition resulting from three toast levels: light (115°C–125°C on the wood surface), medium (200°C–210°C), and heavy (220°C–230°C). Four barrels of each toast level were produced at a commercial cooperage "...according to the standards defined by the coopers..." Wood shavings from the top 6 mm were taken from 10 staves from each barrel. For many of the volatiles reported by Chatonnet et al. (1999), an increase was seen from light toast to medium toast, followed by a decrease for heavy toast. Only the lactones and the furfurals behaved differently, showing a general increase from light to heavy toast levels.

A study by Campbell et al. (2005) attempted to understand compositional changes from toasting, without introducing the variability of the barrel toasting process. In this case, shavings were made from air seasoned untoasted wood staves procured from local cooperages. The shavings were toasted in constant temperature ovens at 200°C (medium toast) or 235°C (heavy toast) for 1 hour. The shavings of untoasted and toasted wood were then extracted for volatile analysis. The results from these more controlled toasting conditions were at odds with the results reported earlier. In this case, guaiacol and 4-methyl guaiacol were not detected in the untoasted wood and increased in both the 200°C and 235°C treatments. Vanillin was present at a low level in the untoasted wood and increased in both heat treatments. The lactones, however, decreased with heat treatment. These results may vary from the Chatonnet study, at least in part, as the conditions in this study were more closely controlled than would be possible in a commercial cooperage.

Farrell et al. (2015) used a real-time method to analyze the volatile compounds in the process gas stream from a toasting oven to monitor the formation of oak aroma compounds during toasting at three different temperatures (175°C, 200°C, and 225°C). They used sections of oak boards for the toasting trials. The process gas stream containing the volatiles produced during the toasting process was transferred to the inlet of a proton-transfer reaction time-of-flight mass spectrometer allowing the researchers to monitor how the volatile composition of the process stream changed over time during the toasting process. They developed time-intensity profiles for selected oak volatiles that reflected the dynamic generation of the volatiles from the thermal degradation of lignin, hemicellulose, and cellulose. The concentrations of polysaccharide-derived volatiles remained low in the early stages of toasting until 10–15 minutes for the lower temperature toasting trials and then increased for some period before declining. The highest temperature treatment produced a more rapid increase in concentration, but also a more rapid decrease in these volatiles.

Farrell et al. (2015) also found that vanillin and guaiacol increased beginning at 10 minutes for the lower temperatures and continued to increase throughout the toasting process. At 225°C, the profile for these compounds was similar to that for the polysaccharide compounds (i.e., a period of increasing concentration followed by declining concentrations as the toasting continued). They also reported that concentrations of the lignin-derived compounds increased by an order of magnitude for each 25°C temperature increase. At lower temperatures oak lactones, however, showed a rapid increase in concentration initially, followed by a slower decline in concentration from after 10 minutes. At 225°C, there was an initial rapid increase, followed by a less steep increase cure, again with maximum concentrations reached at 30 minutes of the hour-long toasting process. Based on concentrations of 5-methyl furfural, furfural, hydroxyl methyl furfural, eugenol, guaiacol, oak lactone, and vanillin, Farrell et al. were able to use principal component analysis to differentiate among French and American oak toasted at 225°C, and to differentiate the higher temperature treatments from the low temperature treatments. The lower temperature treatments were not differentiated from each other either by oak type or temperature.

Similarly, Collins et al. (2015) collected wood samples from barrels during the toasting process for subsequent analysis using gas chromatography/mass spectrometry (GC/MS) to follow the evolution of several oak aroma volatiles during the toasting process. They also found that compounds derived from hemicellulose and cellulose were present only in low concentrations in the untoasted wood, but increased rapidly as temperature increased during the toasting process before leveling off or declining during the latter stages of toasting. These researchers used thermocouples embedded in the barrel staves to monitor the surface temperature of the wood throughout the toasting process. They also found that the lignin-derived compounds increased rapidly from very low levels in the untoasted wood and generally continued to increase throughout the toasting process. Eugenol, however, increased slightly initially, before declining during the middle stages of toasting. Isoeugenol, which was not present in the untoasted wood, began to increase in concentration as the eugenol declined, suggesting that a heat-related isomerization of the double bond could be responsible for the decline in eugenol as the toasting proceeded. Isoeugenol concentrations continued to increase after the eugenol was depleted, suggesting either an alternate source or that as quickly as eugenol was being produced or transported from deeper layers in the wood, it was isomerized to isoeugenol.

Matricardi and Waterhouse (1999) conducted a study that also used thermocouples embedded in the wood to follow temperature changes during the toasting process. In this case, the wood was toasted in a commercial cooperage and thermocouples were embedded in the wood by drilling a hole through the stave and inserting the thermocouple so that the tip of the probe was flush with the interior surface of the barrel. Thermocouples were placed at 22 cm from either end of the barrel and in the center along the length of the stave. There were four toasting techniques used, in a two-way matrix that included the barrel on a raised stand, which allowed air circulation under the barrel or not, and spraying with water during the toasting process or not. In the work of Matricardi and Waterhouse (1999), analysis of the composition focused primarily on ellagitannins and color. The floor toasting method produced higher temperature in the wood and greater degradation of ellagitannins. Ninety seven and ninety nine percent of the castalagin and vescalagin, respectively, were degraded in the surface layer of the toasted wood. The cooler stand toasting protocol resulted in a 72% degradation of the two ellagitannins. They also analyzed the ellagitannin content in the layer of wood 2–4 mm below the surface and found that 92%–95% of the castalagin and vescalagin was degraded in the floor method, but only 10%–40% was degraded using the stand method. The color intensity of the wood was not indicative of the ellagitannin breakdown, however, because the stand toasted wood was darker, but had higher levels of ellagitannin (i.e., less degradation). They further reported that the ellagic acid concentration in the wood increased proportionately with the degradation of the ellagitannins.

As outlined before, there have been numerous studies investigating the volatile and nonvolatile chemistry of both the source wood for barrel making and the wood after it has been processed into barrels. However, there has been little research focused on the variability in the coopering processes in a commercial setting. In most of the studies cited earlier, the toasting process is taken as a given, without critical evaluation of the variability associated with the process. Taken as a whole, the results of these studies have often been contradictory or inconclusive—in part, because the definitions of toasting levels are not consistent from cooper to cooper and, in part, because the toasting protocols as applied have sufficient variability to affect the internal consistency within the study.

9.4 EXTRACTION OF AROMA AND FLAVOR COMPOUNDS

Finally, the extraction of oak aroma and flavor compounds from the wood by wine or spirits has been the subject of many investigations, which have focused on the relative extraction rates and amounts of various oak-derived aroma compounds present in the wood (Towey and Waterhouse 1999; Garde-Cerdan et al. 2010), as well as the impact of multiple extractions of the same wood (Towey and Waterhouse 1996). Some studies have investigated properties associated with the wood, such as grain tightness or density, or of the process, such as toast level, on the extraction. Similarly, other studies have looked at properties of the wine or spirit, such as alcohol content or varietal, on the extraction characteristics.

Piggott et al. (1993) evaluated the impact of using ex-bourbon barrels of varying ages on the composition of a pair of single malt Scotch whisky at two different proofs. The whisky was aged for 36 months in three ex-bourbon barrels not previously used for maturation of Scotch whisky, in three ex-bourbon barrels used once for Scotch whisky, and three ex-bourbon barrels used multiple times for maturation of Scotch whisky. A composite sample was taken from each set of three casks at 3-month intervals for the first year and 6-month intervals out to 3 years of aging. The whiskies were analyzed using GC/MS and high-performance liquid chromatography (HPLC)/MS for a selection of volatile and nonvolatile compounds. They reported a general decline in the concentration of wood-related compounds with increasing age of the casks. Ellagitannins declined from 3.7 mg/L in the first refill cask to 1.1 mg/L and 0.7 mg/L in the second refill and multirefill casks, respectively. Similarly, vanillin declined from 3.1 mg/L to 1.6 mg/L and 1.6 mg/L, respectively. Ellagic acid was only found in the first refill cask, at 0.6 mg/L, but not in the older casks.

In a companion study to the Piggott et al. (1993) study, Clyne et al. (1993) investigated the impact of charring the interior of the cask on the composition of the resulting spirits. In this study, a Highlands single malt whisky was aged for 36 months in sets of three new charred American oak casks, at either 67.5% ethanol (v/v) or

63.4% (v/v) ethanol. The resulting spirits were then subjected to chemical and sensory analyses. Clyne et al. (1993) found increased levels of vanillin, syringaldehyde, and syringic acid in the whiskies aged in charred casks, but reduced levels of coniferaldehyde, sinapaldehyde, and ellagitannins in charred-cask aged whiskies. The whiskies aged in charred casks were more associated with the "maturity" sensory descriptor, while the whiskies aged in uncharred casks were more associated with descriptors associated with immaturity. The character of the whisky aged in the charred barrels, however, was more like that of a bourbon than a Scotch whisky, while the whisky aged in the uncharred casks more closely resembled an immature Scotch whisky.

Fujieda et al. (2008) reported the isolation of three new compounds, named whisky tannins A and B and carboxyl ellagic acid, from a commercial Japanese whisky. Whisky tannins A and B were oxidation products of castalagin, one of the primary ellagitannins in oak wood. Castalagin itself was not found in this whisky. They investigated the pyrolysis products of castalagin to see whether the new compounds were formed during the barrel charring process or were the product of oxidation occurring during the maturation process. As the newly found whisky tannins were not found among the pyrolysis products, they concluded the new tannins were formed from oxidation during the aging process. Fujieda et al. also reported the presence of gallic acid, ellagic acid, three galloyl glucoses, and castacrenin B.

Similarly, MacNamara et al. (2011) used GC/MS and liquid chromatography (LC)/MS to evaluate the composition of pot still whiskies aged for up to 20 years in new, lightly toasted 500 L casks made from either *Q. robur* or *Q. alba* wood. They reported the presence of 32 compounds of low volatility from the LC/MS analysis, including coniferyl and sinapyl alcohol, sinapic acid, syringic acid, vanillin, coniferaldehyde, ferulic acid, and ellagic acid. MacNamara et al. (2011) also found lignans and their glycosides, including lyoniresinol and five and six carbon glycosides of lyoniresinol. Several of the unidentified compounds reported have subsequently been demonstrated to be triterpenoid saponins and their glycosides.

Arramon et al. (2003), developed an HPLC-electrospray ionization MS method for the analysis of triterpenoids in oak wood and wines and spirits aged in wood. They used the method to evaluate the content of four triterpenes in a range of oak woods used for cooperage, including *Q. petraea* and *Q. robur*, both from several different French forests, and *Q. alba* from Oregon. Only one of the triterpenes was found in *Q. alba* and then only at a concentration of 10 μg/g. Two of the four triterpenes were found in all the French oak samples, at concentrations ranging from 64 to 540 μg/g. There was a trend towards higher concentrations of these triterpenes in *Q. petraea*. One of the triterpenes was not found in any of the oak samples, but was found in Armagnac brandies aged in oak casks. The masses of two of the triterpenes in this study matched the masses of unknown triterpenes reported by MacNamara et al.

Marchal et al. (2016), reported an HPLC/MS method for the analysis of four taste-active triterpenoids and their glycosides in wines and in oak wood. Three of the triterpenoids were perceived as sweet in wine, while the fourth was perceived as bitter. They found that *Q. petraea* had somewhat higher levels than *Q. robur* of the sweet triterpenoids and less of the bitter triterpenoid, although there was a high degree of variation from tree to tree within each species. These sweet triterpenoids had been previously reported by Marchal et al. (2011) using taste guide partitioning to isolate the sweet compounds. These were determined to be oleanane triterpenoids, substituted with glucosyl and galloyl moieties. Marchal et al. named these previously unreported compounds Quercotriterpenoside I and II.

These example studies have introduced the range of wood-derived compounds present in aged spirits. The toasting or charring process produces a wide range of thermal degradation products of the structural polymers of wood and makes these compounds more readily extractable through the degradation of the wood surface, allowing better penetration of the spirit into the wood. As the Fujieda et al. (2008) study demonstrated, the compounds extracted from the wood can be further modified by oxidation or through subsequent reaction with other components in the spirit. Finally, while much of the focus has been on understanding the role that the furanic and phenolic degradation products, as well as the ellagitannins, play in the aroma and flavor of spirits, more recently, an interest has developed in the contribution of the triterpenes in oak wood to the flavor of oak aged spirits.

REFERENCES

Arramon, G., Saucier, C., Tijou, S. and Y. Glories. 2003. Estimation of triterpenes in wines, spirits and oak heartwoods by LC-MS. *LCGC North America*. September: 910–918.

Bosso, A., Petrozziello, M., Santini, D., Motta, S., Guaita, M. and C. Marulli. 2008. Effect of grain type and toasting conditions of barrels on the concentration of the volatile substances released by the wood and on the sensory characteristics of Montepulciano d'Abruzzo. *Journal of Food Science*. 73: S373–S382.

Cadahia, E., De Simon, B. F. and J. Jalocha. 2003. Volatile compounds in Spanish, French, and American oak woods after natural seasoning and toasting. *Journal of Agricultural and Food Chemistry*. 51: 5923–5932.

Cadahia, E., Munoz L., De Simon, B. F. and M. C. Garcia-Vallejo. 2001. Changes in low molecular weight phenolic compounds in Spanish, French, and American oak woods during natural seasoning and toasting. *Journal of Agricultural and Food Chemistry*. 49: 1790–1798.

Campbell, J. L., Sykes, M. Sefton, M. A. and P. A. Pollnitz. 2005. The effects of size, temperature and air contact on the outcome of heating oak fragments. *Australian Journal of Grape and Wine Research*. 11: 348–354.

Chatonnet, P., and J. N. Boidron. 1989. The incidence of thermal treatment of oak wood on its chemical composition 1st part. Definition of the thermal parameters for the heating of barrels in the cooperage. *Connaissance de la Vigne et du Vin*. 23: 77.

Chatonnet, P., Cutzach, I., Pons, M. and D. Dubordieu. 1999. Monitoring toasting intensity of barrels by chromatographic analysis of volatile compounds from toasted oak wood. *Journal of Agricultural and Food Chemistry*. 47: 4310–4318.

Chatonnet, P. and D. Dubourdieu. 1998. Comparative study of the characteristics of American white oak (*Quercus alba*) and European oak (*Quercus robur*) for production of barrels used in the barrel aging of wines. *American Journal of Enology and Viticulture*. 49: 79–85.

Clyne, J., Conner, J. M., Paterson, A. and J. R. Piggott. 1993. The effect of cask charring on Scotch whisky maturation. *International Journal of Food Science and Technology*. 28: 69–81.

Collins, T. S., Miles, J. L., Boulton, R. B. and S. E. Ebeler. 2015. Targeted volatile composition of oak wood samples taken during toasting at a commercial cooperage. *Tetrahedron*. 20: 2971–2982.

De Simon, B. F., Cadahia, E., Conde, E. and M. C. Garcia-Vallejo. 1999. Ellagitannins in woods of Spanish, French and American oaks. *Holzforschung*. 53: 147–150.

De Simon, B. F., Sanz, M., Cadahia, E., Poveda, P. and M. Broto. 2006. Chemical characterization of oak heartwood from Spanish forests of *Quercus pyrenaica* (Wild.) Ellagitannins, low molecular weight phenolic and volatile compounds. *Journal of Agricultural and Food Chemistry*. 54: 8314–8321.

Doussot, F., De Jeso, B., Quideau, S. and P. Pardon. 2002. Extractive content in cooperage oak wood during natural seasoning and toasting; influence of tree species, geographical location, and single-tree effects. *Journal of Agricultural and Food Chemistry*. 50: 5955–5961.

Farrell, R. R., Wellinger, M., Gloess, A. N., Nichols, D. S., Breadmore, M. C., Shellie, R. A. and C. Yertzian. 2015. Real-time mass spectrometry monitoring of oak wood toasting: Elucidating aroma development relevant to oak-aged wine quality. *Scientific Reports*. 5: 17334.

Fujieda, M., Tanaka, T., Suwa, Y., Koshimizu, S. and I. Kouno. 2008. Isolation and structure of whiskey polyphenols produced by oxidation of oak wood ellagitannins. *Journal of Agricultural and Food Chemistry*. 56: 7305–7310.

Garde-Cerdan, T., Lorenzo, C. and J. M. Carot. 2010. Effects of composition, storage time, geographic origin and oak type on the accumulation of some volatile oak compounds and ethylphenols in wines. *Food Chemistry*. 12: 1076–1082.

Gimenez Martinez, R., Lopes De La Serrana, H., Villalon Mir, M., Quesda Granados, J. and M. C. Lopez Martinez. 1996. Influence of wood heat treatment, temperature and maceration time on vanillin, syringaldehyde, and gallic acid contents in oak wood and wine spirit mixtures. *American Journal of Enology and Viticulture*. 47: 441–446.

Guchu, E., Diaz-Maroto, M. C., Perez-Coello, M. S., Gonzales-Vinas, M. A., and M. D. C. Ibanez. 2006. Volatile composition and sensory characteristics of Chardonnay wines treated with American and Hungarian oak chips. *Food Chemistry*. 99: 350–359.

Guichard, E., Fournier, N., Masson, G. and J. L. Puech. 1995. Stereoisomers of beta-methyl-gamma-octalactone. 1. Quantification in brandies as a function of wood origin and treatment of the barrels. *American Journal of Enology and Viticulture*. 46: 419–423.

Hale, M. D., McCafferty, K., Larmie, E., Newton, J. and J. S. Swan. 1999. The influence of oak seasoning and toasting parameters on the composition and quality of wine. *American Journal of Enology and Viticulture*. 50: 495–502.

MacNamara, K., Dabrowska, D., Baden, M. and N. Helle. 2011. Advances in the ageing chemistry of distilled spirits matured in oak barrels. *LCGC Europe*. September: 448–466.

Marchal, A., Prida, A. and D. Dubourdieu. 2016. New approach for differentiating sessile and pedunculate oak: Development of a LC-HRMS method to quantitate triterpenoids in wood. *Journal of Agricultural and Food Chemistry*. 64: 618–626.

Marchal, A., Waffo-Teguo, P., Genin, E., Merilon, J. M. and D. Dubourdieu. 2011. Identification of new natural sweet compounds in wine using centrifugal partition chromatography—Gustatometry and Fourier transform mass spectrometry. *Analytical Chemistry*. 83: 9629–9637.

Masson, G., Baumes, R. Moutounet, M. and J. L. Puech. 2000. The effect of kiln-drying on the levels of ellagitannins and volatile compounds of European oak (*Quercus petraea* Liebl.) stave wood. *American Journal of Enology and Viticulture*. 51: 201–214.

Masson, G., Moutounet, M. and J. L. Puech. 1995. Ellagitannin content of oak wood as a function of species and of sampling position in the tree. *American Journal of Enology and Viticulture*. 46: 262–268.

Masson, G., Puech, J. L. and M. Moutounet. 1994. Localization of the ellagitannins in the tissues of *Quercus robur* and *Quercus petraea* woods. *Phytochemistry*. 37: 1245–1249.

Matricardi, L. and A. L. Waterhouse. 1999. Influence of toasting technique on color and ellagitannins of oak wood during barrel making. *American Journal of Enology and Viticulture*. 50: 519–526.

Miller, D. P., Howel, G. S., Michaelis, C. S., and D. I. Dickman. 1992. The content of phenolic acid and aldehyde flavor components of white oak as affected by site and species. *American Journal of Enology and Viticulture*. 43: 333–338.

Mosedale, J. R., Feuillat, F., Baumes, R. Dupouey, J. L. and J. L. Puech. 1998. Variability of wood extractives among *Quercus robur* and *Quercus petraea* trees from mixed stands and their relationship to wood anatomy and leaf morphology. *Canadian Journal of Forest Research-Revue Canadienne De Recherche Forestiere*. 28: 994–1006.

Piggott, J. R., Conner, J. M., Paterson, A. and J. Clyne. 1993. Effects on Scotch whisky composition and flavor of maturation in oak casks with varying histories. *International Journal of Food Science and Technology*. 28: 303–318.

Prida, A., Boulet, J. C., Ducousso, A., Nepveu, G. and J. L. Puech. 2006a. Effect of species and ecological conditions on ellagitannin content in oak wood from an even-aged and mixed stand of *Quercus robur* L. and *Quercus petraea*. *Annals of Forest Science*. 63: 415–424.

Prida, A., Ducousso, A., Petit, R. J., Nepveu, G. and J. L. Puech. 2007. Variation in wood volatile compounds in a mixed oak stand: Strong species and spatial differences in whisky-lactone content. *Annals of Forest Science*. 64: 313–320.

Prida, A. and J. L. Puech. 2006b. Influence of geographical origin and botanical species on the content of extractives in American, French, and East European oak woods. *Journal of Agricultural and Food Chemistry*. 54: 8115–8126.

Sanz, M., Cadahia, E., Esteruelas, E., Munoz, A., De Simon, B. F., Hernandez, T. and I. Estrella. 2010. Phenolic compounds in chestnut (*Castanea sativa* Mill.) heartwood. Effect of toasting at cooperage. *Journal of Agricultural and Food Chemistry*. 58: 9631–9640.

Sanz, M., De Simon, B. F., Esteruelas, E., Munoz, A. M., Cadahia, E., Hernandez, T., Estrella., I. and E. Pinto. 2011. Effect of toasting intensity at cooperage on phenolic compounds in Acacia (*Robinia pseudoacacia*) heartwood. *Journal of Agricultural and Food Chemistry*. 59: 3135–3145.

Towey, J. P. and A. L. Waterhouse. 1996. Barrel-to-barrel variation of volatile oak extractives in barrel fermented Chardonnay. *American Journal of Enology and Viticulture*. 47: 17–20.

Towey, J. P. and A. L. Waterhouse. 1999. The extraction of volatile compounds from French and American oak barrels in Chardonnay during three successive vintages. *American Journal of Enology and Viticulture* 47: 163–172.

10 Marketing Beer, Wine, Spirits, and Vinegar: Who Are the Consumers? What Do They Want? How to Talk to Them so They Hear You

Julie Kellershohn and Inge Russell

CONTENTS

10.1 OVERVIEW

Marketers have been influencing consumers for thousands of years. In ancient times, posters in Egypt were created on papyrus to advertise new products. Over the past 100 years, traditional marketing methods expanded beyond print to include billboard, radio, and then television advertisements. As technology became more prevalent and more cost effective, retailers had many more options for approaches to identify and target niche consumer groups. The marketing world is continuously evolving with the generation of new technology-enhanced approaches, and therefore how beer, wine, spirits, and vinegar are marketed is evolving as well.

10.2 MILLENNIAL CONSUMERS AND THEIR BEHAVIORS

The millennial generation is referred to as the generation that came into early adulthood around the year 2000. While there is no precise starting date for this generation, the millennial generation is often referred to as those born between the early 1980s to around 2000. By mid-2016, the millennial generation had overtaken the Baby Boomer generation as the largest living American generation (Pew Research 2016a) and the majority of them are now of legal drinking age in America. As a result of this shift in population demographics, the habits and behaviors of the Millennials are changing the shape of the retail industry, as retailers evolve their offerings to meet latest consumer demands. Millennial consumers tend to mistrust traditional approaches to advertising and their views of national news media have grown more negative over the past 5 years (Pew Research 2016b). While they are critical of traditional media, Millennials are very open to online advertising, and they rely on sources such as online marketing, blogs, social media, and recommendations from friends when making purchase decisions. Unlike prior generations, Millennials have a connection to gathering information from the digital world, with an inherent belief that the role of digital technology is to make their lives easier. Indeed, more than one-third of Millennials turn to a blog, online website, or social networking site for information prior to making an important purchase (Elite Daily 2015). They expect that brands will publish content online and rank authenticity of information as more important than the content itself. Millennials, as the most connected consumer generation in history, are highly connected to the digital world, and the majority use two to three tech devices daily (Elite Daily 2015).

Their relationship with technology has shifted how they buy products. Using technology to stay connected while shopping, they can easily reach out to family and friends for opinions, read online product reviews, comparison shop for price, and instantly share their own experiences online to influence others. They have an expectation that brands will engage with them on social media, their preferred communication medium. They seek out brands that will value their opinions, as demonstrated by brands that actively seek consumer feedback on new product development. While previous generations may have focused on convenience and price point when making a purchase decision, Millennials interweave the value of the experience into their decisions, including alcohol consumption decisions (Fona International 2016). A 2014 study by Anheuser-Busch found that 44% of drinkers aged 21 to 27 had never tried their Budweiser beer, which in 1990 was the best selling beer in the world (*Wall Street Journal* 2014). Millennial drinkers are thought to be the driving force behind the craft beer industry, doubling the industry's market share from 2011 to 2014, due to this group's focus on flavor exploration and a desire for unique experiences.

Their exploratory approach to beverages is not limited to beer. Millennials have the most varied set of tastes of any wine drinker in history (*Wine Spectator* 2016) and are 60% more likely than all other US wine drinkers to want to try as many different wines as they can (*Wall Street Journal* 2015). In addition, 86% of millennial wine drinkers report that two to three times a month they buy a glass or bottle of wine they have not tried before (Mintel 2014). With an openness to variety and an interest in creating experiences worthy of sharing through social media, this generation is more receptive to both a wide range of varietals, as well as alternative outlets for consuming wine. This is reflected by the growth of bookstores, movie theatres, and even carwashes that serve wine. With a focus on taste over price, Millennials are three times more likely to have purchased bottles of wine costing more than $20 in the past month, compared to Baby Boomer wine purchases. According to research done by the Wine Market Council, women make up a slightly larger share of US wine purchasers than men. However, among Millennials, this gap is much larger, with women much more involved than men in wine purchase decisions (*Wine Spectator* 2016).

Millennials are also active spirit consumers and their behaviors of taste exploration continue in terms of how Millennials consume spirits. While past generations may have favored one or two spirits as their "go-to" drink choices, 20% of Millennials claim to routinely consume four or more different types of spirits (Nielsen 2014) and the interest in taste exploration has also driven the growth in flavored spirits, especially in the flavored vodka category. While older generations are more likely to drink spirits for themselves, Millennials are much more likely to associate spirit consumption with social events, or use spirits to create the theatre aspects of a cocktail, feeding into their desire for experiential events. For Millennials, alcoholic beverages are about entertainment and self-expression. Consumer demand for experiential beverages with a unique twist has driven not only the craft beer boom, but also a similar phenomenon, starting in the early 2000s in the craft distilling market. For example, in the United States in 2016, there were over 1,300 craft distilleries producing product and most of these small distilleries were less than 10 years old. Their products continue to take more shelf space and are posing competition to the well known brand names in a similar manner to what has happened in the craft beer industry.

Unlike Millennials, the Baby Boomer consumer has a different set of consumer needs. While Millennials are willing to spend a premium for a beverage with unique ingredients, or a compelling brand story as a part of a social drinking experience, Baby Boomers are apt to purchase fine wines or spirits to drink for the simple enjoyment of drinking. Baby Boomers spend less overall as they age and drink less as they age. However, this is not unique to this generation (PYMTS 2016); rather, this tapering of alcohol consumption is a typical consumer pattern with age, with volume of alcohol consumption typically peaking during a generation's 20s. Baby Boomers also have a less exploratory approach to beverage consumption. Craft beers have found a way to appeal to almost half of millennial drinkers through offering options for flavor exploration, while less than a third of the Baby Boomer segment finds appeal in craft beers (Craft Brewing Business 2013). While Millennials routinely have a range of three to four spirits that they drink, Baby Boomers more typically have one or two spirits that become staples of their drinking routines.

10.3 INCREASED CONSUMER FOCUS ON QUALITY

Throughout retail, there has been a rise in consumer interest in product quality over the past several years. The interest in quality is not exclusive to the millennial consumer, as Baby Boomers also have a high appreciation for quality products. However, the millennial generation prides itself on being unique, creative, and purpose-driven and on buying quality products with positive origin stories that align with millennial consumer identities. In addition, the majority of Millennials believe that they will not be able to purchase a house in light of the current economic instability of this generation, and, without the large expenses associated with home ownership (Elite Daily 2015), they have a higher level of discretionary income relative to previous generations.

Millennials are shifting their spending into higher quality products in terms of beverages and beverage experiences. Vocal and active, the millennial consumer is interested in understanding the products that he or she is consuming. With several paths to quality perceptions, Millennials seek out differentiated product offerings, looking for high-quality ingredients, artisanal production, and traceability of ingredients. To earn brand loyalty and brand advocacy with a millennial consumer, product quality is critical, as Millennials describe the number one reason they share information about a brand or product online is to share their opinions on product quality (Elite Daily 2015). As the generation that grew up with Starbucks, which uses customization to offer over 87,000 beverage options, Millennials view customization of products as an expected option, not as a unique and distinguishing attribute. They seek out brands with "just-for-us" appeal, and brands that support their individual approaches to life. Quality is viewed in terms of ingredients that have names that the consumers can recognize. The ability to customize or seek out unique flavors is important. Transparency, which allows the consumer to trace ingredients back to their source, is expected. Is the Millennial's desire for transparency of ingredients leading many major beer companies to updating their labels to include all ingredients? This label transparency matches the "wholesome halo" that craft brewers established and is very popular with millennial consumers.

It is not surprising that the marketing of craft spirits is following a similar path to that which we have seen in the marketing of craft beers, as the demographics and customer interests that they are appealing to are similar. Again, it is "farm to table" or "grain to glass" and the traceability and the provenance of a small distillery that attracts the large millennial consumer segment looking for a unique product. Many spirit producers are even growing their own grains as part of the product presentation story. Craft whisky took over from vodka as the largest craft spirit category early in 2010 and today over 60% of craft distilleries sell as least one type of whisky product. Gin is especially popular with craft distillers as it allows them to reach the market much more quickly because the production process can be as fast as 10 days from "grain to glass" since a long aging period is not required. Gin allows the distiller to be very creative with unique flavors and ingredients. As the government rules for craft distilling continue to change in various countries, a surge can be seen in new openings. For example, in the United Kingdom licensed boutique gin distilleries increased in number from 116 gin distilleries in 2000 to 233 in 2015.

10.4 HEALTH CLAIMS DRIVE A SHIFTING CONSUMPTION PATTERN AS MILLENNIALS AGE

With wellness a key driver of consumer spending, the Millennials and Baby Boomers share the common trait of an interest in and concern for health and welfare. Consumers are using food trends to address some of their health concerns, with Millennials willing to put their financial resources behind their dietary interests. Surveys indicate that 42% are willing to pay for foods that are made with all-natural ingredients, 36% are willing to pay more for foods that are organic, and 33% feel that it is very important to eat foods with little or no artificial ingredients. This results in an interest in health and wellness fueled by "real" beverages. Millennials view the top barriers to their being healthy as a lack of time and difficulty in finding ways to incorporate healthier choices (Allidura 2014). A beverage choice marketed with this in mind offers an appealing and convenient option.

As Millennials age and have an increased focus on health, this is reflected in their alcohol consumption habits as they shift away from beer and more into wine and spirits. Of the Millennials that self-report as consuming less beer, opting for a healthier lifestyle is the number one reason they cite, while 43% say they are choosing wine more often as they believe it is better for their health (Liquor Control Board of Ontario 2016).

10.5 MARKETING AND CONSUMER TRENDS IN VINEGAR

Many of the trends seen in alcoholic beverages are seen throughout the food and beverage industry. Changing consumer demands and the influential role of the millennial consumer have also shaped many of the recent trends in vinegar. In 2017, the global vinegar market was expected to rise as consumers embraced vinegar for medicinal usages. The therapeutic benefits of vinegars were recently reviewed by Samad et al. (2016). There are many articles in the popular press touting the benefits of various types of vinegar based on folk medicine reports, as well as some recent scientific studies. Apple cider vinegar may help regulate blood sugar and boost digestive health, but there is still controversy as to other numerous health benefit claims (Chen et al. 2016). With the current consumer interest in healthy products, vinegar is undergoing a newfound popularity with many options for artisanal products. Apple cider vinegar is being listed by the Whole Foods grocery chain as a top trend in 2017, while plum vinegars are making the move from Japanese cuisine to mainstream American pantries.

10.5.1 Health Claims and Vinegar

Consumers are looking for options with ingredients that they recognize and that they believe will also provide them with health benefits, resulting in the addition of vinegar to health products ranging from beverages to organic cough syrups. While apple cider vinegar has been positioned for its health benefits for many years, the belief in the therapeutic effects of vinegar and vinegar-based drinks for health-conscious consumers continues to grow. Approaches to healthy vinegar beverages have expanded to include drinking vinegars such as ginger-tumeric vinegar and coconut vinegar from Suja Juice (San Diego, California) and the addition of vinegar as a healthy additive to more traditional beverages, such as the blueberry lemonade, sold by Pressery, a Boulder, Colorado, company. There are apple cider cleansing probiotic vinegar tonics marketed by KeVita (a California company now wholly owned as a Pepsi subsidiary with the goal of increasing sales in the functional beverage area) and products such as chili ginger lime, which is claimed to aid in digestion. There is a combination product that is a vinegar concentrate with soda water by Pok Pok Som (Portland, Oregon), marketed as an alternative to soft drinks, which would be viewed by some consumers as a healthier alternative and on trend in terms of the lower sugar content, available through retailers such as Whole Foods. This product and similar ones are also on trend as they are marketed as a "craft" product, with a story to tell, like the marketing strategy of many small craft breweries.

10.5.2 Differentiating Vinegars through Ingredients and Flavors

Artisanal products are an opportunity for vinegars to distinguish themselves. An interest in understanding locally sourced ingredients, combined with the millennial interest in unique experiences, has led to a growing range of syrups and cocktail mixes that contain vinegar, often referred to as "shrubs." Shrubs are sweetened vinegar-based syrups that can be added to boost the taste of beverages and dressings. For example, False Ox, a Toronto, Canada, company, combines herbs, sugar, and vinegar to create concentrated syrups for cocktails, in flavors such as beet and ginger, celery jalapeno, or grapefruit rosemary. Using local and seasonal ingredients, the products appeal to consumers that are looking for unique experiences. Locally sourced, small-batch, vinegar-based shrubs often use experimental flavors, and the local sourcing aspect is key. These vinegars are freshly made, lasting up to a week, adding to the fresh consumer connotations. For consumers who are interested in creating their own unique vinegars, there are growing numbers of DIY instructional videos on how consumers can create their own, at-home, herb-infused vinegars.

10.5.3 Limited Editions and Tastings

Consumers interested in a true vinegar tasting experience can now find that option in a number of locations. For example, the winery Round Pond in Napa Valley, California, in addition to traditional wine tastings, offers vinegar tastings to set itself apart from competitive local wineries. As a part of the Round Pond experience, visitors are invited to taste vinegars alongside a selection of olive oils, gourmet breads, and cheeses. Vinegar is one of only a few acidic condiments, and thus it holds a unique place in the world of food. Whether produced from grains or fruits, the basis is an alcoholic fermentation where the alcohol is then consumed by vinegar bacteria to form the acid and then, like fine wine, it is aged and treated further to produce a range of products. From the expensive aged balsamic vinegars made from cooked grape juice and marketed as an artisanal product from Modena Italy to the New Age functional vinegar drinks, there is clearly a product available for every taste and wallet. The consumer desire for unique experiences extends not only to the product but also to its packaging—ranging from iconic limited-edition bottles, such as the Sarson's Malt Vinegar Special Chip Lovers Bottle, which celebrates the relationship between fish and chips, to alternative bottle designs such as those of Fox Run, which hold the oil and vinegar together in a single bottle, to ornate bottles that are hand-blown double dramatic glass vessels, truly elevating the product when it is placed on a dining table.

10.6 TRENDS IN PACKAGING

As technology is enabling more innovation in design, several key trends are emerging in packaging, including environmental packaging, ingredient-inspired packaging, experiential packaging. and technology-enabled packaging.

10.6.1 Environmental Packaging

Today, sustainability and an environmental approach to packaging are a key differentiator for many brands. The approach to addresses this can range from supply chain standards to production approaches to how the final products are packaged. Different brands are trying different techniques to address the consumer demand for an environmental approach to packaging. Cráneo Organic Mezcal is an example of a spirit producer using this approach in its packaging. The bottle is made of hand-blown glass, which uses 100% recycled glass, a wooden cap, natural cork, recycled paper for the label, and soy-based ink; the labels are attached with a vegetable-based glue (*The Dieline* 2016). Another example of how environmental efforts of interest to the millennial consumer are being addressed is by companies such as the 'Farmer Willie's' craft beer company in the United States. It addresses environmental concerns by promising to leave no carbon footprint by offsetting its emissions with reforestation and restoration projects. Recently, the company committed to planting one tree for every 10 cases of beer sold, using this as a "green" marketing tool (*The Valley Breeze* 2017).

10.6.2 Ingredient-Inspired Packaging

Targeting the consumer desire for increased transparency, brands are showcasing ingredient lists to enhance trust. Listing product ingredients demonstrates an openness to the consumer, and clear bottles allow brands to use their packaging design to build consumer confidence. With more consumers educating themselves about premium alcohol, there is a shift in focus towards alcohol brands that promote regionally sourced ingredients and authentic craftsmanship. Millennials in particular are seeking to align themselves with brands that are genuine in their offerings. The Volcan de mi Tierra tequila line, which loosely translates to "land of the volcano," is produced at the Volcan distillery in the Mexican lowlands. This super-premium brand highlights the authenticity and sourcing of ingredients, noting that the production occurs at the base of the Tequila Volcano, thought to provide the ideal soil for producing the blue agave needed for tequila (*The Spirit Business* 2016). Gin produced by the House of Elrick also uses company packaging to highlight the sourcing of their ingredients, noting that the ingredients include hand-picked botanicals, the product is produced in small batches, and filtered water sourced from Loch Ness, the legendary Scottish lake, is used (House of Elrick 2017).

10.6.3 Experiential Packaging

Packaging designed to play a more active role in the drinking experience can resonate strongly with the consumer base that seeks out beverages based on creating memorable experiences. Champagne makers have been creating a variety of packaging options to enhance the champagne experience, ranging from a Krug champagne amplifier (a Limoges porcelain shell) that sits atop the bottle and amplifies the sound of the popping champagne bubbles (Luxury Launches 2013) to the Veuve Clicquot sheath that can transform from being a compact champagne sleeve into a dramatic origami-shaped waterproof ice bucket that can be used to keep the champagne cold (Gizmodo 2013). Understanding how consumers use the product and finding a way to have the packaging contribute to the experience can be a way to uniquely differentiate oneself from the competition.

10.6.4 Technology-Enabled Packaging

With the price of technology on the decline and the continued growth of the "internet of things," technology-enabled packaging is becoming more mainstream. While early versions of technology friendly packaging included labels with QR codes or that were scannable, we are now starting to see technology built directly into packaging, such as the wine bottle by Kuvée (2017). The Kuvée wine bottle is Wi-Fi enabled with a touch screen built into the bottle. The screen reveals the story behind each wine, suggests food pairings, offers tips and tricks, and recommends new wines to try. It also lets you order more wine and rate the wines you drink so that recommendations can be made based on your tastes. While some brands are incorporating technology into all of their packages, other brands are using technology-enabled packaging in limited runs to generate consumer buzz. For example, Beck's created a tiny camera called a "CapCam" that fits on the top of a beer bottle, allowing consumers to take 360° photos, and then automatically connects with a Facebook app. This addresses consumer interest in both a technology- and social-media–connected lifestyle and connecting the consumption of the beverages to a social moment with friends (*Adweek* 2016).

10.7 NEXT-GENERATION MARKETING

Traditional marketing relies on time-tested approaches, such as print advertisements, billboards, and television advertisements. These types of marketing approaches often require large budgets and target a broad base of potential customers. However, with a growing number of nontraditional marketing options available, marketers are expanding their toolboxes and embracing new options with mixed success.

10.7.1 Stunt Marketing

Stunt marketing is the use of a promotional gimmick or publicity event designed to attract the public's attention. Brands may release a false or comedic press release or advertisement to gain media coverage and public attention. Alternatively, stunt marketing may be an organized event, arranged primarily for media coverage and to drive consumer awareness of a product. An example of stunt marketing is when Kraken Rum created an event using a live octopus as a matchmaker in the days leading up to Valentine's Day in 2017. Consumers were encouraged to tweet a photo of themselves to the Kraken Rum twitter account. Presented with photos of a randomly selected short list of contestants, "Ken, the Octopus" selected two winners and they were sent on a Kraken Rum-themed blind date together. Consumers could watch a live streaming video as "Ken" made his decision. For those customers keen to join the Kraken Valentine's experience but not to participate in the blind date, Kraken Rum also set up several pop-up "Kraken and Love" themed bars in 10 locations in London for Valentine's Day.

10.7.2 Experiential Marketing

With both craft beers and craft distilling, a key marketing venue is visitor centers and road trips. These have long been popular in locations such as Scotland for whisky distillery tours and California for winery and tasting tours. The popular road trip route for Kentucky's Bourbon Trail was founded 18 years ago. The impact is significant for the 20 bourbon producers that participate in it. In 2016, the route hosted over one million visitors. Following up on the growth in craft distilleries, there is now also a Kentucky Bourbon Trail Craft Tour, which saw 177,228 visits in 2016 among its 11 participating distilleries, an increase of 187% since 2013, its first full year in existence (Kentucky Bourbon Trail 2017). These visits to tour a number of producers and sample their products on such a route is an increasingly popular pastime for consumers, especially Millennials. New make spirit is often sold at visitor centers (e.g., high-strength unaged whisky straight from the stills or gin products). For the more affluent consumer, entire barrels of unaged whisky are sold as an investment. The marketing emphasis is on the experience of seeing how the product is made and on tasting and learning about unusual or interesting products. Not to be forgotten is how large breweries and distilleries effectively use their visitor centers as an educational marketing tool. A good example is the Guinness Storehouse visitor center in Dublin, which in 2016 had over 1.6 million visitors. Clearly, a very effective vehicle to tell a brand's product story to a large number of interested consumers.

10.7.3 Marketing through Smartphones

Technology has become an integral part of how the millennial consumer shops; providing easily accessible, mobile-friendly information to a tech-savvy consumer base can pay off. While conventional thinking has noted that price and recommendations from friends/family have been the most important factors influencing beer purchase decisions in the past, there is a growing role for online information to become one of the major influencing factors. The growth of mobile shopping apps is also changing how consumers shop. For a small fee, many apps now not only deliver products, such as alcohol, to the location of the consumer, but they also recommend products based on what cocktails or recipes the consumer may want to make. Delivery apps such as "Drizly" or "Minibar" connect traditional retailers with predominantly younger drinkers, and their role as an influential resource is key, as over half of millennial wine drinkers consider wine reviews to be very important in their wine purchase decision process (*Wine Spectator* 2016). Because of their affection for technology, Millennials are generally more highly engaged in prestore buzz than older generations. Consumers are looking for easy-to-read content that can be read on the go that enables them to engage with the product and make quick, informed purchase decisions.

10.7.4 Digital Marketing and Social Media

During the early years of social media, many brands avoided social media, unsure of how to navigate the challenges in ensuring posts did not break strict regulations meant to prevent pitching alcohol to underage consumers. However, as the technology-enabled ability to more narrowly target specific consumer groups continues to improve and the millennial set has almost all hit the 21-year-old and over age demographic, companies have responded by increasing their budgets and focusing on digital marketing and social media marketing opportunities. Millennial consumers are more trusting of digital media than the general population, with an estimated 66% of Millennials trusting search engines as a source for news and information (Edelman 2016). Largely evolved past the traditional marketing approach of celebrity-endorsed products, social media campaigns are now more focused on creating an online buzz through interactive opportunities and the pursuit of campaigns so successful that they go viral, an oft sought but rarely achieved status. In the digital marketing space, consumers share their personal experiences with their audiences on Facebook, Twitter, and Instagram, as well as other social media platforms. With an average of 3 hours or more spent per day on social media, the millennial consumer is tuned into social media messaging. More than half of millennial wine drinkers say they talk about drinking wine on Facebook, and nearly a third share their wine experiences on YouTube, Twitter, or Instagram (*Wine Spectator* 2016). The brands do not control these individual consumer-led brand communications, and yet they have the potential to be both beneficial and detrimental to brands.

10.8 THE IMPORTANCE OF CONSUMER RESEARCH IN MARKETING

Understanding the needs and wants of the consumer is a critical component to marketing plans. By understanding what the consumer wants, marketing campaigns can be created that connect to the consumer's interests. As a brand, this keeps the effort focused on the elements most essential to the target consumer and saves money, time, and energy by not draining resources supporting elements that the consumer does not deem to be important. Consumer research can be used to customize strategies and find a brand's unique position in the marketplace. Including customers in the product development process is an approach that many craft brewers have leveraged with great success. For example, at the Dogfish Head Brewery in Delaware, customers have the chance to sample one-off brewing experiments, and if the consensus is positive enough, the company considers rolling those beers out nationally. Using research to understand not only what consumers want to drink but also how they want to drink it can lead to product innovation and addressing unmet, as well as unarticulated, consumer needs.

For example, it was important to understand that one of the consumer barriers to ordering wine through the mail was the challenge associated with accepting the wine delivery. Consumers were frustrated with needing to be home to accept the delivery or missing deliveries and having to follow up with depot-collection cards. In response to this consumer insight, Garçon Wines created a flattened 750 mL wine bottle, designed to fit through a typical mailbox. The bottles, made from a tough glass-like plastic, come packaged in a cardboard box to avoid any spillages once it comes through the letterbox (*The Telegraph* 2017). In a time-pressured environment, there may be urgency expressed to shave time lines by eliminating consumer research, but using consumer research to ensure that there is a clear understanding of your target consumer's needs and how your brand's product resonates with that consumer is critical to enabling long-term marketing success in today's rapidly changing technology-enhanced environment.

REFERENCES

Adweek (2016). Beck's created a 360-degree camera that fits into a beer bottle cap. Syncs with Facebook to upload clips. June 9, 2016. Available from: http://www.adweek.com/digital/becks-created-360-degree-camera-fits-beer-bottle-cap-171918/

Allidura (2014). The Millennial mindset: The worried well. Available from: http://www.allidura.com/downloads/141020_millennial_report.pdf

Chen, H., Chen, T., Giudici, P., and Chen, F. (2016). Vinegar functions on health: Constituents, sources, and formation mechanisms. *Comprehensive Reviews in Food Science and Food Safety*, 15:1124–1138.

Craft Brewing Business (2013). Craft beer consumer stats: How will they affect your business plan? May 20, 2013. Available from: https://www.craftbrewingbusiness.com/business-marketing/craft-beer-consumer-stats-how-will-affect-your-business-plan/

Dieline, The (2016). Celebrate authentic Mexican craftsmanship with Cráneo Organic Mezcal. Dec. 27, 2016. Available from: http://www.thedieline.com/blog/2016/12/16/celebrate-authentic-mexican-craftsmanship-with-crneo-organic-mezcal

Edelman (2016). The 2016 Edelman trust barometer—Slide 45. Available from: http://www.edelman.com/insights/intellectual-property/2016-edelman-trust-barometer/global-results/

Elite Daily (2015). Elite Daily Millennial Consumer Study 2015. Accessed from: http://elitedaily.com/news/business/elite-daily-millennial-consumer-survey-2015/902145/

Fona International (2016). The 2016 trend insight report. Millennials: Alcoholic beverages/spirits/beers. Accessed from: http://www.fona.com/sites/default/files/Millennials_alcoholic%20beverages_0116.pdf

Gizmodo (2013). Veuve Clicquot's new packaging transforms into its own bottle chiller. Nov. 13, 2013. Available from: http://gizmodo.com/veuve-clicquots-new-packaging-transforms-into-its-own-1463754439

House of Elrick (2017). Beautiful crafted artisan gin. https://houseofelrick.co.uk/

Kentucky Bourbon Trail (2017). Kentucky Bourbon Trail barrels past one million visits in 2016. Jan. 25, 2017. Available from: http://kybourbontrail.com/kentucky-bourbon-trail-barrels-past-1-million-visits-2016/

Kuvée (2017). Uncork possibilities. Available from: https://www.kuvee.com/

Liquor Control Board of Ontario (2016). Wine trends and sales. Available from: https://wgao.ca/wp-content/uploads/2016/03/Shari_MogkEdwards_Presentation.pdf

Luxury Launches (2013). Krug unveils an audio device for listening to the sweet sound popping champagne bubbles. Dec. 3, 2013. Available from: http://luxurylaunches.com/other_stuff/krug-audio-device.php

Mintel (2014). Mintel—Wine Overview 2014. London.

Nielsen (2014). Generations on tap: Beverage alcohol purchases vary by age group. Accessed from: http://www.nielsen.com/us/en/insights/news/2014/generations-on-tap-beverage-alcohol-purchases-vary-by-age-group.html

Pew Research (2016a). Millennials overtake baby boomers. Accessed from: http://www.pewresearch.org/fact-tank/2016/04/25/millennials-overtake-baby-boomers/

Pew Research (2016b). Millennials' views of news media, religious organizations grow more negative. Accessed from: http://www.pewresearch.org/fact-tank/2016/01/04/millennials-views-of-news-media-religious-organizations-grow-more-negative/

PYMTS (2016). Millennials vs. baby boomers: Raise a glass to purchasing preferences. Available from: http://www.pymnts.com/consumer-insights/boomers-vs-millennials/2016/millennials-vs-baby-boomers-raise-a-glass-to-purchasing-preferences/

Samad, A., Azlan, A., and Ismail, A. (2016). Therapeutic effects of vinegar: A review. *Current Opinion in Food Science*, 8:56–61.

Spirit Business, The (2016). Moët Hennessy enters high-end tequila sector. Oct. 31, 2016. Available from: http://www.thespiritsbusiness.com/2016/10/moet-hennessy-enters-high-end-tequila-sector/

Telegraph, The (2017). Say cheers to the ingenious flat wine bottle that gets delivered straight through your letterbox. Jan. 12, 2017. Available from: http://www.telegraph.co.uk/food-and-drink/news/say-cheers-ingenious-flat-wine-bottle-gets-deliveredstraight/

Valley Breeze, The (2017). Two new breweries bring Pawtucket's total to seven... and counting. March 7, 2017. Available from: http://www.valleybreeze.com/2017-03-07/pawtucket/two-new-breweries-bring-pawtucket-s-total-seven-and-counting#.WMXoInnD_3g

Wall Street Journal (2014). Bud crowded out by craft beer craze. Business section, Nov. 23, 2014.

Wall Street Journal (2015). How millennials are changing wine. Nov. 5, 2015. Accessed from: https://www.wsj.com/articles/how-millennials-are-changing-wine-1446748945

Wine Spectator (2016). The youngest millennial just turned 21. What does that mean for wine? Accessed from: http://www.winespectator.com/webfeature/show/id/52689

Index

C

N

O

P

Q

R

Y